Nigeria:
Its Petroleum Geology, Resources and Potential

Volume 1

Nigeria:
Its Petroleum Geology, Resources and Potential
Volume 1

Arthur Whiteman

Professor of Geology,
Petroleum Exploration Studies,
University of Aberdeen, Scotland

Graham & Trotman

First published in 1982 by

Graham & Trotman Ltd
Sterling House
66 Wilton Road
London SW1V 1DE

© A. Whiteman
Softcover reprint of the hardcover 1st edition 1982
Additional material to this book can be downloaded from http://extras.springer.com

ISBN 978-94-009-7363-3   ISBN 978-94-009-7361-9 (eBook)
DOI 10.1007/978-94-009-7361-9

Typeset in Great Britain by Input Typesetting Ltd, London SW19 8DR

CONTENTS

Volume 1

CONTENTS

Volume 2

"So geographers [geologists], in Afric maps,
With savage pictures fill their gaps;
And o'er uninhabitable downs
Place elephants for want of towns."

Jonathan Swift
On Poetry

1

INTRODUCTION

At attempt is made here to provide a comprehensive account in book form of the Petroleum Geology of Nigeria, a country which in 1979 was the world's sixth largest oil producer and rated the twelfth giant petroleum province of the world by Ivanhoe (1980) in terms of known recoverable resources (cumulative production + proven + probable reserves) of oil and gas.

Nigeria, which has been an independent sovereign country since 1960, faces the Atlantic Ocean on the south, is bounded by the Peoples' Republic of Benin (ex-Dahomey) on the west, by the Republic of Niger and by the Sahara on the north, the Republic of Chad on the northeast, and is bounded by the United Republic of Cameroun on the east. It now consists of 19 states organized in a federation and, largely because of oil and gas, Nigeria has become one of Africa's richer and more influential nations with a Gross National Product second only to the Republic of South Africa. The distribution of these states is shown in Figure 1.

A regional geological map of Nigeria and adjacent states is presented in Figure 2 and outline maps showing the general geology and main sedimentary basins of Nigeria and adjacent countries are given as Figures 7 and 9. References are made to the Petroleum Geology of adjacent coastal basins in Ghana, Togo, Benin (Dahomey) and Cameroun and to the geology and prospects of parts of the Iullemmeden Basin of Mali and Niger and of parts of the Chad Basin of Chad and Northern Cameroun, but none of these areas are dealt with systematically in this account.

Three main areas of Basin Complex have been mapped in Nigeria:

1. WEST AFRICAN MASSIF, eastern end.
2. NORTHERN NIGERIAN MASSIF centred on Kaduna.
3. EASTERN NIGERIAN MASSIF consisting of the Oban Hills, the Adamawa, Sardauna and Cameroun areas.

The Basement "massifs" roughly delimit the main sedimentary areas of the:

1. ABAKALIKI, BENUE, GONGOLA AND YOLA TROUGHS
2. BIDA OR MIDDLE NIGER BASIN
3. SOKOTO EMBAYMENT OF THE IULLEM-MEDEN BASIN
4. BORNU–CHAD BASIN
5. DAHOMEY BASIN

These basins and troughs, taken together with the onshore part of the Nigeria Delta Complex, occupy about 178 000 square miles, half the total area of Nigeria. Figure 3 shows the area of Nigeria in comparison to areas of other well known petroleum provinces and units, such as the Gulf Coast of the United States, North Sea etc. Generalized physiography, location of main towns etc. are shown in Figures 1 and 4.

Not all the sedimentary areas in Nigeria are considered prospective for hydrocarbons. Grossling (1976) suggested that around 157 000 square miles, about 88% of the onshore basins, are of prospective interest, but this certainly is an overestimate because large parts of the Bida Basin, the Sokoto Basin and the folded pre-Santonian sediments of the Benue Trough, and parts of the onshore Niger Delta Complex itself for that matter, included by Grossling, can be shown to be non-prospective or to be of marginal prospective interest for good geological reasons. Coury, Hendricks and Tyler (1979) in the Miscellaneous Field Studies Map MF 1044B, Europe, West Asia and Africa list the Niger Basin, Benue Trough (including parts of the Anambra Basin and Abakaliki Trough although they did not depict these units), Dahomey Coastal Basin, part of the Iullemmeden Basin and Chad Basin as areas known for oil and gas or which are favourable for their occurrence. Probably less than 20% of the Niger Delta Complex (total onshore-offshore area exceeding 100 000 square miles) is of first order prospective interest and, of

1

Figure 1. Map showing political boundaries of the Federal Republic of Nigeria. New state boundaries and state capitals shown.

course, productive and potentially productive acreage is much less. Most of the Gulf of Guinea Miogeocline (Nigerian Section) is of prospective interest both east and west of the delta complex, and oil and gas have been found in the Benin and Cameroun sections and in Ghana and Côte d'Ivoire.

In a little over 20 years the Niger Delta Province, Nigeria's only productive region, has yielded more than 8.04 billion (10^9) barrels of oil (up to 1 July 1980) and over 4.7 trillion (10^{12}) cubic feet (1977) of associated gas, small perhaps by some Middle Eastern and Saudi Arabian standards but nevertheless impressive in terms of remaining free world and of African production. These remarkable developments took place mainly in the late 1960s and early 1970s and came about largely because of the skill and enterprise of Shell and British Petroleum, Gulf, Texaco, Mobil, Phillips, AGIP and other companies and to some extent because of the liberal nature of Nigerian petroleum law adopted in the early days of exploration and development which encouraged oil finders to invest in the region.

Systematic undiscovered reserve–resource estimates do not appear to have been published officially for Nigeria. According to conservative estimates recoverable proven reserves (up to 1 January 1980) of oil standing at 16.7×10^9bbl and of gas totalling 41×10^{12}cuf are said to have been established for Nigerian sedimentary basins (*Oil and Gas Journal* 29 December 1980). Grossling (1976) put Nigeria's proven reserves at the end of 1975 at 20.2×10^6bbl and 44.3×10^{12}cuf of gas (Figure 5). Nehring (1978) put Nigeria's reserves (1975) at 13.4×10^9bbl. The *Financial Times* (1 October 1979) placed oil reserves at around 20×10^9bbl but did not disclose the source of the data and may will have been rounding out Grossling's figures. The reserve figures quoted above for proven oil are probably underestimates, and the proven gas reserve figures are certainly far too low. Unfortunately without inside information it is not possible to assess the value of various published undiscovered resource figures for oil and gas except in a general way (see Section 5.2). *The Petroleum Economist* (January 1980) states "Nigeria's gas reserves are

2

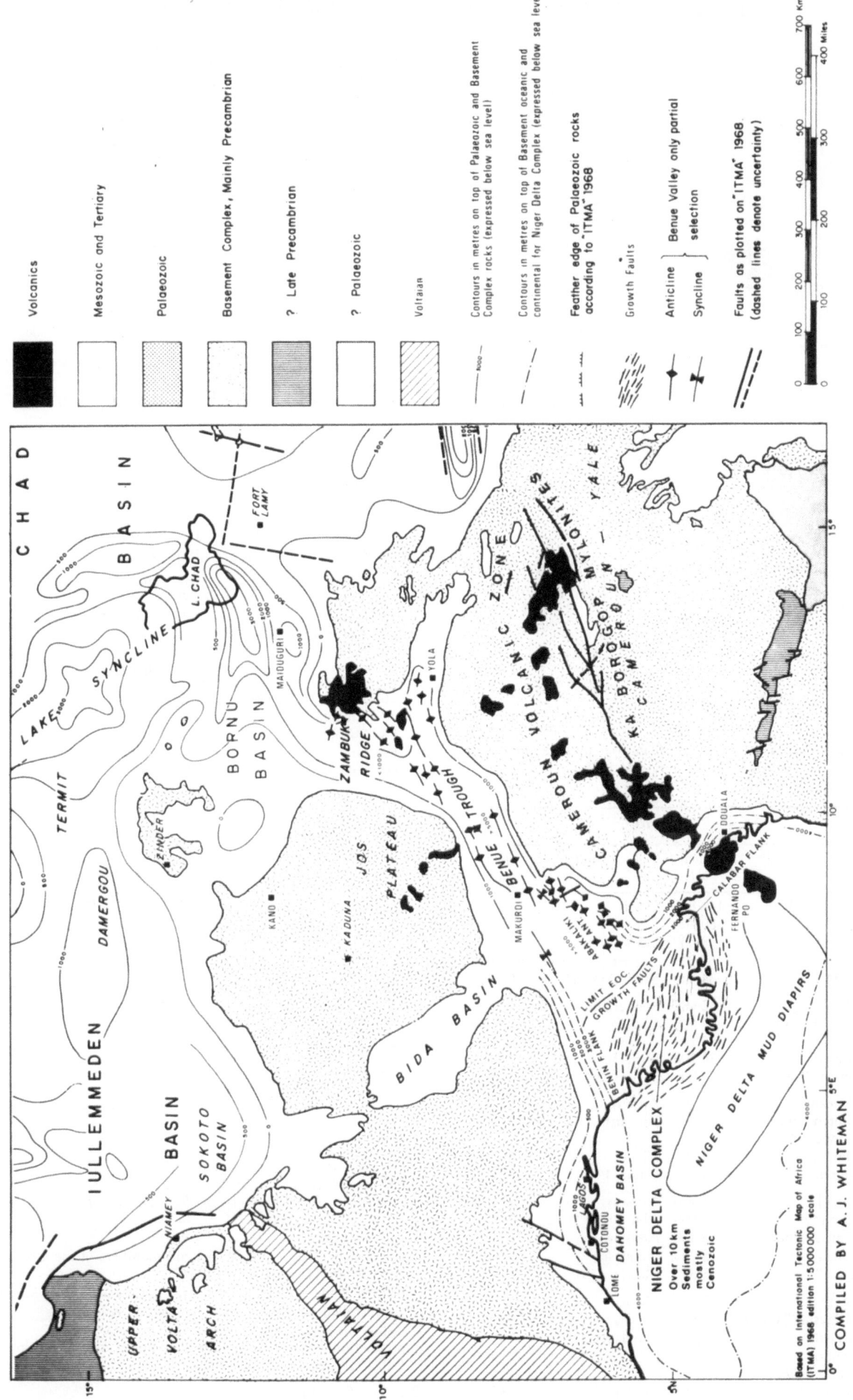

Figure 2. Regional geological map of Cameroun, Nigeria, Benin, Togo, Ghana, Chad, Niger showing distribution of main sedimentary basins, limited and generalized thickness data and selected structural features.

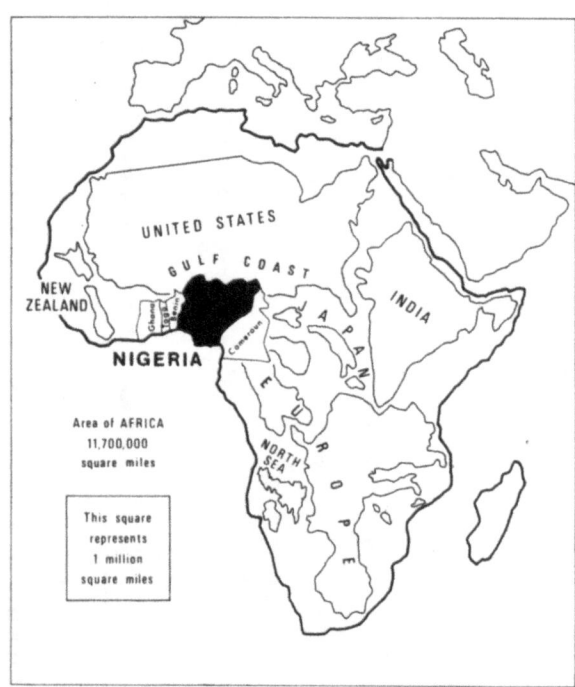

Figure 3. Sketch map showing the area of Nigeria (370 000 square miles) compared with well known geographic units and producing areas.

Figure 5

Ranking of Petroleum Producing Countries by Recoverable Proven Reserves at End of 1975

Rank	OIL[1]			GAS[1]	
	Country	million bbl		Country	billion cu ft
1	Saudi Arabia★	148 800		USSR	800 000
2	USSR	80 400		Iran★	329 500
3	Kuwait★	68 000		USA[2]	228 200
4	Iran★	64 500		Algeria★	126 000
5	Iraq★	34 300		Saudi Arabia★	103 000
6	USA[2]	32 682		Netherlands	70 000
7	Abu Dhabi[6]★	29 500		Canada	53 400
8	Libya★	26 100		United Kingdom	50 000
9	**Nigeria★**	**20 200**		**Nigeria★**	**44 300**
10	China, P.R.	20 000		Venezuela★	42 000
11	Venezuela★	17 700		Australia	32 500
12	United Kingdom	16 000		Kuwait★	31 800
13	Indonesia★	14 000		Iraq★	27 100
14	Mexico	9 500		Libya★	26 300
15	Algeria★	7 370		China, P.R.	25 000
16	Canada	7 100		Norway	25 000
17	Norway	7 000		Abu Dhabi[6]★	20 000
18	Neutral Zone[7]★	6 400		Pakistan	16 400
19	Oman	5 900		Indonesia	15 000
20	Qatar★	5 850		Malaysia (Sarawak)	15 000
21	Egypt	3 900		Mexico	12 000
22	Malaysia (Sarawak)	2 500		Bolivia	10 800
23	Argentina	2 465		Bangladesh	10 000
24	Congo (Brazzaville)	2 450		Italy	10 000
25	Ecuador★	2 450		Brunei	8 700
26	Syria	2 240		Germany, F.R.	8 300
27	Gabon[8]★	2 200		Greece	8 000
28	Brunei	2 000		Neutral Zone[7]★	7 500
29	Australia	1 700		Qatar★	7 500
30	Romania (1974)[3]	1 700		Argentina	7 200
31	Dubai[6]★	1 350		Romania (1974)[4]	6 800
32	Sharjah[6]★	1 350		Bahrain	5 500
33	Angola	1 300		France	5 300
34	Tunisia	1 065		Ecuador★	5 000
35	India	913		New Zealand	5 000
36	Brazil	780		Poland (1974)[4]	4 790
37	Peru	770		Colombia	4 000
38	Italy	700		Egypt	4 000
39	Trinidad and Tobago	700		Trinidad and Tobago	4 000
40	Colombia	556		Afghanistan	3 500
41	Zaire	500		Germany, D.R. (1974)[4]	3 400
42	Yugoslavia	375		Hungary (1974)[4]	3 040
43	Germany, F.R.	373		Gabon[8]★	2 500
44	Bahrain	312		India	2 400
45	Netherlands	250		Chile	2 300
46	Spain	250		Peru	2 260
47	Denmark	245		Oman	2 000
48	Bolivia	235		Japan	1 900
49	Hungary (1974)[4]	195		Angola	1 500
50	Chile	190		Dubai[6]★	1 500
51	Austria	167		Yugoslavia	1 500
52	Turkey	107		Tunisia	1 480
53	France	88		Sharjah[6]★	1 450
54	Afghanistan	85		Syria	1 240
55	New Zealand	75		Congo (Brazzaville)	1 000
56	Burma	70		Ireland	1 000
57	Albania (1974)[4]	66		Taiwan	1 000
58	Poland (1974)[4]	42		Brazil	927
59	Greece	40		Austria	775
60	Pakistan	26		Turkey	538
61	Japan	25		Denmark	500
62	Barbados	21		Spain	500
63	Czechoslovakia (1974)[4]	18		Bulgaria (1974)[4]	490
64	Bulgaria (1974)[4]	17		Czechoslovakia (1974)[4]	470
65	Taiwan	15		Albania (1974)[4]	430
66	Germany, D.R. (1974)[4]	12		Burma	160
67	Cuba[5]	9		Zaire	50
68	Ghana	7.5		Morocco	22
69	Israel	1.47		Israel	20
70	Thailand	0.23			
71	Morocco	0.22			
	TOTAL WORLD	658 007			2 254 742

★ OPEC Member.

1 *Oil and Gas Journal*, 29 December 1975, p. 86.
2 *Oil and Gas Journal*, 5 April 1976, p. 82.
3 *International Petroleum Encyclopedia*, 1975, p. 297.
4 *World Oil*, 15 August 1975, p. 44.
5 BFG estimate.
6 Abu Dhabi, Dubai and Sharjah are members of the United Arab Emirates which is in OPEC.
7 Neutral Zone is included in OPEC aggregates in this study. It is controlled jointly by Saudi Arabia and Kuwait.
8 Gabon became a member of OPEC in 1975, but its production and reserve figures have not been included in OPEC aggregates in this study.

Based on Grossling (1976).

very large" and reported that the Nigerian National Petroleum Corporation (NNCP) in December 1978 put Nigeria's gas "reserves" at 75×10^{12}cuf. Associated gas production is placed at 2×10^9 cuf/d with all but 1/10th of this being flared off. The export potential for gas on the face of it is enormous. Egbogah and Oronsaye (1979) put gas "reserves–resources" of Nigeria between $317–353 \times 10^{12}$cuf (Section 5.2). A selection of reserve–resource guesstimates for oil and gas is presented in Figure 294B.

Ranked in terms of recoverable proven oil reserves (to end 1975), Nigeria stood ninth on the world scale: and second after Libya for oil and gas reserves on an African scale (Figure 5). At the end of 1978, Nigeria still stood ninth on the world scale of oil producers (*Oil and Gas Journal*, December 1978). Recently World Gas Report (14th September, 1981) in a special survey dealing with Nigerian gas placed remaining proven reserves of gas at between $90 – 140 \times 10^{12}$ cuf (attributed to J. B. Owokalu, Deputy Manager,

Petroleum Inspectorate, Nigeria). This ranks Nigeria behind the USSR, USA, Algeria and Saudi Arabia and equal with Canada. The reserves are said to be split two ways in favour of non-associated gas. The Petroleum Inspectorate have suggested that there are 57.6×10^{12} cuf NAG and 30.4×10^{12} cuf AG. There is a feeling that Nigeria is encouraging independent exploration for gas in abandoned structures in proven oil producing areas, unconventional structures and traps, plays in the deep offshore and in the Cretaceous basins of Chad and the Benue Trough.

Oil production data (daily annual and cumulative) and estimated ultimate recovery from existing fields for the period 1957–1978 are presented in Figure 6A–C.

In January 1979 production exceeded 2.4×10^6 b/d, having risen from around 1.5×10^6 b/d in March 1978. Production was expected to stabilize around 2.2×10^6b/d as the Nigerian National Petroleum Corporation (NNPC) have recently ordered

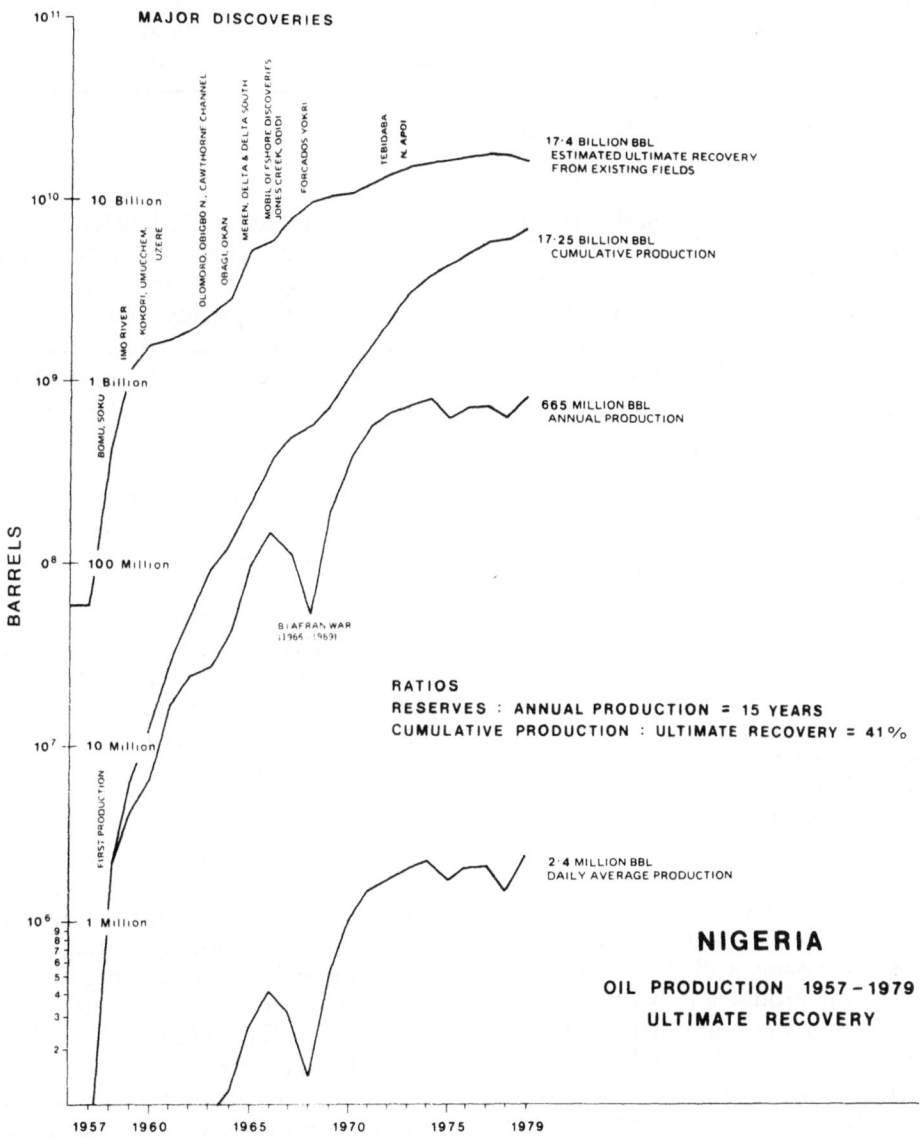

Figure 6A. Graphs (single log scale) showing oil production 1957–78 and estimated ultimate recovery in barrels. Based on Petroconsultants SA.

Figure 6B
Oil Production and (Proven and Probable) Reserves Nigeria to End 1978

Year	Daily average production (thousand bbl)	Annual production (thousand bbl)	Cumulative production (thousand bbl)	Estimated reserves (million bbl)	Ultimate recovery (million bbl)
1960	17	6 367	12 351	1 640	1 650
1961	46	16 802	29 153	1 690	1 720
1962	67	24 624	53 777	2 870	1 920
1963	76	27 613	81 391	2 300	2 400
1964	120	43 997	125 387	2 900	3 000
1965	272	99 354	224 741	5 200	5 400
1966	418	152 425	377 166	5 600	6 000
1967	325	118 481	495 648	7 500	8 000
1968	142	51 906	547 554	9 500	10 000
1969	541	197 537	745 091	10 300	11 000
1970	1 084	395 761	1 140 851	10 400	11 500
1971	1 533	559 606	1 700 458	11 200	12 900
1972	1 821	666 623	2 367 080	12 100	14 500
1973	2 050	748 363	3 115 443	12 700	15 800
1974	2 255	823 104	3 938 547	12 800	16 700
1975	1 785	651 387	4 589 934	13 000	17 600
1976	2 068	756 797	5 346 731	12 800	18 100
1977	2 905	764 547	6 111 278	12 600	18 700
1978	1 671	697 277	6 416 807	18 200	—
1979	2 416	—	7 253 493	17 200	—

Source: Petroconsultants S.A. Geneva and *Oil and Gas Journal*.

a reduction in production for "technical reasons". Average production for 1978 was put at 1.8×10^6 b/d by the *International Petroleum Encyclopedia* 1979. N.B. Actual production figures vary considerably from source to source and date of estimate or announcement. Using firmer 1977 figures:

Reserves: Annual Production, Nigeria = 16 years
Cumulative Production: Ultimate Recovery, Nigeria = 33%

Oil and Gas Journal (31 December 1979) put estimated 1979 production at 2.37×10^6 b/d so placing Nigeria sixth on the world scale of producers. At the end of August 1981 exports of crude are reported to have sagged to a low of 0.6×10^6 b/d from a reported 2.1×10^6 b/d.

In 1976 oil and gas provided about three quarters of the Nigerian Government's revenue and all but 5% of the country's export earnings. The dependence of Nigeria's economy on oil and gas is brought out by the fact that the Federal Government's revenues rose from Naira (N)1.7×10^9 in 1973–74 to N8×10^9 in 1977–78, an increase of 370% in four years (£1 sterling = N1.255 October 1979). Oil is estimated to have contributed N6.6×10^9 (10.5 $\times$ 10^9 US$) to Colonel Obasanjo's 1977–78 Budget.

In 1977 Nigerian oil production began to turn down and the economy plunged into sharp recession. Slacker world demand and a glut of low-sulphur crude coupled with Nigerian Government's misreading of the oil market were the main causes of why this came about. By March 1978 daily production was down by 32% on the previous year and stood at 1.52×10^6 bbl. In April 1978 the Federal Govern-

ment was forced to slash federal expenditure by 30% and imposed rigorous import controls as their balance of payments sagged into deficit. However a more realistic pricing policy for Nigerian crude (Bonny Light marker crude reduced from $14.61 to $14.10 bbl and reduction in other crude prices) and the Iranian oil crises stimulated overall demand and enabled oil production to climb back to 2.44×10^6 b/d in January 1979. This coupled with further OPEC price increases brought a very welcome windfall gain to the economy. Nevertheless it remains to be seen whether there will be an overall long term change in Nigeria's oil fortunes and whether the "new look" exploration policy recently announced will improve the reserve base and offset the damage to confidence done by the nationalization of British Petroleum's Nigerian interests (August 1979) and by other government activities, which currently in some people's views discriminate against foreign companies.

It has been clear for some time that oil production of around 2.0×10^6 bbl could not be far short of installed capacity and that production could not be maintained for more than a few years at high 1979 monthly rates without damaging some reservoirs and without new fields being brought on stream. If oil production is not to fall off in the long term then new exploration and production developments must be undertaken very soon.

In 1978 exploration for oil was well down, and in all but low risk-low cost areas, exploration had vir-

Figure 4. Map showing generalized physiography, locations of selected towns, main rivers.

Additional material from *Nigeria: Its Petroleum Geology, Resources and Potential*
ISBN 978-94-009-7363-3 (978-94-009-7363-3_OSFO1),
is available at http://extras.springer.com

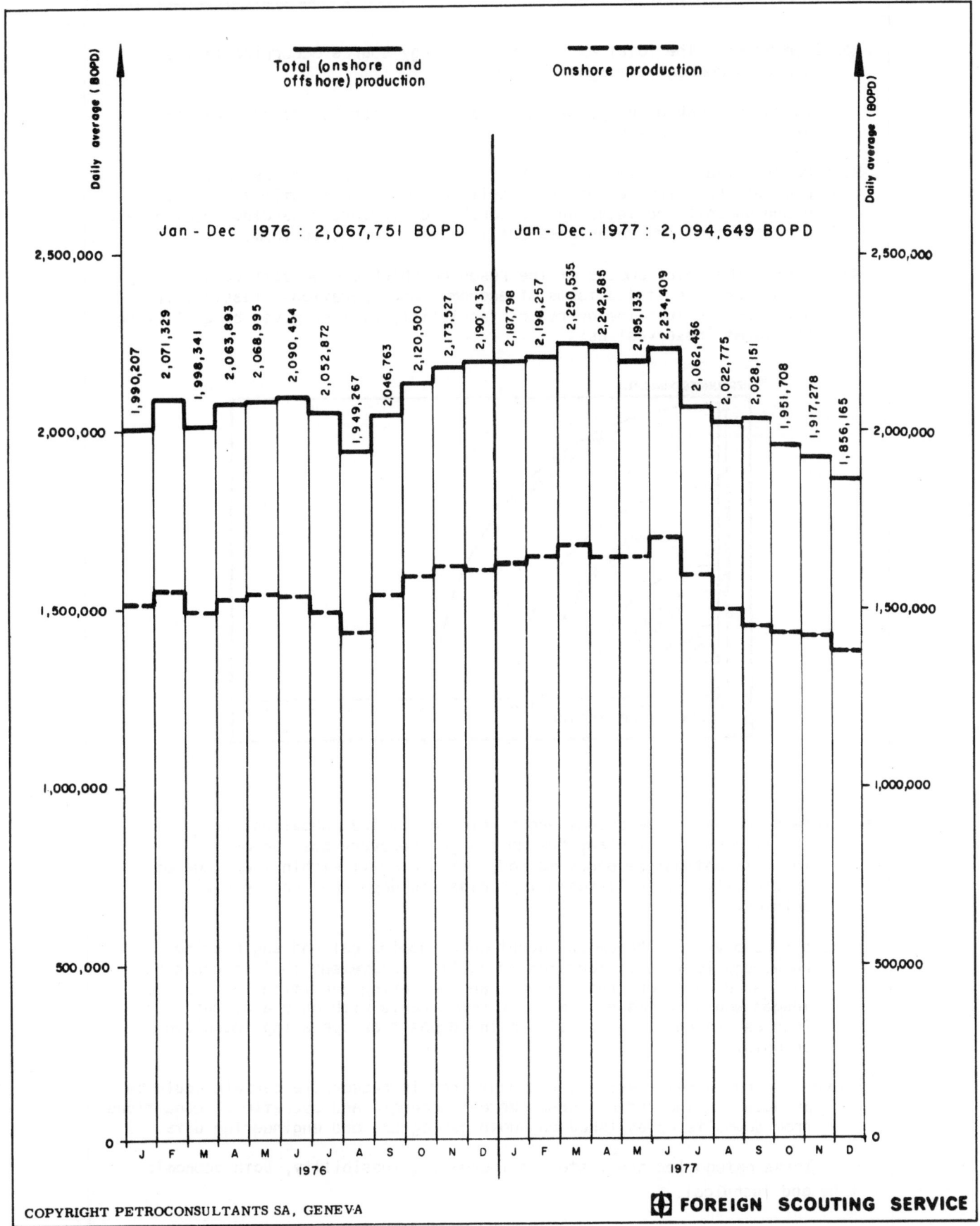

Figure 6C. Bar graphs showing crude oil production 1976–79, Nigeria. Based on Petroconsultants SA.

Resource base: The total amount of the energy source occurring in the world in commonly recognizable form.

In the renewable energy resources a time factor has to be added to allow quantification.

Resources: The total amount of the resource base which is estimated to be probably recoverable for the benefit of man. This estimate will be based on both knowledge and reasonable conjecture regarding location and probable recovery techniques but is a very imprecise term.

Reserves: The total amount of the resource which can be defined as recoverable in stated terms of economic and operational feasibility. Whenever possible the degree of feasibility will be given by qualifying the terms 'reserves' as:

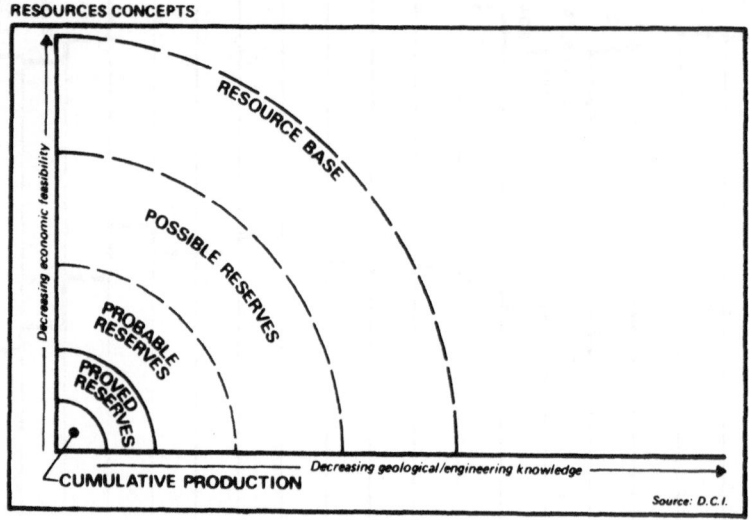

Possible reserves: The amount about which geological knowledge is insufficient to give any but most vague recovery costing or indicate optimum recovery method, yet are still within the range of possibility. This is again imprecise and dependent on individual opinion.

Probable reserves: The amount about which geological and engineering knowledge is insufficient for an explicit statement that it could be recovered under current economic and operating conditions but can be judged would become economically recoverable with only a slight increase in knowledge of either the deposit or operating techniques or both.

Proved or Proven reserves: The amount that is reasonable certain could be produced in the future under current economic and operational conditions from deposits established on known geological and engineering data.

These categories are listed in increasing feasibility, both economic and technical.

Based on: D.C. ION 1975   Figure 6D

Figure 6D. Diagram showing reserve-resource concepts. Based on Ion 1975.

tually stopped. In February 1977, there were 13 drilling rigs operating in Nigeria compared with 28 in April 1975 and in 1977 more exploratory wells were drilled in adjacent Cameroun (32) than in Nigeria (24), although the position had reversed by end 1978 (Cameroun 21, Nigeria 35 Exploratory Wells).

In examining a possible build up of reserves and production as a result of new exploration it should not be forgotten that Nigeria's average oil productivity per well for the Niger Delta Province was 1500 bbl for 1975 (cf. 16.46 bbl/well USA; 15 730 bbl/well Iraq; 1653 bbl/well Libya and 9226 bbl/well for Saudi Arabia) and that up to 1975 some 23 250 000 ft of hole had been drilled to achieve the current reserve/production situation (Figure 6). By international standards Nigeria's delta complex oil fields on average are small; billion barrel reserve fields are unknown in Nigeria (see Halbouty 1970; *Oil and Gas Journal* August 1977; "Petroleum 2000", *Oil and Gas Journal* 1978) and, unless the discovery size pattern changes dramatically, reserves will continue to decline rapidly at the present high rates of production demanded by both the external and internal aspects of the Nigerian economy. The contribution that giant oilfields (i.e. fields with an ultimate recovery of 500 million barrels or more) make to total recovery is the smallest for any major producing country. (See Section 5.2.)

By 1988 internal demand could rise to as much as 500 000 b/d according to NNPC and, without further increases in reserves, this would have to be met at the expense of crude exports and loss of revenue. The seriousness of the situation can be seen from the fact that in 1976 oil and gas provided about three quarters of the Nigerian Government's revenue and all but 5% of the country's export earnings and that the ratio Cumulative Production : Ultimate Recovery changed from 33% in 1977 to 41% in 1979.

Answers to the exploration and production problems posed above not only have internal importance but have considerable external importance, the more so because the biggest single buyer of Nigerian crude is the apparently insatiable United States of America which imports a large part of Nigerian production; next comes the Nederlands which not only supplies its own needs but feeds into the EEC hinterland and then Britain whose purchases have decreased as North Sea oil came on stream and gas finds developed. The politico–economic consequences of failure to stimulate new exploration and expand production could have long lasting effects on the Nigerian economy and Nigeria's position of political influence in Black Africa could change fundamentally. Without increasing oil and gas exports those economists who have called Nigeria the "Brazil of Africa" could easily be proved wrong.

General assessments of hydrocarbon potential for each Nigerian basin are given by formation or major grouping in Chapter 2 which deals with Stratigraphy and in Section 5.3 where they are given in summary form. Unfortunately the assessments are brief and in some respects are cursory; in most cases they cannot be otherwise with data available publicly. Details of the current concession and well situation are dealt with in Chapter 7.

The stratigraphy and structure of Nigeria's six major sedimentary "basins" are described systematically in Chapters 2 and 3, and data is marshalled largely from the point of view of the petroleum explorationist. Stratigraphy is dealt with under regional headings in Chapter 2, as is Structure in Chapter 3. The book does not attempt to provide an all embracing account of the stratigraphy and structure of Nigeria and adjacent regions; this is the purview of the forthcoming edition of the *International Stratigraphic Lexicon* and future publications of the Geological Survey of Nigeria and other studies. Topics such as the Basement Complex, which consists largely of crystalline and metamorphic non-productive rocks, are dealt with briefly and systematic formational descriptions, synonyms etc. are dealt with in summary form only in Section 2.2. Palaeontology is not discussed in any detail, as this has been dealt with in numerous publications by Reyment (1965 etc.), Adegoke (1969 etc.), Fayose (1970 etc.) and others (See References, Chapter 8).

Plate tectonic studies, important for basin evaluation work, are dealt with largely in Chapter 3 Structure.

Chapter 4 is concerned with Oil and Gas Occurrences in the Cenozoic Niger Delta Complex; Source, Reservoir, Trap, Migration data and Case Histories. Chapter 5 contains Production and Reserve data and general comments on Prospects; and Chapter 6 is a chronological review of selected exploration and production developments of Nigeria's petroleum industry 1908–1979. Concession maps and Exploration data are presented in Chapter 7.

Basic geological information for this book was collected from 1960 to 1968 when the writer worked in West Africa utilizing Ford Foundation and Rockefeller Foundation research grants; from 1968 to 1972 when he was Oil Producer's Trade Section, Lagos Chamber of Commerce and Industry Professor of Petroleum Geology at Ibadan University, Nigeria; and from 1972 to 1980 when engaged on research and consulting assignments in West Africa for international agencies and oil companies. The book is based mainly on published or readily available non-proprietary information but it also incorporates information from an "open report" (prepared 1971–72) entitled:

Whiteman, A. J. 1973
Geology and Hydrocarbon Prospects of Nigeria
Vols. 1 & 2 pp. 247, Figs. 1–96,
Tables 1–38 Exploration Consultants Ltd,
Marlowe, England

A great deal of proprietary and confidential data exists in files of oil companies, service companies, consultants, geological surveys, international agencies and of official government departments responsible for the petroleum affairs of Nigeria and adjacent countries but obviously much of this data is not available for inclusion in a book of this kind, nor is it likely that much of it will be made available in the near future because a considerable part of this proprietary data still has commercial and strategic value.

However much of this type of information is highly specific and deals with detailed local situations and, if made available, really would not materially alter some of the conclusions and generalizations included in regional studies, because it should not be forgotten that overviews already published by Short and Stauble (1967); Frankl and Cordry (1967) Murat (1972); Evamy *et al.* (1978) and others have deep roots in detailed proprietary information in company files dealing with well logs, reports, seismic investigations etc. Obviously there are areas of knowledge where the release of oil and service company and government held data would have significant impact and improve our understanding of Nigerian Petroleum Geology and Hydrocarbon Prospects and there are people in oil companies and in government who are better placed to write a more comprehensive study than the one made here.

Selected references collected up to the first quarter of 1980 are presented in Chapter 8. Much of this material is on file in the Petroleum Exploration Studies Library at Aberdeen University, Scotland.

Acknowledgements

I should like to thank Mr J. B. Fulton, Draughtsman, Petroleum Exploration Studies Group, Aberdeen University, Scotland, who with great skill and patience either drew or prepared for presentation the maps or diagrams used in this book. PW Design, Maidstone prepared many of the original diagrams used in the Exploration Consultants Ltd Open Report referred to above (Whiteman 1973), some of these are used in this book. Diagrams from a variety of other sources have been used also. This has resulted in a heterogeneous standard of presentation. To redraw many of these diagrams was beyond our resources and the decision to include them in untouched or slightly modified form is justified in that the book gains in coverage and interest, if not in some cases in artistic presentation and uniformity.

In a compilation and assessment study of this type one naturally has to draw heavily on observations, ideas and conclusions made by those who have gone before, and I trust that due acknowledgement has been made in the text and references. Sources and credits used in constructing maps or diagrams, are given in the captions to illustrations or on the body of maps.

I should like to acknowledge discussions I had with colleagues at the University of Ibadan, Nigeria especially Professor Kevin Burke, Dr Joop Dessauvagie, Dr Heinz Dieter Ludwig, Dr L. H. Schatzl and Professor M. O. Oyawoye; with Professor A. S. Adegoke of the University of Ife; with Dr Anilo Ajakaiye of the University of Zaria, Nigeria; and colleagues in oil and service companies in Nigeria, especially Dr Roland Murat and Mr Graham Davies then with Shell-BP Nigeria; Messrs Bob Johnston, Al Habarta and Lyn Hazel of Gulf Oil Nigeria, Stan Pearson of Gulf Eastern; Art Hawley of Great Basins Petroleum and many others; and with Chief M.O. Feyide who was then Chief Petroleum Engineer, Federal Ministry of Mines and Power.

2

STRATIGRAPHY

2.1 GENERAL GEOLOGY AND BASIN SUMMARIES

The oldest sedimentary rocks dated at surface in Nigeria (excluding the Basement Complex) and of direct interest to the hydrocarbon prospector are Early Cretaceous (Albian) in age and the youngest sediments are the Present Day Niger Delta Complex deposits which are being laid down at the present day by the Niger–Benue System distributaries in southern coastal Nigeria. Sedimentary rocks of Barremian and Neocomian (Early Cretaceous) age have been reported (Murat 1970 and others) but palaeontological proof of age has not been presented.

Sedimentary rocks and sediments are disposed in six major basins which onshore occupy about 178 000 square miles, roughly half the area of onshore Nigeria. These basins are briefly described in this section. Oil and gas have been produced from the Niger Delta Basin and only shows or shut in wells are known from other basins. Four of these basins are of prime interest to explorers for oil and gas. The main Nigerian sedimentary basins (Figure 7) include those described in the following subsections.

1. **Abakaliki and Benue Troughs:** These troughs contain folded and unfolded mainly Cretaceous sediments which were deposited in two Late Mesozoic "Failed Arms" or Rift Troughs which formed the third arm of the Niger Delta Triple Junction. This junction may have had its centre within a triangle formed by a line joining Warri, Port Harcourt and the subaerial delta nose (Figure 7). The Benue Arm also formed the third arm of the Chum Trilete Junction (rrr) consisting of the Benue, Yola and Gongola arm (Figure 7). More than 15 000 ft of the pre-Middle Santonian sediments were deposited in various parts of the Abakaliki and Benue Troughs. Parts of these sediments were folded, intruded, mineralized (Pb–Zn) and some were metamorphosed, especially

in the Abakaliki Fold Belt. Because of this, sediments in parts of the Abakaliki and Benue Troughs are thought to be poor prospects for oil and gas because high geothermal gradients must have prevailed above the zone of modified continental lithosphere and crust which appears to underlie the Benue Trough and above the zone of narrow "oceanic" crust which may occur beneath the Abakaliki Trough. Areas of prospective interest, especially for gas, may remain in the Benue Trough marginal basins (Figures 169 and 170). In post-Santonian basins marginal to the Abakaliki and Benue Troughs such as the Anambra, Wase-Gombe and Wukari – Mutum Biya basins, unfolded or gently tilted, mainly Cretaceous post-Santonian sediments accumulated to great thickness. In the Anambra Basin such sediments exceed 30 000 ft and may exceed 15 000 ft in other basins marginal to the fold belts. Prospects in pre-Santonian rocks are limited but good prospects probably exist for gas in post-Santonian basins, e.g. in the Anambra Basin (Figures 2, 7 etc.).

2. **Dahomey or Benin Basin:** This basin forms the onshore part of the West African Miogeocline in Eastern Ghana, Togo, Benin and Western Nigeria. Sediments encountered in the basin onshore at surface and in wells range in age from Late Cretaceous to Recent and exceed 7000 ft at the coast in Nigeria. They thicken markedly into the offshore and then thin down beneath the deep water area (Figures 234, 235 etc). Onshore basin fill thins westward into the Volta Delta region of Ghana and eastwards and northwards onto the Basement Complex of the Okitipupa High and the metamorphic Basement Complex of Western Nigeria (Figures 2 and 7). Fair to good prospects exist in the Dahomey Miogeocline which is bounded on its seaward side by, and in parts overlaps, the Romanche Fracture Zone. This is essentially a rift margin basin, as in the Abidjan Basin to the west; in contrast to the Ghana offshore basin (west of the Volta Delta) which is a basin whose boundaries parallel the Romanche Fracture Zone.

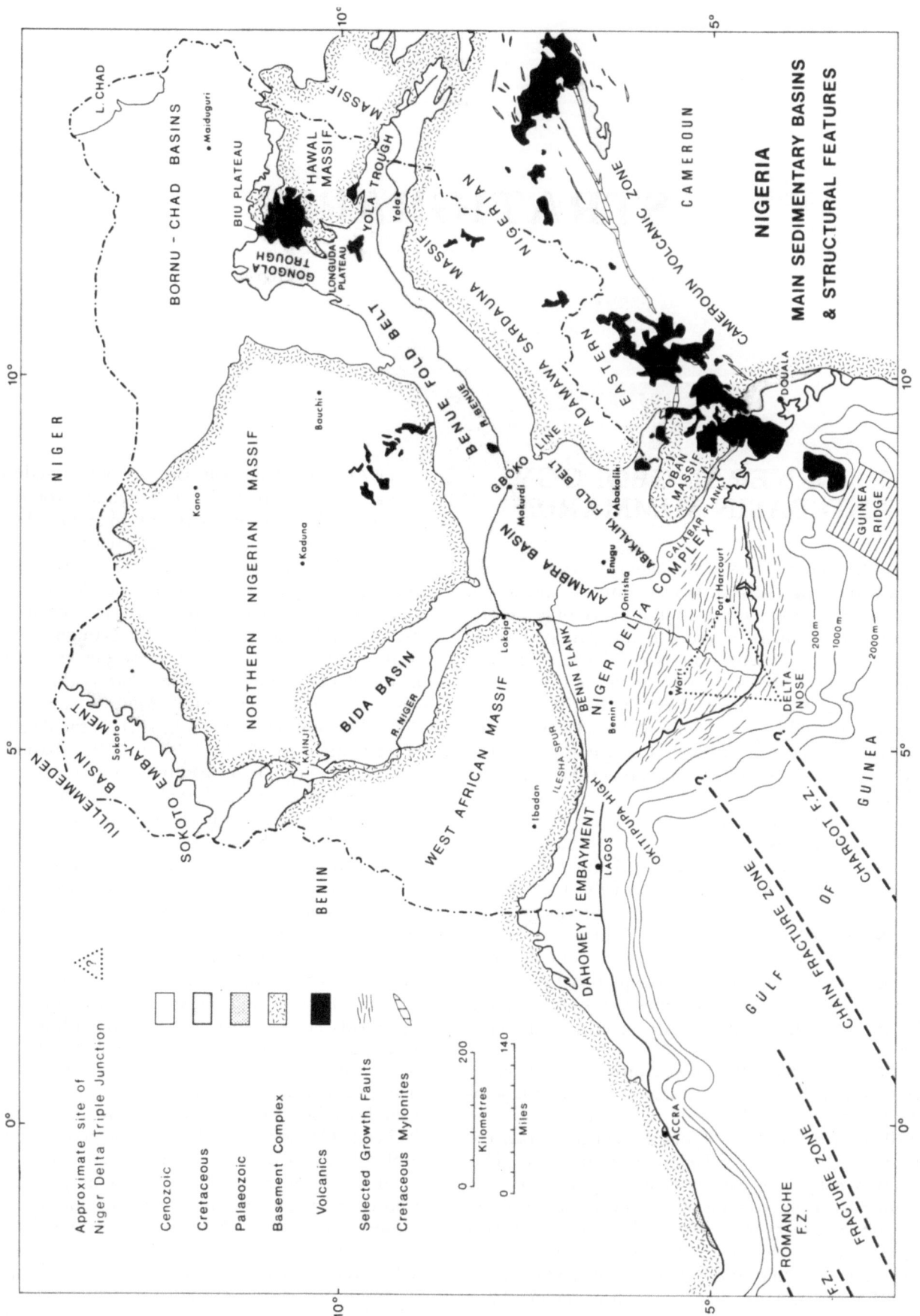

Figure 7. General geological map showing main sedimentary basins of Nigeria, selected growth faults and general structural data.

The small Seme Field is under development in the Benin section of the Dahomey Basin.

3. **Niger Delta Complex:** Consists of Cenozoic formations deposited in a high energy constructive deltaic environment and differentiated into continental Benin, paralic Agbada and pro-delta marine Akata facies. Early Eocene delta complexes formed in three separate depocentres and elongate delta systems may have existed. By Oligocene and certainly by Miocene times the deltas had prograded and united to form a high energy constructive arcuate – lobate delta system which then rapidly prograded into deep water and extended onto transitional continental – oceanic crust. By Oligocene – Miocene times the delta complex had prograded onto oceanic crust generated as the African and South American lithospheric plates had spread apart on the Gulf of Guinea and South Atlantic Ridge – Transform Systems. Sediments may well exceed 35 000 ft beneath the upper and lower deltaic plains, located generally over the triple junction and the Abakaliki "failed arm". Growth faults and extensive mud diapirism disturb the deep water marine and paralic successions and indeed diapiric bulges and oceanward directed mass movements (Delta Toe Thrust Zone) of Akata mud/shale have enabled successive deltas to prograde since Oligocene time. Without the large scale diapirism and the delta toe thrusts the delta complex could not have developed as it has (Figures 2, 7, 246). Nearly all Nigerian oil and gas produced to date comes from Agbada sand reservoirs and is located mainly in rollover anticlines and related structures situated high in the delta prism. Because of the small number of growth faults remaining to be discovered in the Niger Delta Complex and because good Agbada reservoir sands are largely restricted to within a belt roughly 25 miles from the Present Shore and within the Plio-Pleistocene Shore, undiscovered petroleum resources are probably largely limited to within the onshore-shallow offshore part of the Niger Delta Complex; although large undiscovered resources of gas may exist in both the onshore and offshore. "Official" reserve estimates of 41.4×10^{12} cuf proven (*Oil and Gas Journal* 31 December 1979) may well be out by a factor greater than $\times$ 3 or 4 or even more. Reserve data for individual fields have not been published systematically by governmental agencies in Nigeria (See Chapter 5). "Unconventional" gas resources in overpressured zones (i.e. methane dissolved in water) must be very large indeed but such resources must be classed as uneconomic at the present time (Figure 6).

4. **Bida or Middle Niger Embayment or Nupe Basin:** Contains mainly Upper Cretaceous (Maestrichtian) rocks near surface. Sedimentary fill has been put at 3000 – 6500 ft thick using gravity data (Ojo and Ajakaiye 1976) or over 10 000 ft thick using aeromagnetic data (Great Basins Petroleum, A. Hawley, Personal Communication). Prospects are considered to be poor in this basin, and if hydrocarbons are present they may well be gas (Figures 208–217).

5. **Sokoto "Basin" or Embayment, Southern Part of the Iullemmeden Basin:** Contains around 4000 ft of Upper Cretaceous – Lower Eocene – ? Pleistocene sediments which thin southward onto the Basement Complex Massif centred on Kaduna. Prospects are said to be poor in the Sokoto "Basin" of Nigeria. Greater sediment thicknesses occur to the north in Niger where prospects are thought to improve (Figures 208–217).

6. **Bornu Basin and adjacent parts of the Southern Chad Basin:** The Bornu Basin appears to consist of a "two horned" basin at its western end (based on a published reconnaissance gravity interpretation). The basin appears to extend eastwards into the Lake Chad area where Cenozoic and Mesozoic rocks are thought to exceed 10 000 ft (gravity based estimate) in a well developed but poorly described series of pre-Late Cretaceous rift troughs. Oil has been discovered by Conoco in Mesozoic "Continental Intercalaire" sediments in wells in Chad so increasing prospective interest in adjacent areas of Northern Nigeria in the Bornu Basin. Access to tide water and political instability limits interest in these prospects. The distributions of these basins are shown in Figures 2 and 7.

Cretaceous and Cenozoic rocks in Nigeria were laid down within two separate and distinctive tectonic frameworks:

1. Pre-Santonian Tectonic Framework; and
2. Post-Santonian Tectonic Framework.

Maps portraying the Albian (Early Cretaceous), Santonian (Late Cretaceous) megatectonic framework and the Campanian – Maestrichtian – Early Cenozoic megatectonic framework are presented in Figures 22, 153 and 200. Cretaceous and Early Cenozoic seas transgressed and regressed within these frameworks and on their flanks. Sediments were laid down in a complex set of environments and are thickest within the Abakaliki and Benue Troughs.

The Niger Delta Complex constitutes a separate deltaic unit which has developed since Eocene time when sediments over-spread the Cretaceous continental rift margin which had evolved as the Atlantic Ocean Basin extended. The distribution of these Cenozoic rocks and associated Cretaceous sediments and the complex patterns of stratigraphy which evolved are portrayed cartoon-like in Figures 25, 28, 29, 32, 33, 34, 57, 59.

Unfortunately, an up to date official list of stratigraphic units in use in Nigeria has not been published by the Geological Survey of Nigeria. The list included in the Stratigraphic Lexicon (Furon 1956a) is outdated but is in process of being revised. However, an "unofficial" list of stratigraphic units in current use in Nigeria is available and is presented in Figure 8. This was compiled by Dr T. F. J. Dessauvagie, and in part by the writer, to accompany the black and white Geological Map of Nigeria Scale 1 : 1 000 000 made by Dessauvagie in 1972 (in Whiteman 1973) and published in colour by the Nigerian Mining, Geological and Metallurgical Society in

STRATIGRAPHIC UNITS USED IN THIS AC

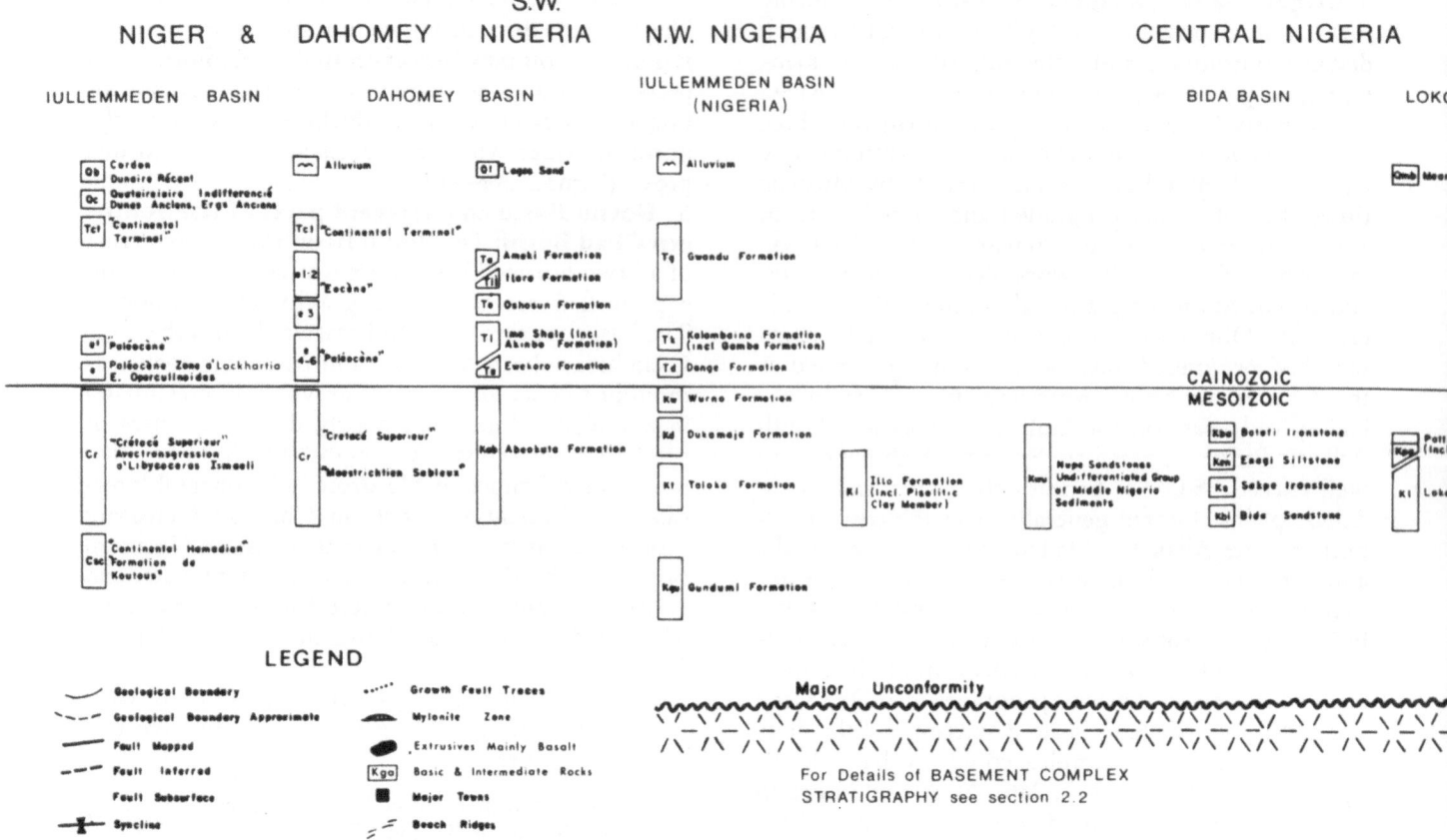

Figure 8. List of stratigraphic units used in this account for Nigeria. Compiled by Dessauvagie and Whiteman (1972). A similar list is included on Dessauvagie's (1974) 1 : 1 000 000 Scale Geological Map of Nigeria.

1974. Formational details are given for some formations in the *International Stratigraphic Lexicon* (Furon 1956b). Short and Stauble (1965) designated and described the subsurface Akata, Agbada and Benin formations etc. of the Niger Delta Complex.

The list presented in Figure 8 was compiled using maps and memoirs of the Geological Survey of Nigeria, the *International Stratigraphic Lexicon* for Nigeria (Furon 1956b); the Shell-BP Geological Map of the Southern Nigerian Sedimentary Basin (Murat 1972); Reyment and Barber (1956 etc.); Dessauvagie, Burke and Whiteman (1971, 1972); Whiteman (1973); official maps and memoirs of the geological surveys of adjacent countries and memoirs of the Bureau de Recherches Géologiques et Minière (BRGM), France; and Dessauvagie (MS 1973 and 1974). Dessauvagie (1974) provided short systematic descriptions, synonymies etc. of lithostratigraphic units to accompany the map. A map showing the distribution of major formational units is presented in Figure 9. It is based on the black and white version of Dessauvagie's 1 : 1 000 000 Geological Map of Nigeria (1972), a coloured version of which has now been published. For simplicity Basement Complex formational units have been omitted on Figure 9. The general distribution of Basement Complex formations and structural trends is shown in Figures 11–15 and discussed in Section 2.2. Mesozoic and Cenozoic sedimentary units are described below (Sections 2.4 and 2.5), ordered first on a regional basis and second systematically. The stratigraphic account dealing with the sedimentary rocks is first subdivided into:

2. POST SANTONIAN UPPER CRETACEOUS
   ROCKS
   ——————UNCONFORMITY——————
   FOLDING, INTRUSION, MINERALIZATION AND
   METAMORPHISM
 1. PRE-SANTONIAN LOWER AND UPPER CRE-
   TACEOUS ROCKS.

2.2 BASEMENT COMPLEX

Basement Complex rocks are normally of little direct interest to the petroleum geologist, but they hold some interest in that they constitute the container for the various basin fills; because they have acted as source areas for sediments now found as rocks in basins in which hydrocarbons occur; and because deep seated structural phenomena affecting the Base-

JNT FOR NIGERIA AND ADJACENT AREAS

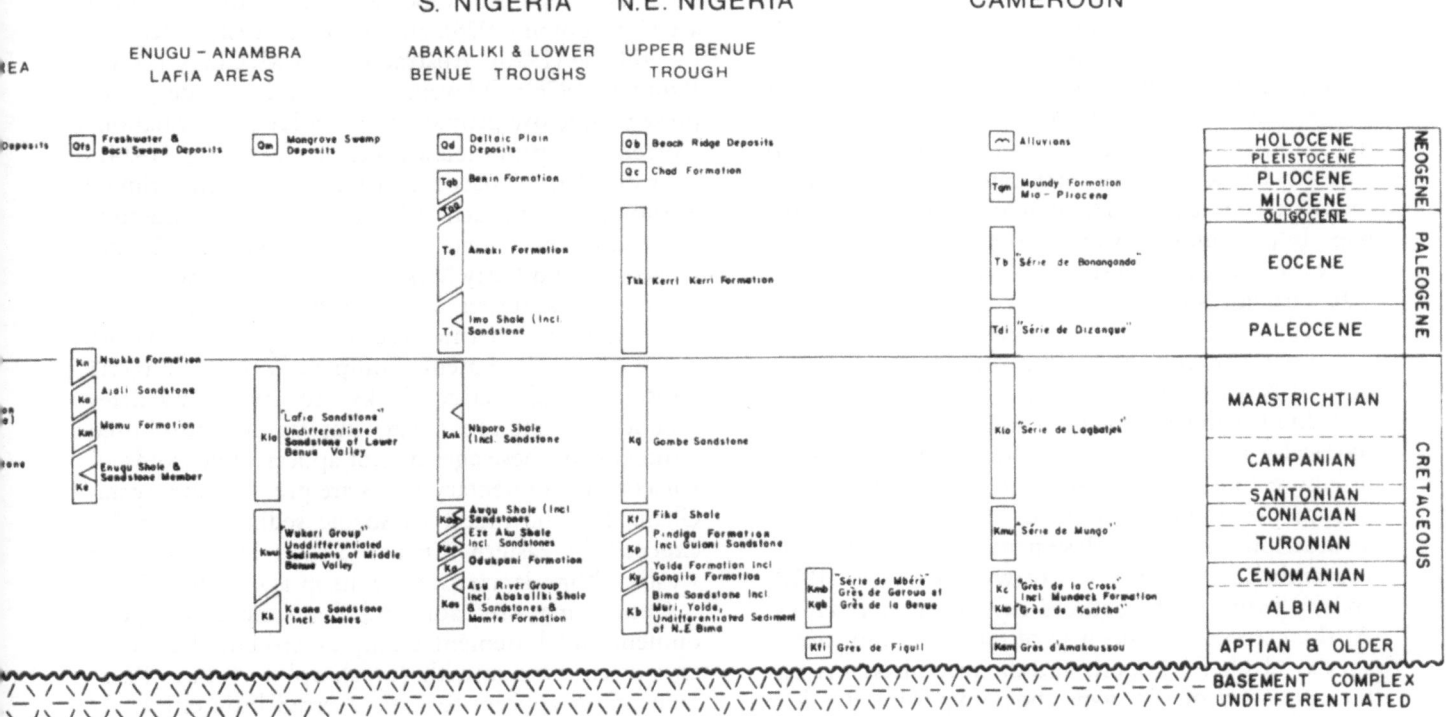

ment Complex and the lithosphere as a whole have exerted strong broad architectural depositional controls on the evolution of the six sedimentary basins which developed in Nigeria in Late Mesozoic and Cenozoic time.

Basement Complex igneous and metamorphic rocks, mainly of Precambrian age, cover about 50% of Nigeria, i.e. approximately 178 000 square miles (Figures 2 and 8). The petroleum prospects of these vast areas are nil and therefore some 50% of the surface area of Nigeria can be written off for hydrocarbon prospecting on first principles.

The main areas of outcrop and general characteristics of the Basement Complex are described below. Classifications, distributions and trends are shown in Figures 10–15. The largest area of Basement Complex is centred on Zaria and consists mainly of igneous and metamorphic rocks of Precambrian age (Figures 9 and 15). Elevations in this region exceed 1640 ft. Around Jos, elevations exceed 3280 ft (Figure 4) where thermal uplift has taken place related to the intrusion of Jurassic Ring Complexes and to young volcanic activity (Figures 15 and 16).

Igneous and metamorphic rocks in Western Nigeria are separated from the northern area by continuous outcrops of sedimentary rocks in the Bida Basin (Middle Niger Embayment) and in the Sokoto Basin (Figure 2). This area of Basement Complex rocks is lower on average than the northern one and

large areas do not exceed 1640 ft and lie between 984 and 1640 ft. The highest outcrops of Basement Complex in this area are situated southeast of Offa, in Kwara State where elevations exceed 1640 ft (Figures 4, 7 and 9). The area extends westwards into Benin, Togo and Ghana. The third main area of Basement Complex extends from the Calabar Flank of the Niger Delta Complex to Sardauna where it disappears beneath the sediments of the Chad Basin. It is broken across by the Mamfé and Yola Depressions which contain Cretaceous and Cenozoic sediments. The Basement outcrops continue into Cameroun and into Adamawa and Sardauna in Nigeria where there are extensive areas between 3280 and 6560 ft.

All these areas, because of uplift in the past and because of active uplift in eastern Nigeria and adjacent Cameroun have acted, or are acting, as important sources of clastic sediments. This is especially true of the Sardauna, Adamawa and Western Cameroun areas of Basement Complex. These areas provided the enormous quanities of sediment which were swept down the proto-Benue and Cross River drainage systems and which now form part of the Niger Delta Complex (Burke and Whiteman 1973). The position of the present day Benue marginal escarpment is a measure of the extent to which the Basement Complex formations have been eroded. Great pediplains, escarpments and inselbergs cut into Basement formations are witnesses to the vast

15

amount of erosion which has taken place. In sediment volume and source terms the Niger Delta Complex would have been better named the Benue Delta Complex.

Oyawoye (1972), Rahman (1972, 1971 and 1976), McCurry (1976) and Ajibade (1976) have provided reviews dealing with Basement Complex rocks of Nigeria, although very little has been written about the basement Complex of Sardauna, Adamawa and adjacent Cameroun, the main Niger Delta Complex sediment source area. The Nigerian Basement Complex is predominantly gneissose and granitic, but there are extensive schist belts west of a line trending 020° through Kaduna. Interesting areas of rocks occur such as charnockites and highly potassic syenites. Oyawoye (1972) in reviewing the Basement Complex of Nigeria followed Hockey and others (1966) and divided the Basement Complex into four lithological groups (Figure 10). The classification given in Figure 10 is purely a lithological one and an order of superposition is not implied. Detailed petrographic descriptions of the Basement Complex (mainly Precambrian) are given in Oyawoye's review (1972) and by many of the authors cited in his references. At the present time because of the few radio-metric ages available for Nigeria and the reconnaissance nature of many of the maps, poor exposure etc. it is difficult to erect a comprehensive order of superposition and a sequence of events for the Basement Complex, although Burke and Dewey (1972) and Burke and Whiteman (1973) attempted to build up plate pictures. The oldest rocks proven in Nigeria appear to be the banded gneisses which were intruded by the Ibadan granite gneiss (Grant *et al.* 1969).

Figure 10

Classification of the Basement Complex Rocks of Nigeria Following Hockey and Others (1965) and Oyawoye (1972)

1. OLDER GRANITES
   (a) Porphyroblastic Older Granite
   (b) Granite Gneiss or Gneissose Granites

2. MIGMATIC COMPLEX
   (a) Banded Gneisses
   (b) Migmatic Gneisses
   (c) Transition Gneisses
   (d) Augen Gneisses
   (e) Pegmatites

3. METASEDIMENTARY "SERIES"
   (a) Metasedimentary schists
   (b) Quartzites and quartz schist
   (c) Marble
   (d) Calc-silicates
   (e) Metaconglomerate
   (f) Amphibolite
   (g) Metamorphic Iron beds

4. MISCELLANEOUS ROCK TYPES
   (a) Bauchites
   (b) Charnockites and pyroxene diorites
   (c) Gabbro and metagabbro
   (d) Epidiorite
   (e) Potassic syenite
   (f) Dolerite and other dyke rocks

The Basement Complex rocks: of Northern Nigeria were reviewed by McCurry (1976); of Southwestern Nigeria by Rahman (1976 etc.); of the Mid-West region of Obeyemi (1976); of the Bauchi area by Eborall (1976); and of Northwestern Nigeria by Ajibade (1976). Abstracts from McCurry (1976); Rahman (1976); Odeyemi (1976) and Ajibade (1976) presented below provide basic regional information. Correlations proposed are shown in Figures 11–14.

From Early Palaeozoic to Late Mesozoic times, erosion affected much of Nigeria and over thousands of square miles, extending as far west as Ghana, Palaeozoic and Early Mesozoic rocks are now absent. In Late Precambrian and Early Palaeozoic times the Pan African Thermo-Tectonic Episode affected many of the Basement Complex rocks in Nigeria resetting "radio-active clocks" to give many dates around 480 million years, but unmetamorphosed sediments of these ages do not appear to be exposed. Palaeozoic sedimentary rocks are present near Accra, Ghana but the oldest Mesozoic sedimentary rocks exposed in eastern Ghana, Togo, Dahomey and western Nigeria are Cretaceous in age (Figure 2).

Concerning Mesozoic and Cenozoic basin development and Basement Complex structural control, as is often the case with "break up" tectonics, Basement grain appears to have exerted little effect. As is visible in Figure 154 the continental margin flexures of the Dahomey Miogeocline cut charply across the predominantly meridional trends of Nigeria, Benin, Togo and Ghana. The sub-surface Okitipupa Ridge, related to the trace of the Chain Fracture Zone, which limits the Niger Delta Complex along the Benin Flank, cuts obliquely across the Basement Complex Trend. On the eastern side of the Niger Delta Complex on the Calabar Flank the continental margin flexure, again cuts sharply across the Basement trend of the Oban Hills (Figures 2 and 9). The Benue Depression cuts obliquely across the trend of the Late Precambrian–Early Palaeozoic Pan African Mobile Belt, as do the Yola and Gongola Troughs, the Bida Basin and the Mamfé Embayment. The Benue Trough also cuts across Jos Plateau–Aïr Newer Granite (Jurassic) trends and the margins of the Bornu and Sokoto Basins truncate Basement trends. As in the Kenyan and Tanzanian sections of the Karroo and Cenozoic Eastern African Rift Systems, Basement structural trends have certainly exerted some control in the development of marginal downwarps and faults bounding the troughs and basins but these are mainly local and the overall effect is that the rift troughs and basins cross cut Basement grain.

Abstracts – Basement Complex Papers, Nigeria. From Kogbe, C.A. (Ed.) 1976:

The Geology of the Precambrian to Lower Palaeozoic rocks of Northern Nigeria – A review (P. McCurry 1976)

Abstract

Northern Nigeria is underlain by gneisses, migmatites and metasediments of Precambrian age which have been intruded by a series of granitic rocks of late Precambrian to lower Palaeozoic age. The oldest rocks are represented by a series of Older Metasediments and gneisses believed to be of Birrimian age and older. These rocks have been variably metamorphosed and granitized through at least two tectono-metamorphic cycles so that they have been largely converted to migmatites and granite-gneiss. Younger Metasediments, believed to be upper Proterozoic in age, were deposited on this granitized basement and folded along with it during the Pan-African orogeny. They are of low metamorphic grade and are now represented as synclinal troughs among older rocks in north-west Nigeria. Intrusive into both the basement rocks and the younger supra-crustal cover is a series of basic, intermediate and acid plutonic rocks known as the Older Granites. The youngest rocks in the area belong to a suite of volcanic rocks intruded into Older Granite bodies during lower Palaeozoic epeirogenic uplift following the Pan-African orogeny.

At least two phases of tight isoclinal folding have been recognized in both the Younger Metasediments and the basement gneisses. These deformational episodes were accompanied by progressive regional metamorphism, and separated and followed by phases of static metamorphism. P-T conditions remained essentially constant throughout both deformations. Accompanying migmatization and granitization of the basement gneisses resulted in intrusion of a suite

Figure 11

Classification of Basement Complex Stratigraphic Units and Events, According to Rahaman (Unpublished Thesis 1972)

Events	Bauchi and Western Railway	Niger, Sokoto and Zaria Area	Niger, Sokoto and Zaria Areas	Nigeria General	South Western Nigeria	Western Nigeria two sheet 60	Western Nigeria Akure Sheet 61	Western Nigeria Iseyin Area		Orogenic and Thermotectonic Episodes
Intrusion of Dykes								Dolerite	Dykes (480 my) Ibadan	
Deposition					Newer Sediments	Effon Psammitic Formation	Effon Psammitic Formation			Pan African Thermo-tectonic events 600 ± 150 my
Older Granite Cycle	Younger intrusive series or older granites, older intrusive series	Older granites	Older granites	Gneisses migmatites older granites	Older granites	Older granites	Older granites	Older granites	Peamatites, Apilites, Quartz Veins Porphyroblastic Granite (Metosomatic) Okeiho Potassic Syenite 'intrusive'	
Deposition and Igneous activity	Older intrusive series	Para-schists	Birnin Gwari schist formation Kushaka schist formation Kusheriki psammitic formation					Paraschists and meta intrusives	Biotite schist. Biotite garnet schist. Biotite garnet staurolite schist. Quartzites, Amphobilites, Talc-chloride, Schists	
		Meta-gabbros			Charnockitic intrusives gneissic complex	Charnockitic meta-intrusives	Charnockitic meta-intrusives (? 1000 my)			?
Ancient Granite Cycle		Gneissic complex				Gneissic complex	Migmatite gneiss complex	Migmatite gneiss complex (2000 my)	Granite Gneiss bonded Gneiss Quartzites (Ibadan) Calc-sulicate rocks Bio-Hornblende schist. Intercolated amphibolites	Eburnean orogeny c. 2000 my
Deposition	Metamorphism of ancient sediments Ancient meta-sediments	Ancient meta-sediments		Ancient meta-sediments	Ancient meta-sediments	Ancient Meta-sediments	Ancient meta-sediments	Meta-sediments and volcanics	Gneiss migmatites amphibolites	
	Bain (1926) (Bauchi) Wilson (1922) (Western Railway)	Russ (1957) King and de Swardt (1949)	Truswell and Cope (1963)	Oyawaye (1964)	Jones and Hockey (1964)	De Swardt (1963) Hubbard (1968)	Dempster (1968)	Rahaman (1973)		

of syn- to late-tectonic granites. The closing stages of orogeny were marked by cooling, uplift and fracturing, and by the intrusion of high level volcanic rocks. Rock types and the sequence of events during the upper Proterozoic in northern Nigeria can be correlated with similar sequences in Ghana, Togo and Dahomey to the west and with the Ahaggar in the north. Available evidence suggests that in these areas the Pan-African orogeny was the result of the opening and closing of a small ocean comparable in size to the Red Sea.

Review of the Basement Geology of South-Western Nigeria (M. A. Rahman 1976)

Abstract

The Precambrian of Southwestern Nigeria is a part of the Nigerian Basement Complex. Five major groups of rocks have been recognized:

1. Migmatite-gneiss complex which comprises biotite and biotite hornblende gneisses, quartzites and quartz schist and small lenses of calc-silicate rocks.
2. Slightly migmatized to unmigmatized paraschists and metaigneous rocks which consist of pelitic schists, quartzites and amphibolites, talcose rocks, metaconglometates, marbles and calc-silicate rocks.
3. Charnockitic rocks.
4. Older Granites which comprise rocks varying in composition from granodiorite to true granites and potassic syenite.
5. Unmetamorphosed dolerite dykes believed to be the youngest.

Two phases of folding believed to be related to the Older Granite Orogeny (Pan African) have been described affecting the gneiss-migmatite quartzite complex and the slightly migmatized to unmigmatized schists and metaigneous rocks. The first folds (F1) are reclined in style with axial planes trending E–W whereas the second folds (F2) have forms that are variable from open to isoclinal with steep to vertical N–S trending axial planes.

A Barrovian type metamorphism has affected the area and metamorphic grade is from green schist to amphibolite facies. Three episodes of metamorphism have been recognized in the pelitic schists. The first (M1) is related to F1 deformation; the second (M2) metamorphism developed during a static phase following F1 and the third (M3) metamorphism which reached a climax in the amphibolite facies started during F3 deformation and outlasted it.

The migmatite-gneiss complex is thought to have resulted from a complex association of deformative, shearing and folding and granitization and migmatization processes. The slightly migmatized to unmigmatized paraschists represent a sedimentary cover on the gneiss-migmatite complex.

The association of amphibolites (meta-igneous and volcanic rocks) and the Effon Psammite Formation east of Ile-Ife and the association of carbonates (marbles and calc-silicate rocks) and argillaceous rocks

(schists) in the Igara area further to the east are thought to represent an eugeosynclinal and miogeosynclinal sequence of sediments respectively.

The charnockitic rocks are mainly of magmatic origin and not the result of high grade metamorphism in the granulite facies. They belong to two generations, the first preceeding the emplacement of the Older Granites and the second emplaced in the last stages of the Older Granite Orogeny. A magmatic origin is suggested for most of the Older Granite bodies.

Radiometric data from rocks within the Nigerian Basement complex indicate at least two orogenic episodes, The Eburnean Orogeny (1950 ± 250 my). The second orogenic episode, the Older Granite Cycle (Pan African Orogeny 600 ± 150 my) culminated in the emplacement of the Older Granite and is responsible for the widespread amphibolite facies metamorphism of the Basement Complex. The significance of a Kibaran age 1150 ± 140 my for granite gneisses from near Ile-Ife has not been estimated.

Preliminary report on the field relationships of the basement complex rocks around Igarra, Mid-West (I. B. Odeyemi 1976)

Abstract

The region is a quarter degree sheet and is underlain by rocks of the Nigerian Basement Complex. About four main groups of rocks have been noticed. These are the migmatite-gneiss complex, the metasediments composed of schists, calc-gneisses, quartzites and metaconglomerates, the porphyritic Older Granites and the late discordant non-metamorphosed syenite dykes.

The migmatite-gneiss complex is presumably the oldest rock group in this region and might be the source rocks for the metasediments. Xenolithic inclusions of quartzitic bands might be relics of an older ancient metasedimentary sequence.

The calc-gneisses, marbles, schists and metaconglomerates form a lineally disposed metasedimentary unit and were probably laid down in a miogeosynclinal trough, underlain by rocks of the migmatite-gneiss complex. This sequence was affected by the Pan African Thermotectonic episode during which it was folded and metamorphosed.

The porphyritic granites are found as intrusive rocks in migmatites and metasediments. They often show cross cutting contacts in places with the country rocks. Xenolithic inclusions of country rocks are common in these granites. These granites are thought to be Synorogenic and appear to have been emplaced in a magmatic phase probably during the Pan African Thermotectonic event. The porphyroblastic gneisses—often a gneissic equivalent of the porphyritic granite—is always found in association with biotite gneisses and porphyritic granites into which it often grades.

The emplacement of the non-metamorphosed syenite with associated biotite pyroxenite is probably one of the last events to affect this region. It is thought to be a composite dyke emplaced discordantly within the schists.

The rocks show a general NW–SE foliation trend which seems to lie parallel to the limbs of major folds. The folded band of calc-gneiss and metaconglomerate forms the single most continuous structurally deformed unit. This shows a big antiformal fold, open to close in style and over-turned to the East. Minor folds are commonest in calc-gneisses, schists and gneisses. These are tight isoclinal folds with axes parallel to foliation planes. Most folds plunge to the north, though the over turned folds in the calc-gneisses plunge south as well.

Observations in the field show that rocks in this region have suffered at least two phases of structural deformation.

Provisional classification and correlation of the schist belts in North-Western Nigeria (A. C. Ajibade 1976)

Abstract

The schists of north-western Nigeria occur in linear north-south trending belts which are more or less continuous for a distance of about 400 km along the strike. There are five main belts separated by migmatite-gneiss complex and intruded by granitic rocks of the Older Granites suite. Occasional small isolated bands and lenses of the schists occur in the migmatite-gneiss complex.

The schist belts consist of pelitic, semi-pelitic and psammatic rocks of various lithologies. The rocks are classified into lithostratigraphic units and six formations are distinguished. Each formation is considered to have been laid down contemporaneously on the migmatite-gneiss complex.

Figure 12
Classification and Sequence of Events for Basement Complex Rocks of Southern Nigeria

Events	Stratigraphic Units 1 Western Railway	Osi Area	1:25 000 sheet 59 Ibadan	1:250,000 sheet 60 Iwo	Stratigraphic Units 1:250 000 sheet 62 Lokoja	1:250 000 sheet 61 Akure	Iseyin Area Nigeria	Nigeria	Orogenic Episode
Intrusion of dykes			Dolerite dykes		Unmetamorphosed charnockitic and syenitic intrusives and dykes. Dolerite dykes	Dolerite dykes			
Younger Deformation cycle				Effon Psammite Formation	Effon Psammite Formation			Newer metasediments	
Deposition				Sandstones, siltstones, mudstones etc.	Sandstones, siltstones, mudstones etc.			Sandstones, siltstones mudstones etc.	
Older Granite Cycle	Younger intrusive series or Older Granite	Older Granites 3. Late leucocratic Granit 2. Coarse Porphyritic Granite 1. Early Granite	Older Granites 3. Late Phase 2. Main Phase 1. Early Phase Migmatites Gnesses	Older Granites	Older Granites Younger Gneiss-Schist Complex	Older Granites Igarra and Kabba Jakura Formations	Older Granite 3. Post Kinematic 2. Late Kinematic Potassic syenites and main phase granites 1. Early phase Unmitigatized–slightly migmatized, paraschists and metaigneous rocks	Older Granite Migmatites Gneisses	
Deposition and igneous activity					Sandstones, calcareous rocks, mudstones, siltstones etc. Intrusion of Basic to Ultramafic rock	Conglomerates, sandstones, siltstones, calcareous rocks, mudstones, limestone etc.	Sandstones, mudstones, siltstones, minor calcareous rocks, emplacement of minor basic to ultramafic rocks		
Intrusive Cycle		Gabbros now metagabbros	Charnockitic intrusives	Charnockitic intrusives now metamorphosed	UNCONFORMITY Charnockitic intrusives	Charnockitic intrusives	Charnockitic intrusives	Charnockitic intrusives	? 1000m.
Ancient Granite Cycle	Older intrusive series	Gneiss Complex		Gneiss Complex	Migmatite-Gneiss Complex	Migmatite-Gneiss Complex	Migmatite-Gneiss Complex		Eburnean Orogeny 2000±250 my
Deposition igneous activity and metamorphism	Ancient metasediments	Ancient metasediments	Ancient metasediments	Ancient metasediments	Ancient metasediments	Ancient metasediments and minor volcanics	Ancient metasediments and minor volanics	Ancient metasediments	
Authors	Wilson (1922)	King and De Swardt (1949)	Jones and Hockey (1964)	De Swardt (1953), Hubbard et al. (1966), Hubbard (1968)	Dempster (1967)	Odeyemi and Rahaman (1973)	Rahaman (1973)	Oyawoye (1964, 1972)	

PAN AFRICAN OROGENY 600 ± 150 my

Based on Rahaman (1976)

Figure 13
Proposed Evolutionary Scheme for Basement Complex Rocks of Southwestern Nigeria

Events	Stratigraphic Units	Tectonics	Metamorphic Episode	Process
Intrusion of Dykes	Dolerite dykes e.g. Ibadan dykes 480 + 20 my Grant (1970) Unmetamorphosed charnockitic and syenitic dykes e.g. South of Igarra	Tensional Rebound		
Older Granite Cycle	Pegmatites, Aplites, Quartz veins. Late phase granites e.g. Okeiho porphyroblastic granite Late charnockitic intrusive, e.g. Osuntedo and Oke-Patara charnockites. Potassic syenites e.g. Okeiho syenite Main phase porpyritic Older Granite e.g. Abeokuta Granite Igarra granite Osi granite Early phase granite and granodiorite; charnoekites e.g. Ikerre Ekiti charnockite. Unmigmatized to slightly migmatized paraschists or Newer Metasediments and metaigneous rocks, e.g. Biotite garnet staurolite, biotite garnet and biotite schists, talc-premolite chlorite schists and amphibolites of Iseyin area. 2. Effon psammite and associated epidiorite and amphibolite complex, 3. Igarra and Kabba-Jakura formations.	Post F2 deformation Late F2 deformation Late F2-SynF2 deformation Syn F2 SynF2 main period of E-W stress Static phase F1 main period of N-S stress	M3 M3 M3 M3 M2 of garnet M1	Local migmatisation followed by mobilization Intrusion Intrusion Intrusion Amphibolite facies metamorphism Development of 1st generation garnets in pelitic rocks
Sedimentary and Igneous activity	Deposition of mudstones, quartzites, siltstones, limestones etc. Or Newer sediments and emplacement of basic to ultrabasic rocks. Later converted to unmigmatized paraschists and metaigneous rocks Or Newer Metasediments e.g. (1) Deposition of Effon psammite and emplacement of associated epidiorite and amphibolite complex. (2) Deposition of Igarra and Kabba-Jakura Formations. Charnockitic intrusives e.g. Wasimi charnockite. Gabbro now metagabbros e.g. Osi metagabbros.			Sedimentation and volcanicity Intrusion
Ancient Granite Cycle	Migmatite-Gneiss Complex e.g. Ibadan and Iseyin migmatite-Gneiss complex. Granite gneiss e.g. Ibadan granite gneiss Transition gneiss K feldspar bearing Banded Gneiss Early gneiss, quartzites, calc-silicates and intercalated amphibolites.	Main period of deformation isoclinal folding and shearing.		Widespread migmatization and granitization, locally mobilization and intrusion of granite.
Deposition and Igneous activity	Ancient sediments largely greywacks and minor volcanics largely basalts.			

Based on Rahaman (1976).

2.3 JURASSIC IGNEOUS COMPLEXES

The Late Palaeozoic to Precambrian Basement Complex formations of Northern Nigeria are intruded by ring complexes and associated bodies of Jurassic age. Similar bodies occur in Aïr, Niger Republic. The general distribution of Nigerian ring complexes is shown in Figures 16–18. On Figure 16 they are marked as Younger Granites. The Nigerian and Niger Young Granites form a north south trending zone and the Benue Trough cuts obliquely across the trend. Turner (1976) recently summarized data on their structure and petrology. His abstract is reproduced below. Jacobson et al. (1963) also described the main features of these ring structures and in the intervening years an extensive literature has built up about these rocks (Turner 1976 etc.) which has only peripheral interest to petroleum geologists.

Structure and Petrology of the younger granite ring (D. C. Turner 1976)

Abstract

Ring complexes of Jurassic age—Younger Granites—intrude the late Precambrian to Lower Palaeozoic basement rocks of northern Nigeria in a N–S zone which continues northwards to the Aïr region of Niger Republic. This zone is parallel to the main Pan-African trends in the basement, indicating control by earlier structures. It also lies on a continuation of the African continental margin to the south and possibly formed in a region of crustal arching developed prior to the separation of the African and American plates in the Cretaceous.

The structural and petrological sequence of many of the individual ring complexes follows the same general pattern. The earliest rocks are volcanic, preserved from erosion through cauldron subsidence. A two-fold volcanic sequence is characteristic. Early rhyolites, often ignimbritic and including the peralkaline comendites, form bedded sequences with minor associated basalts and trachytes. Late rhyolites are porphyritic and form thick caldera filling flows and also intrusive stocks, plugs and dykes. They are closely associated with the emplacement of outer ring dykes of granite-porphyry which contain the minerals fayalite, hedenbergite and hastingsite or arfvedsonite. Within these ring dykes are later annular or stock-like intrusions of arfvedsonite and biotite granite.

The northern complexes are exposed at a higher structural level—that is have been less deeply eroded—than those of the Jos Plateau and the south. In the north, abundant rhyolites are associated with fayalite granite-porphyries and peralkaline granites, with only the roof zones of later biotite granites

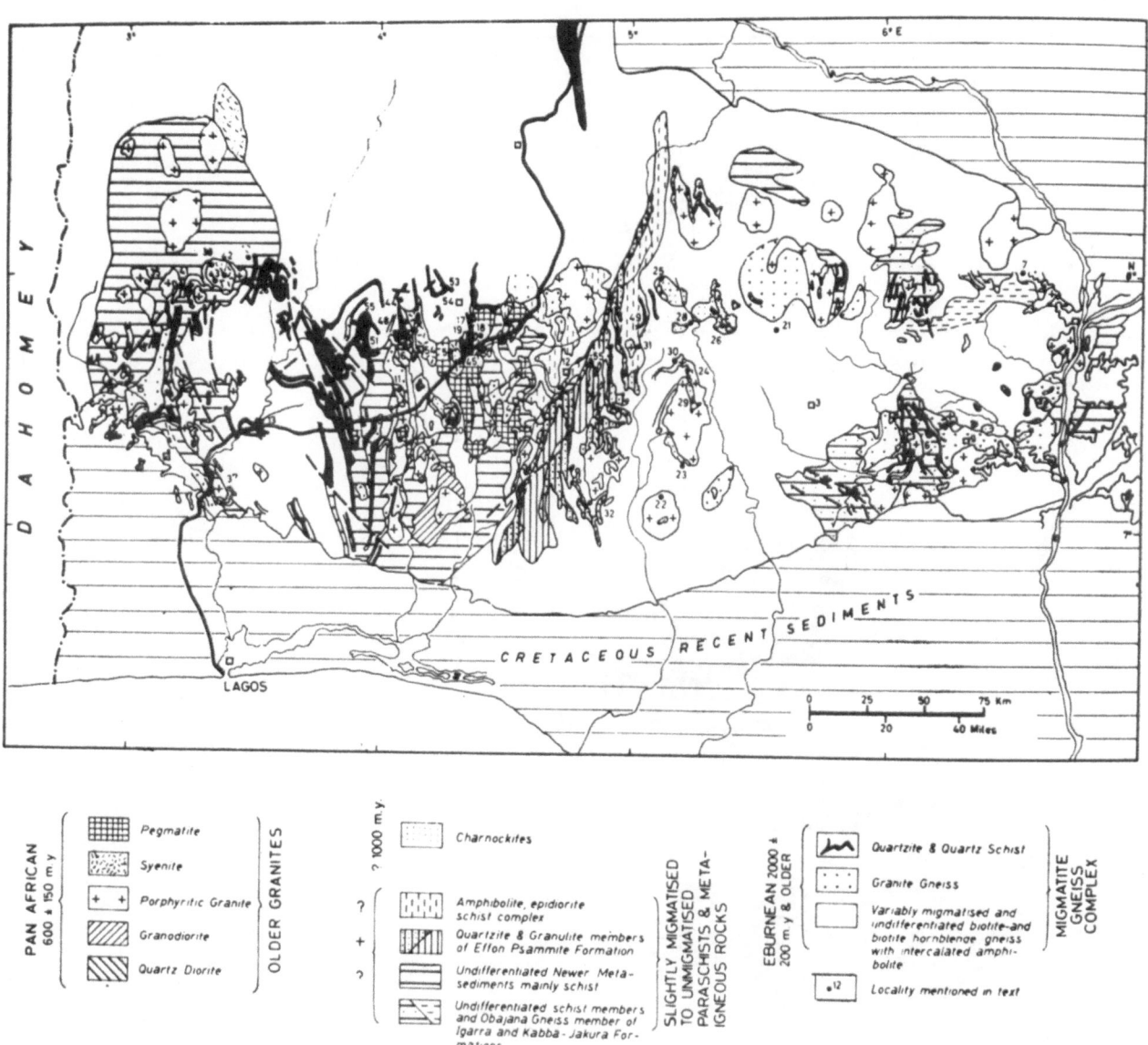

Figure 14. (A) Map showing main lithological units and localities, Basement Complex, Southwestern Nigeria. (B) Locality List. Based on Rahaman (1976).

Figure 14B
Locality List for Figure 14A

1. Ibadan	20. Wasimi	39. Shaki
2. Iseyin	21. Ikole-Ekiti	40. Igboho
3. Ikare	22. Idanre	41. Iganna
4. Ife	23. Akure	42. Ilua
5. Kuta	24. Ado-Ekiti	43. Isemi
6. Igarra	25. Otun	44. Ajawa
7. Jakura	26. Egosi	45. Ikoyi
8. Igbetti	27. Osi	46. Ogbagba
9. Ukpilla	28. Ifaki	47. Ikire-Ile
10. Aiyetoro	29. Ikere	48. Ikonifin
11. Lagun	30. Ewu	49. Ijero-Ekiti
12. Ilesha	31. Aramoko	50. Ede
13. Oloke-Meji	32. Ondo	51. Akinmorin
14. Imala	33. Ore	52. Oyo
15. Oyan	34. Iwo	53. Ola
16. Oke Patara	35. Iwere-Ile	54. Ejigbo
17. Ara	36. Ososo	55. Erinmo
18. Awo	37. Abeokuta	
19. Osuntedo	38. Okeiho	

21

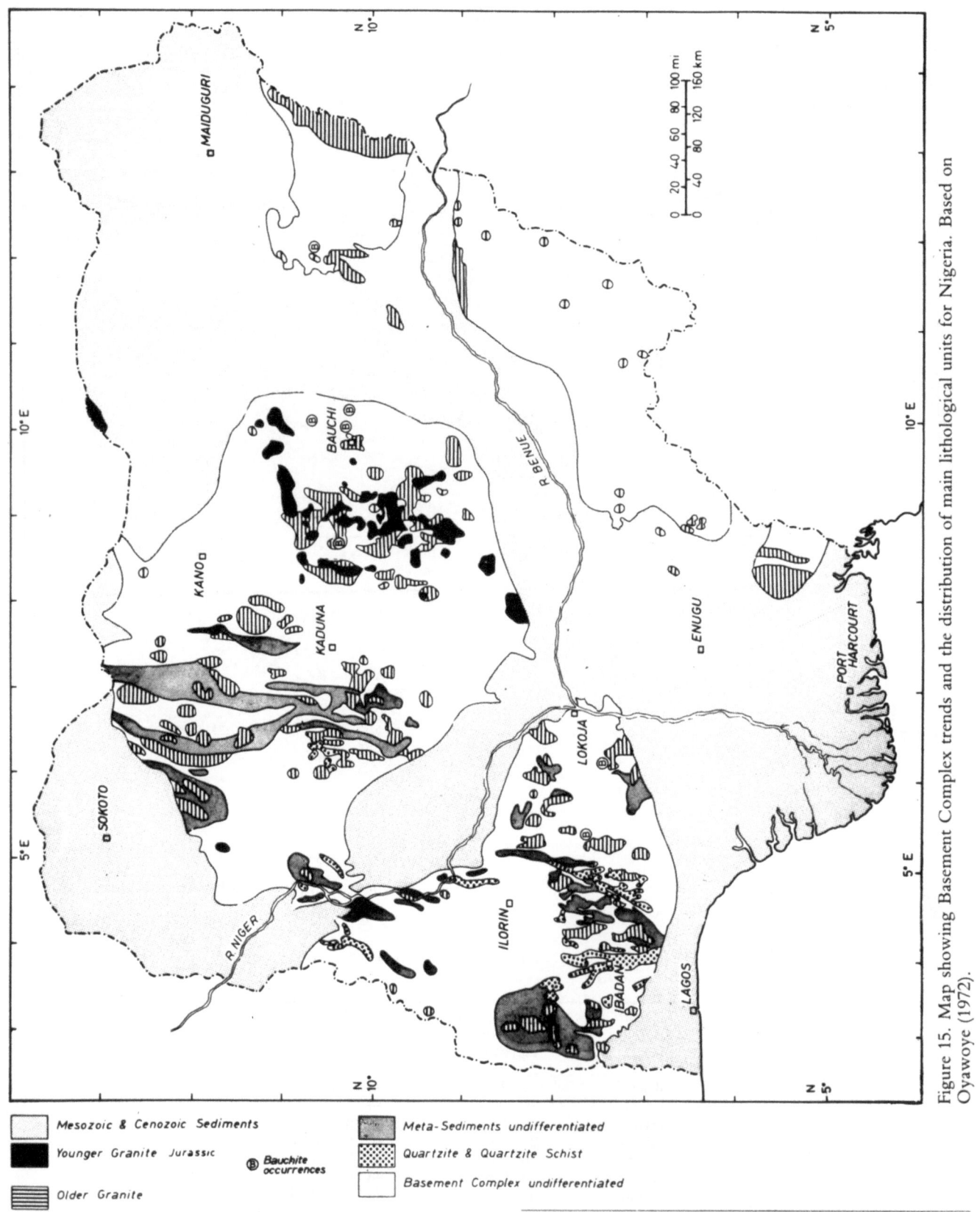

Figure 15. Map showing Basement Complex trends and the distribution of main lithological units for Nigeria. Based on Oyawoye (1972).

Mesozoic & Cenozoic Sediments	Meta-Sediments undifferentiated
Younger Granite Jurassic	Quartzite & Quartzite Schist
Bauchite occurrences	Basement Complex undifferentiated
Older Granite	

Figure 9. Geological Map of Nigeria showing the distribution of sedimentary formations. Compiled by Dessauvagie (1972) included in Whiteman (1973) with modifications (Basement Complex data excluded); and modified 1977 to include data included on Dessauvagie (1974).

Scale 1 : 2 000 000 (approximately).

Additional material from *Nigeria: Its Petroleum Geology, Resources and Potential*
ISBN 978-94-009-7363-3 (978-94-009-7363-3_OSFO2),
is available at http://extras.springer.com

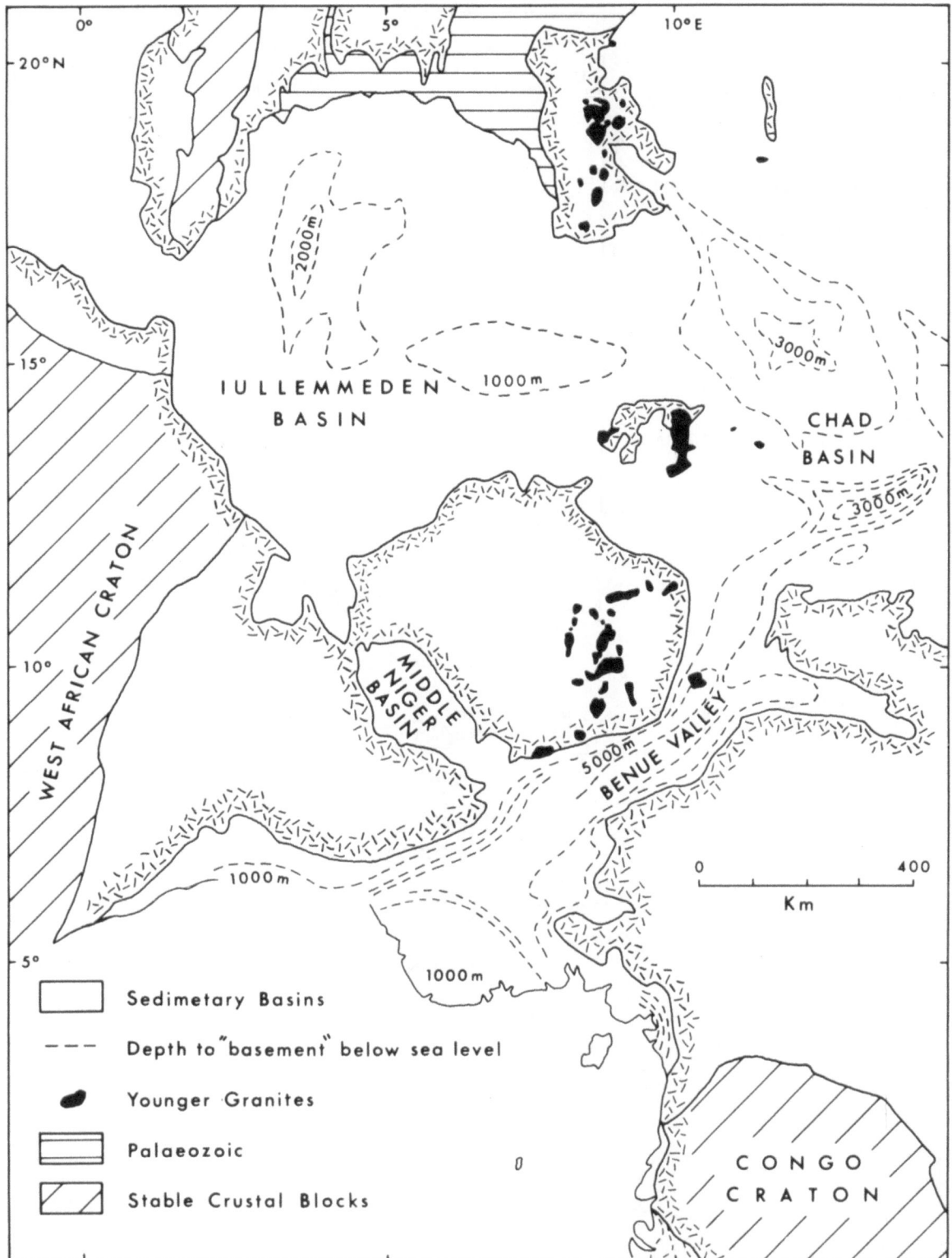

Figure 16. Map showing the general structural setting of the Nigerian and Niger Ring Complex Province. Thickness data shown on sedimentary area is based mainly on the Tectonic Map of Africa 1 : 5 000 000 Scale (1968). Based on Turner (1976).

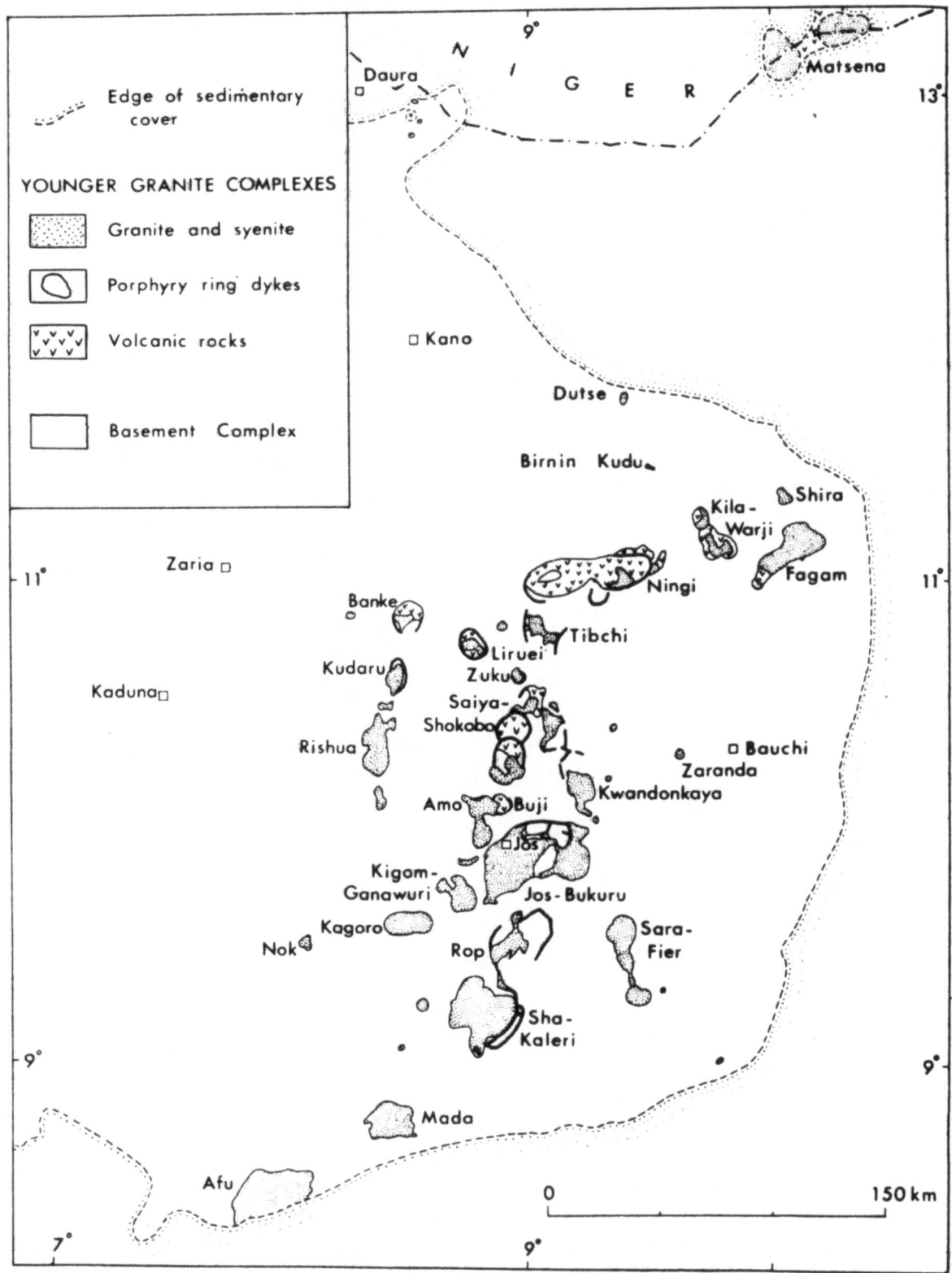

Figure 17. Map showing the general distribution of the Younger Granite Ring Complex of Nigeria. Based on Turner (1976).

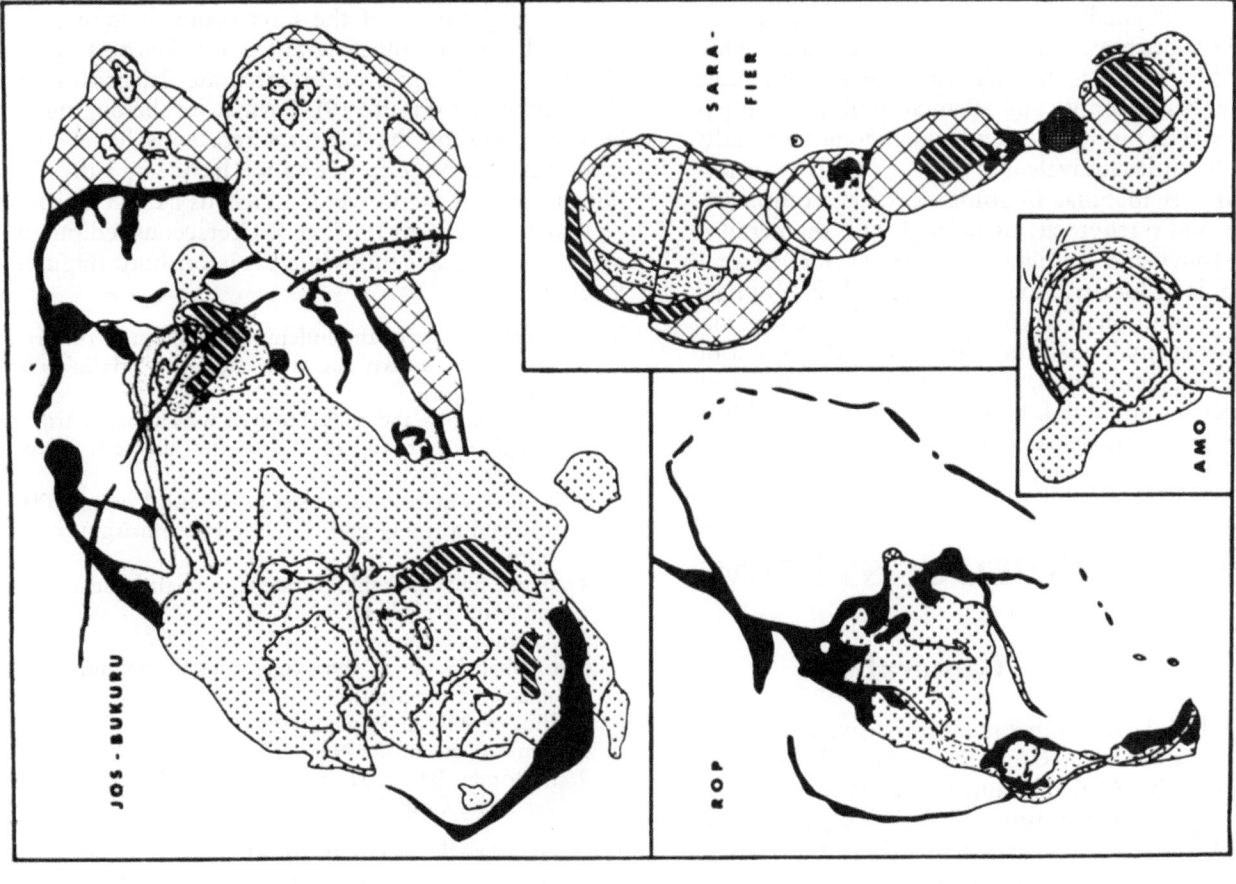

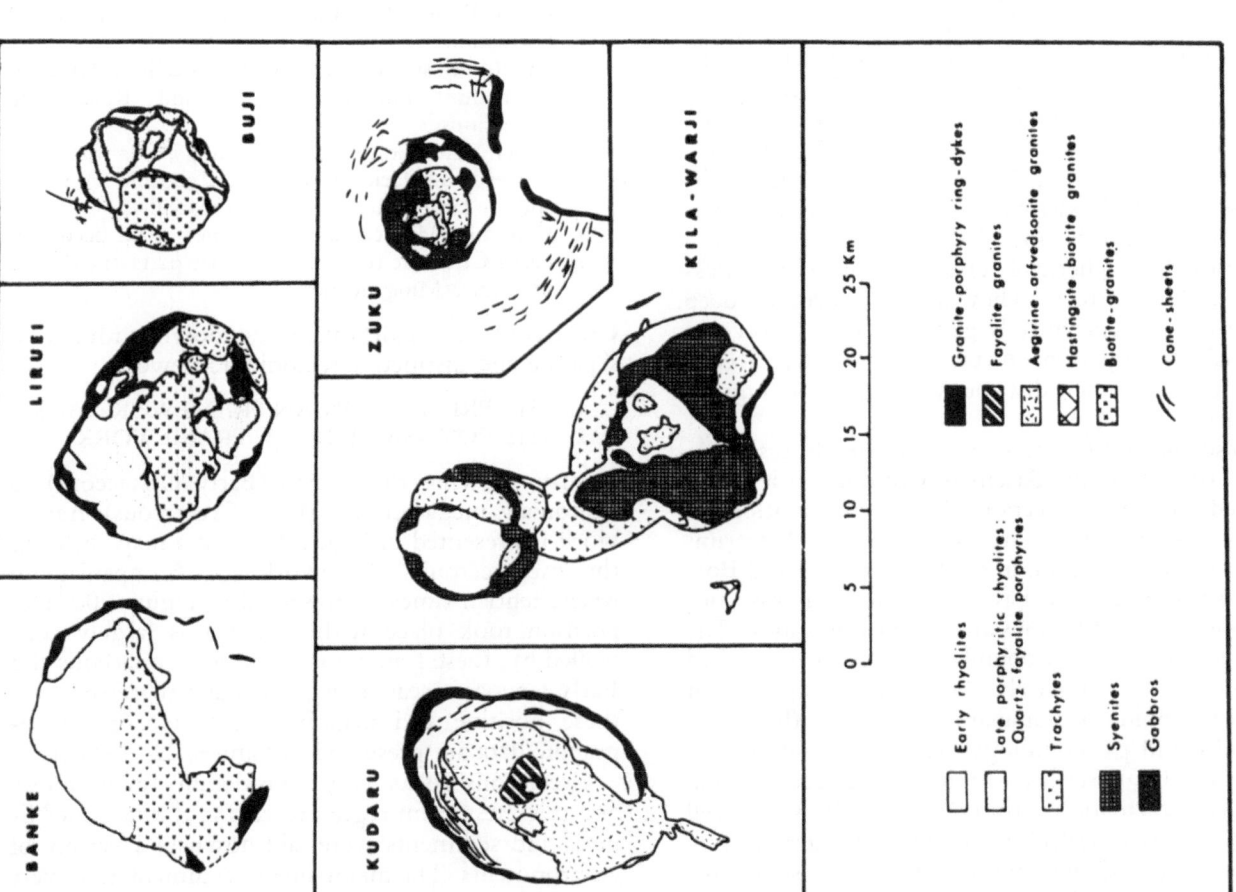

Figure 18. Simplified geological maps of selected ring complexes, Nigeria. (a) Complexes with important volcanic rocks and (b) mainly intrusive complexes. Based on Turner (1976).

exposed. In the centre and south the complexes mainly of massive biotite and hastingsite-biotite granites with smaller, generally peripheral intrusions of fayalite porphyries and peralkaline granites. Thus the earliest magmas crystallized to rocks containing fayalite, hedenbergite and alkali amphiboles and pyroxenes. The biotite granites, which do not generally have rhyolitic equivalents, represent a later, more voluminous magma. In some northern complexes, syenite and trachyte are associated with the peralkaline granites and rhyolites; a quartz syenite magma may, at least locally, have a parental role in Younger Granite evolution. The contribution of basic magma, available in small quantities throughout the evolution of the ring complexes, is less apparent; there is no clear petrographic link between granite and gabbro, and separate sources seem probable.

2.4. CRETACEOUS SYSTEM

2.4.1. General

The oldest dated sedimentary rocks exposed in Nigeria are the Lower Cretaceous (Albian) Asu River Group of the Abakaliki and Lower Benue Troughs and the Bima Sandstone of the Upper Benue Trough. Older formations may exist in the offshore miogeoclinal sedimentary areas and beneath the delta complexes. Older sedimentary rocks have been recognized onshore in adjacent sedimentary areas to the north but not in countries immediately to the west. Claims have been made that sediments older than Albian are present onshore in Nigeria but palaeontological proof has not been given (Murat 1972, Uzuakpunwa 1974 and Adeleye 1976). Such sediments may well exist considering the date of initiation of the break up of the African and South American lithospheric plates and Atlantic Ocean Basin spreading history. They must lie beneath deep water and are currently largely inaccessible to commercial investigation. A tabulation of Cretaceous formations and their distribution is shown in Figures 8 and 19.

Mesozoic and Cenozoic rocks normally rest noncomformably on the Basement Complex, briefly described above in Section 2.2. Sedimentation in south-eastern Nigeria, around the Niger Delta rim, in the Dahomey Basin, in the Iullemmeden and Bornu Basins appears to have been more or less continuous from Cretaceous into Cenozoic times. The Cenozoic/Cretaceous boundary has been penetrated in wells but at surface, because of poor exposure of rocks belonging to boundary formations, their stratigraphy and palaeontology are poorly known. Fayose (1970) described topmost Cretaceous and Cenozoic formations from Mobil's Ofowo-1 well and Kogbe et al. (1976) described the biostratigraphy of the Upper Cretaceous and Cenozoic sediments penetrated in Shell-BP's Gbekebo "B" well. These

appear to be the only two wells for which data has been published.

The subdivisions of the Cretaceous System used in this book are those adopted by Teichert et al. (1964) in the Treatise on Invertebrate Palaeontology (K) Mollusca 3 (Figure 19). Formations belonging to all the stages of the Upper Cretaceous have been recognized in Nigeria but as we have said only the Albian stage of the Lower Cretaceous has been proven to date (Reyment 1965 etc.). Cretaceous sediments are exposed in the following areas onshore (Figures 9, 22 etc.):

1. Southern part of the Iullemmeden Basin in the area generally known as the Sokoto "Basin" or Embayment.
2. The Gongola and Yola Troughs which form a trilete trough system together with the northern Benue Trough.
3. Limited outcrops northeast of the Biu Volcanic Plateau and forming part of the southern margin of the Bornu Basin.
4. Benue and Abakaliki Troughs with the side arm of the Mamfé Embayment.
5. Anambra Basin.
6. Calabar Flank of the Niger Delta extending into adjacent Cameroun.
7. Bida or Middle Niger Basin.
8. Benin Flank of the Niger Delta.
9. Dahomey Basin extending westwards from the Niger Delta Complex into eastern Ghana.

Cretaceous rocks occur in the subsurface onshore in the following areas (Figure 9):

1. Bornu Basin where they subcrop beneath the Chad Formation.
2. Sokoto "Basin" or Embayment beneath the Gwandu Formation and the Dange and Kalambaina formations.
3. Niger Delta Complex where they occur mainly as deeper water facies beneath the Cenozoic sediments and are largely inaccessible to the drill.
4. Dahomey Miogeocline where they again occur beneath Cenozoic rocks and in other parts of the Gulf of Guinea Miogeocline.

Cretaceous sediments were laid down within two separate and distinctive tectonic frameworks:

1. THE PRE-SANTONIAN FRAMEWORK AND
2. THE POST-SANTONIAN FRAMEWORK.

A map showing the Albian (Early Cretaceous) to Santonian megatectonic (Late Cretaceous) framework is presented in Figure 153 and a map showing the megatectonic framework for Campanian to Maestrichtian times is presented in Figure 200. Deposition took place within, and was largely controlled by, these frameworks and the Cretaceous and Early Cenozoic seas transgressed and regressed over these positive and negative elements. Palaeogeographic data are presented in Figures 22, A–H etc.

Cretaceous rocks vary considerably in lithology and thickness from region to region throughout Nigeria and sediments were laid down in a variety of environments. The major sites of sedimentation were rift troughs which eventually failed to develop and

Figure 19
Subdivisions of the Cretaceous, Palaeogene and Neogene Systems

Erathem	System	Series	Stage	Substage
CENOZOIC	NEOGENE	Holocene Pleistocene Pliocene Miocene		
	PALAEOGENE	Oligocene Eocene Palaeocene		
MESOZOIC	CRETACEOUS	UPPER CRETACEOUS	Maestrichtian	
			Senonian	Campanian Santonian Coniacian
			Turonian Cenomanian	
		LOWER CRETACEOUS	Albian Aptian Barremian Hauterivian Valanginian Berriasian	

Based on Teichert et al. (1964)

Figure 20
The Cretaceous and Tertiary Sequence in Southeastern Nigeria

	Stage	Formation	
Neogene	Recent	Marine deltaic deposits, alluvium	
	Miocene-Pleistocene	Benin Formation	
	Oligocene? – Miocene	Ogwashi-Asaba Formation	
Palaeogene	Ledian	Not represented	
	Bartonian	possibly upper part of Ameki Formation	
	Lutetion	Ameki Formation	
	Ypresian	possibily part of Ameki Formation	Nanka Sand
	Palaeocene	Imo Shale	
Upper Cretaceous	Danian	Nsukka Formation	
	Maestrichtian	Ajali Sandstone	
		Mamu Formation	
	Campanian	Enugu Shale / Nkporo Shale	
	Coniacian-Santonian	Awgu Shale	
	Turonian	Eze-Aku Shale	
	Cenomanian	Odukpani Formation	
Lower Cretaceous	Albian	Unnamed Formation Abakaliki Shale	"Asu River Group"

Based on Reyment (1965)

miogeoclines. Transgressions and regressions recognized mainly in the Abakaliki and Lower Benue Troughs by Murat (1970) and adjacent areas of Nigeria are listed in Figures 21 and 23. The Albian (Early Cretaceous)–Santonian (Late Cretaceous) Megatectonic Framework for the Southern Nigeria and Benue - Niger areas developed in response to crustal modifications processes related to Atlantic opening and the post-Santonian framework developed in the south as a consequence of these processes (Figure 22).

As we have already noted the Niger Delta Complex overlies a triple junction consisting of the Gulf of Guinea Translation Zone comprising the Equatorial Fracture Zone Complex bounded by the Romanche and Fernando Po Fracture Zones in the eastern part of the Gulf of Guinea (F); the South Atlantic Spreading Ridge System with minor transforms (R); and the Benue Trough or "Failed Arm" (r).

The trough system which comprises the pre-Santonian Benue and Abakaliki Troughs and the troughs of the Benue and Calabar Flank areas of Southern Nigeria are thought to have evolved largely in a tensional environment until Santonian–Campanian times when the stress field appears to have changed and folds were generated in the troughs. How exactly the Benue–Abakaliki Fold System was produced is still a matter for discussion (See Section 3.3.5) and more basic facts are needed for an acceptable solution to be reached. These areas were also mineralized and intruded by granodioritic bodies,

especially in the southern part of the area in the Abakaliki Trough, where sediments are metamorphosed in some areas. Gravity data presented by Cratchley and Jones (1963) can be interpreted to indicate that thinned crust underlies parts of the Benue Trough (Whiteman 1973; Artsybashev and Kogbe 1975; and Adighije 1979).

The uplifted area within the Late Cretaceous Benue Trough constituting the Benue Fold System was eroded and marginal sedimentary basins, such as the Anambra Basin, the Wase–Gombe and the Wukari–Mutum Biya Basins developed on the flanks of the belt. Sediments in these basins are largely Late Cretaceous in age. In places they are thick and commonly are only gently flexed or are undisturbed. The Anambra Basin certainly has considerable prospective interest and may contain undiscovered gas resources. The rocks in these marginal basins apparently were subjected to a less intense geothermal regime than those in the pre-Santonian basins constituting part of the trough system. In places pre-Santonian trough sediments were intruded by granodioritic bodies and were metamorphosed. As a result of the development of a compressional stress field in the Abakaliki–Benue troughs and adjacent areas, apparently consequent on the formation of

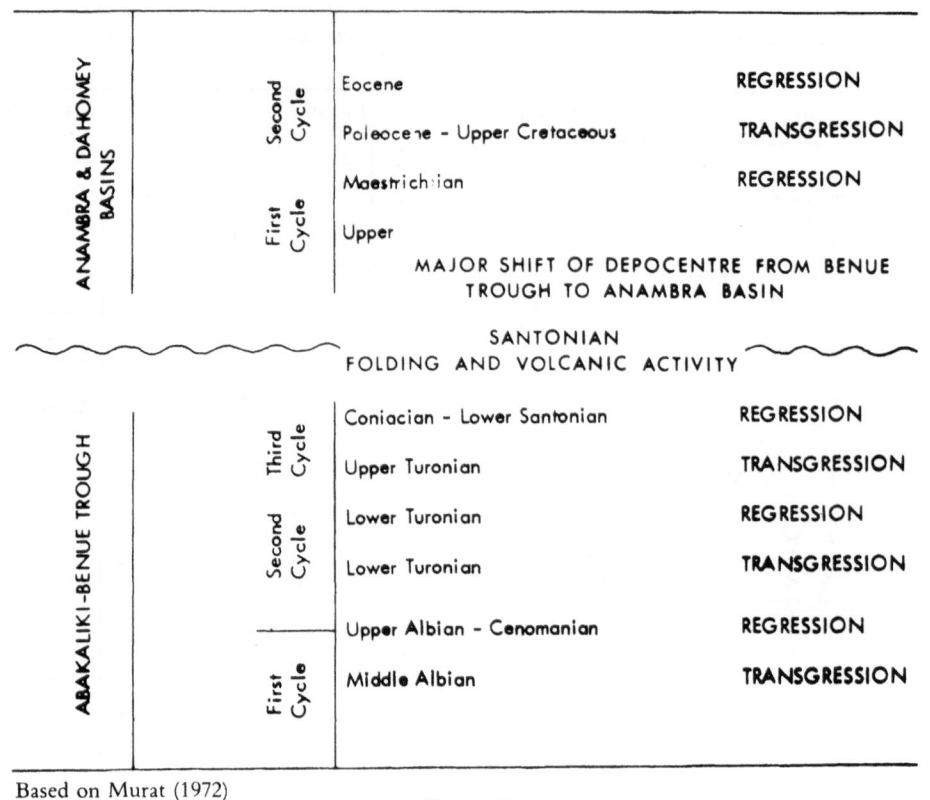

Based on Murat (1972)

Figure 21
Cretaceous and Tertiary Transgressions and Regressions Recognized in the Benue Trough and Adjacent Areas of Nigeria

long ridge–ridge transforms in the Gulf of Guinea region, the two interconnected troughs and their flanking shoulders were uplifted and became major sources of sediment. A large and extensive unconformity developed during Santonian and Campanian times in areas marginal to the Abakaliki–Benue Fold Belt (Figure 22). As we have already said the existence of this unconformity greatly facilitates subdivision of the Cretaceous System in southern and eastern parts of Nigeria.

The post-Santonian Maestrichtian Gombe Sandstone of the Upper Benue Trough is the best known of the marginal deposits and consists of coarse clastics which were widespread within the trough. These deposits resulted from uplift and erosion of the flanks of the depression (Figure 22) and are unfolded. In the Wase–Gombe Basin and the Wukari–Mutum Biya Basin Late Cretaceous (Post Santonian) sequences accumulated (Figure 22). A short lived, but extensive, marine transgression is said to have affected the whole of the Benue Trough in Maestrichtian times (R. A. Reyment, Personal Communication 1970) and extended at least as far north as the Bornu Basin where Maestrichtian fish, chelonian and reptilian fossils occur at Damagum (Carter *et al.* 1963). Because of the extent and elevation of the Benue–Abakaliki Fold Belt the writer holds that this transgression probably reached the area via the Iullemmeden, Chad and Bornu Basins and not via the Benue Depression as Reyment suggested because the Benue and Abakaliki Fold Belts were probably positive features at this time supplying sediment to outside areas (Figure 22). Murat (1972) (Figure 23) does not show a Maes-

trichtian transgression as having effected the Benue Trough.

Down valley, paralic sedimentation developed in post-Santonian time in the Enugu area and the paralic "Lower Coal Measures" and "Upper Coal Measures" were laid down in delta complexes. By the end of Cretaceous time the Benue–Niger–Delta Systems had reached the Onitsha Region at the south end of the Anambra Basin where perhaps as much as 30 000 ft of sediments accumulated. Much of this thickness consists of post-Santonian and pre-Cenozoic sediments. Eventually the delta systems spread down the newly generated continental margin to the south and thick sequences of sediment were deposited on transitional and oceanic crust in Oligocene time (Figure 22).

In the Dahomey Basin during Coniacian times sediments began to accumulate in what is now the onshore part of the basin, although marine sediments must have been accumulating in leptogeoclinal basins (mainly in the offshore) developed on the flanks of the Gulf of Guinea Depression since the African and South American lithospheric plates had begun to separate in Early Cretaceous times.

An Early Palaeocene–Late Cretaceous Transgression has been recognized in the Dahomey Basin and in southern parts of the Anambra Basin but apparently this did not penetrate into the Benue and Abakaliki Fold Belts which were still upstanding and acting as sediment source areas (Figure 22).

In the Bornu Basin the Cretaceous is overlain by horizontal Cenozoic Kerri-Kerri and Chad Formations and the youngest Cretaceous rocks exposed are

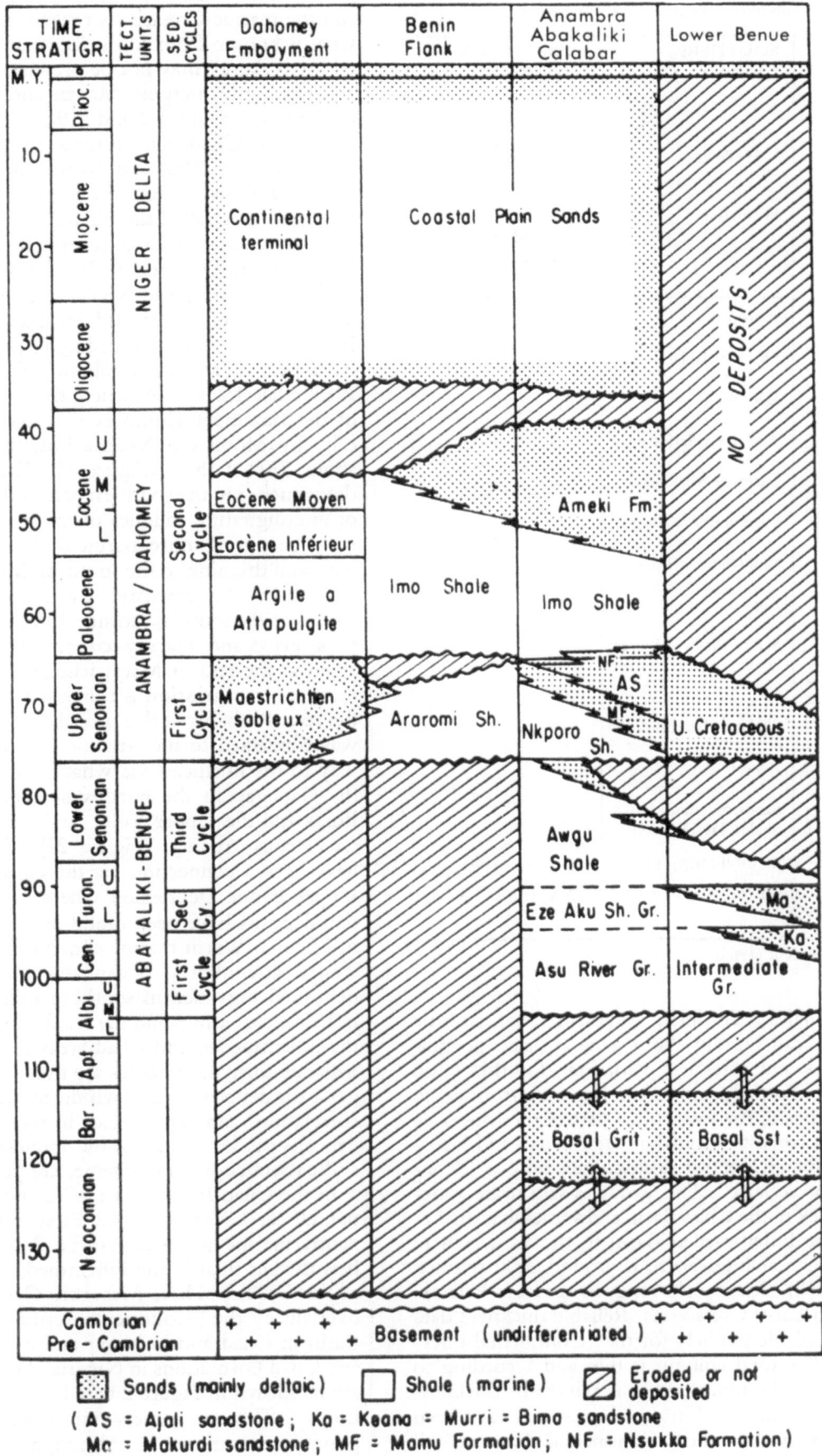

Figure 23. Table showing stratigraphic synopsis for Cretaceous and Tertiary rocks (excluding the Niger Delta Complex) of Southern Nigeria. Based on Murat (1972).

Figure 24
Subdivision of the Asu River Group

			SOUTHERN NIGERIA	NORTHERN NIGERIA
UPPER CRETACEOUS				Muri Sandstones
			Odukpani	Bima Sandstones and Grits?
			Formation	Guyak Sandstones? (Abandoned) Gateri Grits? (Abandoned)
LOWER CRETACEOUS	ALBIAN	UPPER ALBIAN	Zone 5	
			Shales at Ishiago	
			Zone 4	
			Asu River Series	
			Shales at Abakaliki	
			Limestones at Arapu	
			Sandstones at Nwofe	
			Zone 3	
			Sandy shales at Ibenta	
		MIDDLE ALBIAN	Zone 2	
			Shales at Nkpume	
			Zone 1	
			Sandstones and Shales Tiv Division	

---------- UNCONFORMITY ---------------------

PRECAMBRIAN BASEMENT COMPLEX GRANITES
AND GNEISSES

Based on Reyment (1965)

the Fika Shales (?Turonian, Reyment 1965), so little conclusive can be said about general palaeogeography, facies and thickness variations for this basin (Figure 2). To the north of the Nigerian frontier, in the Chad Basin, the lowermost part of the Mesozoic succession consists mainly of "red beds" and lacustrine and deltaic deposits known generally as "Continental Intercalaire". This group is said to be about 6300 ft thick in the Chad Basin and somewhat thicker than in the Iullemmeden Basin to the west, where the Continental Intercalaire is around 2000 ft thick (ITMA 1968). These rocks range in age from Permo–Triassic to Early Cretaceous. Reliable thickness data are not available publicly for the "Continental Intercalaire". The total column is thin and according to Greigert and Pougnet (1967) it ranges in thickness from 300 to 6230 ft. Various formations have been recognized within the "Continental Intercalaire" group in the northern part of the Iullemmeden Basin but how far these descriptions apply to the southern section of the basin is not known because the "Con-

tinental Intercalaire" pinches out under the Tegama and the marine Cretaceous rocks (Figure 219). From what is known from subsurface data the "Continental Intercalaire" may not be present south of latitude 16°N in the Damergou, Mantas and Sokoto sections of the basin (Figures 2 and 219).

The Late Cretaceous in these northern areas, as in many parts of West and North Africa, began with a marine transgression. In the Iullemmeden Basin this episode is called the *Neolobites* transgression and the sea appears to have penetrated from the north (Greigert and Pougnet 1967). The whole of the Upper Cretaceous in the Niger Section of the Iullemmeden Basin is less than 1650 ft and thins marginally. However, in Nigeria according to the section shown on Sheet 1, Tangaza (Geological Survey of Nigeria 1965, Scale 1:250 000), where the position of the top of the Basement Complex was determined seismically by Mobil Oil Nigeria Ltd, the Cretaceous is about 3700 ft thick (Figure 81). Broadly speaking, the sequence in the southern part of the basin consists of interdigitating tongues of marine and continental rocks becoming more marine to the north and northwest and thinning to the south. In Medial Maestrichtian times the extensive *Libycoceras* transgression spread as far south as latitude 13°N. Regression then took place and marshy conditions developed; but again at the end of Maestrichtian times another *Libycoceras* transgression is thought to have penetrated the area. Probably at this time the Iullemmeden Basin was connected to the Atlantic Ocean, not by a seaway to the northeast via what is now Libya but via the Gao gap in the northwest between Adrar des Iforas and the Upper Volta Arch and so to the Late Cretaceous Atlantic (Figure 22). There may well have been a connection via the Chad Basin and the Upper Benue Depression where marine Maestrichtian deposits have been found in the Damagum area of Nigeria. Warm marine conditions in which limestones, attapulgites, phosphates and gypsum were deposited prevailed in southern Iullemmeden Basin in Palaeocene times and the sea finally retreated from the Iullemmeden and Chad areas in Ypresian times. Separate formation names are used in Niger and Nigeria for deposits laid down during these times. The correlation, and the manner in which the Nigerian deposits can be fitted into the above-mentioned general palaeogeographic scheme, are presented below.

In the Chad Basin north of the Nigerian frontier, the Upper Cretaceous and Palaeocene is predominantly marine and is said to be about 3000 ft thick. It is thicker than in the Iullemmeden Basin where it attains 2000 ft. These Mesozoic–Cenozoic rocks are overlain by the "Continental terminal" in the Chad Basin in Chad and in Niger and by the Kerri-Kerri and Chad Formations in Nigeria. The thickness variations shown in Figure 2 based on ITMA (1968) and Louis (1970) are largely interpretations based on gravity measurements. Although wells have been drilled recently into sediments in these basins, formational, thickness and depth data have not been published.

Post-Santonian Cretaceous formations in Southern and Western Nigeria include:

1. IN WESTERN NIGERIA: Nkporo Shale, Araromi Shale, Nsukka Formation, Ajali Sandstone and Abeokuta Formation.

2. IN ANAMBRA BASIN: Nkporo Shale, Mamu Formation, Ajali Sandstone and Nsukka Formation.

3. IN BIDA BASIN AND ADJACENT LOKOJA AREA: Bida Sandstone, Sakpe Ironstone, Enagi Siltstone, Batati Ironstone and the Patti Formation and Lokoja Sandstone.

4. IN NORTHEASTERN NIGERIA: Gombe Sandstone

In northwestern Nigeria Upper Cretaceous (Maestrichtian) rocks outcrop and include:

1. NORTHWESTERN NIGERIA, SOKOTO AREA: The Rima Group which consists of the Taloka, Dukamaje and Wurno Formations.

All these formations are either unfolded or very slightly tilted, unlike the Pre-Santonian rocks.

A stratigraphic summary chart for the Cretaceous and Cenozoic rocks of the Dahomey Embayment, Benin Flank and Middle Benue areas (excluding the Niger Delta Complex) is presented in Figure 23; and a regional stratigraphic correlation chart for Nigeria and adjacent areas of Cameroun is presented in Figure 8.

Because of the considerable differences between the depositional patterns and structural styles in the various parts of Nigeria where Cretaceous rocks have been studied, generalizations and details about these areas are best considered regionally under the following headings:

1. SOUTHERN NIGERIA
   including the Abakaliki Trough, Mamfé Embayment, Calabar Flank, and Anambra Platform.
2. BENUE DEPRESSION
   including the Lower and Upper Benue subdivisions.
3. BORNU AND CHAD BASINS
4. IULLEMMEDEN BASIN
   (Sokoto area)
5. DAHOMEY BASIN

For ease of presentation the Cretaceous and Early Cenozoic stratigraphic history of all these areas is best discussed in two parts:

2. The Post-Santonian or Campanian–Maestrichtian–Early Cenozoic History; and
1. Pre-Santonian or Late Cretaceous–Early Cretaceous history

This scheme is especially useful for description of the sequences in the Benue–Abakaliki Troughs and adjacent areas of Southern Nigeria, where a well developed Late Cretaceous (Santonian) unconformity has been recognized. However, some repetition fol-lows as a consequence of adopting this method of description.

The Cretaceous formations in the areas listed above are described systematically below basin by basin under the headings: (1) General Description (mainly lithology); (2) Thickness Variations; (3) Age; and (4) Conditions of Deposition and Palaeogeography. Such systematization of information is useful for basin evaluation studies.

2.4.2 Southern Nigeria including Anambra Platform, Abakaliki Trough and Calabar Flank

Pre-Santonian Lower and Upper Cretaceous Formations

2.4.2.1. Lower Cretaceous Formations

2.4.2.1.1. General

Reyment's classification of Cretaceous and Cenozoic rocks occurring in Nigeria is presented in Fig. 20. Only Lower Cretaceous rocks belonging to the Albian stage have been recognized with certainty in Nigeria and then only the Middle and Upper divisions (Reyment 1965). He suggested that beds slightly older than Albian might be present in Nigeria because *Douvilleiceras* has already been collected from Gabon (Reyment 1965). However, Murat (1969 and 1972) classed the Basal Grit of the Anambra and Calabar areas as Lower Cretaceous Neocomian and Barremian and Uzuakpunwa (1974) assigned an Aptian age to sub-pyroclastic "Intra-continental sandstone" in the Abakaliki Town area. In both cases palaeontological evidence was not provided and therefore the oldest known sediments for which we have published proof are of Albian age.

Stratigraphic units recognized on the Anambra Platform, in the Abakaliki Trough and on the Calabar Flank areas and assigned an Albian age by Reyment (1965) include:

Asu River Group, Abakaliki Shale, Arufu Limestone, Gboko Limestone, Mamfé Formation and Uomba Formation.

All these units were deposited during the Medial Albian Transgressive Phase and the Late Albian Regressive Phase.

2.4.2.1.2. Asu River Group

General Description: During the early stages of exploration for oil in Nigeria, Shell-D'Arcy (later Shell-BP) geologists named the Asu River Group and established the following sequence in the western tributaries of the Cross River in the so-called Cross River Plain area (Figure 9):

3. AWGU-NDEABOH SHALES (Top)
2. EZE-AKU SHALES
1. ASU RIVER GROUP (Bottom)

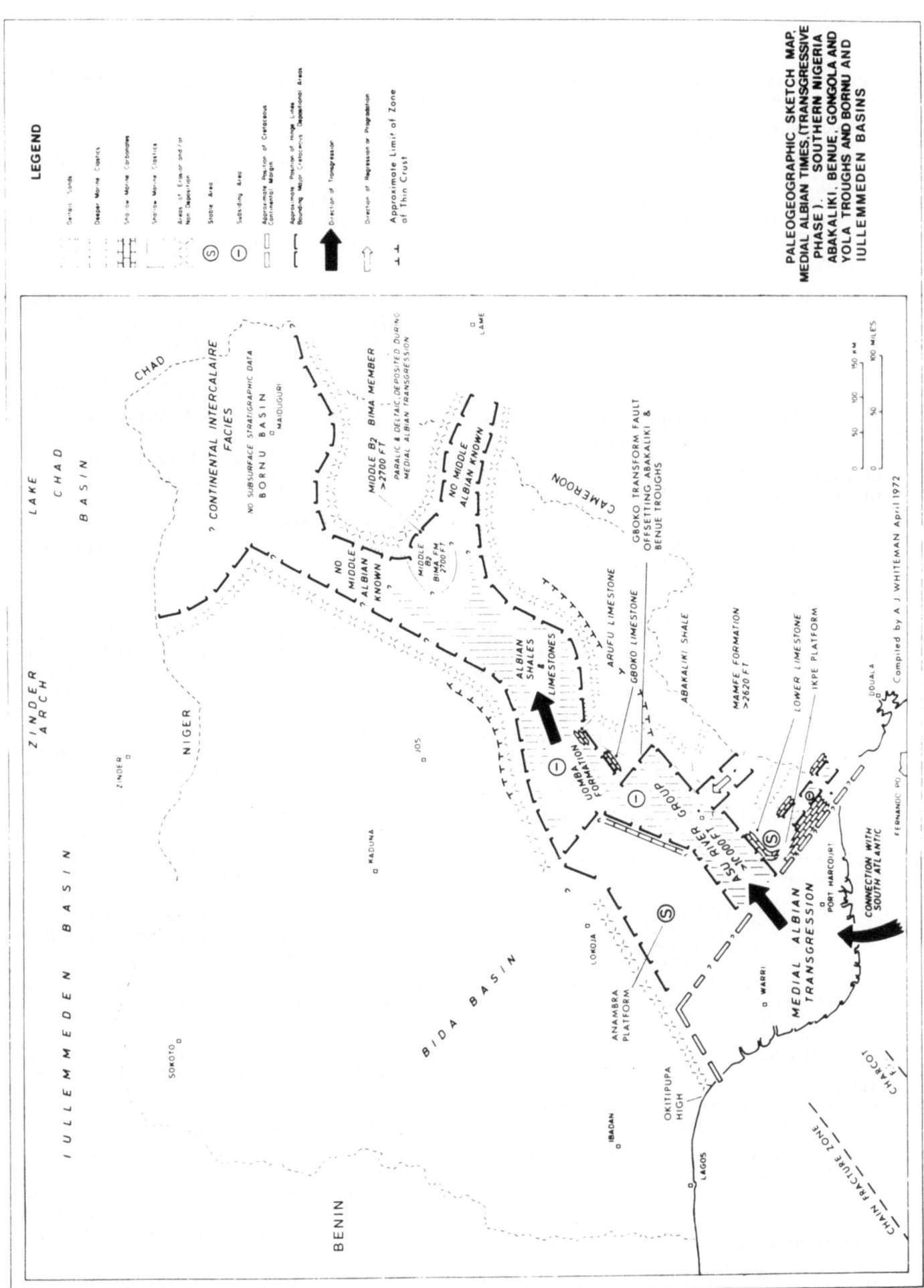

LEGEND

Clastic Sands

Deeper Marine Clastics

Shallow Marine Carbonates

Shallow Marine Clastics

Areas of Erosion and/or Non Deposition

Stable Area

Subsiding Area

Ⓢ Approximate Position of Cretaceous Continental Margin

Ⓘ Approximate Position of Hinge Lines Bounding Major Cretaceous Depositional Areas

Direction of Transgressor

Direction of Regression or Progradation

Approximate Limit of Zone of Thin Crust

PALEOGEOGRAPHIC SKETCH MAP. MEDIAL ALBIAN TIMES, (TRANSGRESSIVE PHASE). SOUTHERN NIGERIA ABAKALIKI, BENUE, GONGOLA AND YOLA TROUGHS AND BORNU AND IULLEMMEDEN BASINS

Compiled by A J WHITEMAN April 1972

Figure 25. Palaeogeographic sketch map, Medial Albian Transgressive Phase for Southern Nigeria, Abakaliki, Benue, Gongola and Yola Troughs and Bornu and Iullemmeden Basins. Based on various sources (see Figure 21) mainly Murat (1972) and Whiteman (1973).

Simpson (1955) basing his comments on reports made by the Shell-D'Arcy geologists gave the following description:

> The Asu River Group consists largely of olive brown sandy shale, fine grained micaceous sandstones and micaceous mudstones. Bluish grey or olive brown shales which weather to a rusty brown colour are also present. The sequence is poorly fossiliferous.

The Survey also mentioned that the group is termed the Asu River Series and that the formation is over 10 000 ft thick.

Reyment (1955) erected the divisions of the Asu River Group shown in Figure 26. It should be remembered however that this sequence is not a continuous one and that some of the localities are separated by tens of kilometres of unmapped, folded and poorly exposed ground. The units which yielded fossils are not part of a continuously exposed sequence.

Figure 26
Correlation of Lithological Units of Albian Age

Time–Stratigaphical Units		
Upper	Unnamed Formation	Arufu Limestone
Albian	Abakaliki Shale	
Middle	Unnamed Formations	Uomba Formation
Albian		

Based on Reyment (1965)

The main outcrop area of the Asu River Group is shown in Figure 9. These rocks, as mentioned above, are all restricted to southeastern Nigeria. The Asu River Group occupies the core of the Abakaliki Fold Belt which is more than 130 miles long and 80 miles wide maximum and extends from the headwaters of the Ivo and Asu River systems, east of Okigwi to the Itaka region, east of the Workum Hills, where the folds end in pitching anticlines which affect the Basement Complex. The fold axes in the Abakaliki anticlinorium trend predominantly NE–SW (Figure 156).

East of Okigwi the Campanian Nkporo Shales rest unconformably on the Asu River Group at the southwestern nose of the fold belt but on the flanks and at the eastern end of the structure the Eze-Aku Shale succeeds conformably and the Asu River Groups rests non-conformably on the Basement Complex. In this account of lithostratigraphic units, Reyment (1965) gave a synonymy for the Asu River Group but did not give a formal lithological description. The Asu River Group is said to include the Lower Shales of Bain (1924, p. 15); Cross River – Benue Shales (in part) (Tattam 1944, p. 28); Asu River Series (Simpson 1955, p. 9). Actually Simpson cited the Asu River rocks as a Group. Farrington (1954) described shales, limestones and sandstones resting unconformably on the Basement Complex (Farrington

1952, Figure 2) and within these rocks he recognized thick folded shales of Albian age which pass under tilted but unfolded sediments at the Awgu escarpment. The folded shale "series" is said to be intruded and mineralized, whereas the overlying "Awgu beds" are not. The Geological Survey of Nigeria (in *International Stratigraphic Lexicon* 1956) described the Asu River Group. The type reference is in Simpson (1954) although the name was first proposed in an unpublished report by Shell-D'Arcy. The group is said to consist of:

> Olive-brown sandy shales, fine grained micaceous sandstones and micaceous mudstones. Bluish-grey or olive-brown shales which weather to a rusty brown colour are also present. Although this sequence is poorly fossiliferous, there are occasional outcrops of thin shelly limestone. Thickness exceeds 5600 ft (Geological Survey of Nigeria 1956). Its age was given as Albian; its type locality as Asu River; and its distribution as occupying the Abakaliki area, Ogoja Province. The foregoing description then stands as the official type reference, with the slight modification made later by the Geological Survey of Nigeria (1957) that the age of the Asu River Group is 'Albian and Pre-Albian?'.

The Asu River Group was intruded by "Minor Basic and Intermediate Intrusives" which outcrop out around Ishiago and to the west; and around Abakaliki northwest of Odomobe. In the Workum Hills the Asu River Group shows contact metamorphism. Lead-zinc mineralization has affected the Abakaliki area (Farrington 1952). The mineralization and igneous activity is discussed in Chapter 3 Structure. Reyment (1965) subdivided the Asu River Group further as shown in Figure 24. The status of Reyment's (1965) subdivisions of the Asu River Group has yet to be decided officially. Presumably they must be thought of as formations (or perhaps as members) within the group (or formation) but more information is needed before they can be given official status.

Abakaliki Shale: This unit was proposed by Reyment (1965). The type locality is given as the area around Abakaliki Town, Abakaliki Province, (Reyment 1965, Plate XVII, Figure 3).

> Exposures occur in ditches and drains; there are no sizeable outcrops. The sediments of the Abakaliki Shale consist of rather poorly bedded shales, occasionally sandy, and there are lenses and sandstone and sandy limestone. One limestone bed attains a local thickness of thirty metres. The approximate distribution of the Abakaliki Shale occurs roughly over radius of twenty-five kilometres. Palaeontologically, it is characterized by species of *Mortoniceras* and *Elobiceras*. The sediments are folded, particularly in the country south of Abakaliki; the fold axes stretch NE–SW. The formation is associated with lead-zinc mineralization. The shales are deeply weathered. Radiolaria occur and echinoids are locally abundant. Pelecypods and gastropods are relatively rare.

Above and below the Abakaliki Shale are "unnamed formations" (Reyment 1965). Pre-Albian Abakaliki Shale pyroclastics have been described from an

Abakaliki Town Quarry by Uzuakpunwa (1974) and beneath these pyroclastics "Intra-continental sandstone" occurs resting on Basement Complex. An Aptian age was assigned to these sediments but no palaeontological proof was given.

The total thickness of the Abakaliki Shale is not known.

Gboko Limestone: The name Gboko Limestone was proposed by Reyment (1965 p. 27). It was described at its type locality as:

> a ten to fifteen metres thick limestone near Yandev-Tiv on the Gboko-Makurdi road, which is probably Albian in age. It contains pelecypod and gastropod shells. Its exact position in the Albian succession has not yet been elucidated and no formal name has yet been proposed.

Farrington (1952) mentioned that Albian fossils were collected from folded shales near to Gboko and in the region between the Egedde Hills and Gboko. The shales (presumably the Asu River Group) are thick, fissile and carbonaceous and contain thin beds and lenses of limestone underlying a thick series of siltstones intercalated with thin beds of sandstone. He also mentioned that limestone rests directly on the Precambrian. Reyment (1965) dated this limestone as Late Albian. The limestone forms part of the Asu River Group outcrops mapped to the west of Gboko (Figure 24). The Asu River Group rests on the Basement Complex in this area and is overlain apparently conformably by the Eze-Aku Shale Group of Turonian-Cenomanian age (Geological Survey of Nigeria Sheet 64 Makurdi). Basal Albian limestone also occurs north of Gboko. Farrington (1952) thinks that the Albian beds in this region are either reduced in thickness or there has been a major lateral change of facies (Figure 25).

Uomba Formation: Reyment (1965) gave the type locality as the Uomba River, Benue Province and described the formation as consisting of:

> mainly of sandstones, shales and sandy shales. They are distributed, over parts of Lafia and Tiv Divisions. The characteristic fossils are oxytropidoceratids and diploceratid ammonites and represent the oldest forms yet identified in the Nigerian Cretaceous succession.

Fossils are figures in Reyment (1955c, 1957b). No thickness data was given.

Arufu Limestone: Reyment (1965) gave the type locality as Arufu, Benue Province. Arufu is situated about 45 miles east of Makurdi in the Benue Depression. The limestone, which is apparently of local distribution (Figure 25) and may not be worth more than the rank of member, is associated with lead-zinc mineralization. It is thought to be in part equivalent to the Abakaliki Shale, as shown by the occurrence of *Elobiceras*. Echinoids and pelecypods are fairly common. Fossils from this unit are figured in Reyment (1957b).

Thickness Variations: Asu River Group: Because of the complexity of the folds and because of the reconnaissance style 1: 250 000 maps available for the area involved it is not possible to provide accurate thicknesses for the Asu River Group and indeed few reliable estimates of thickness have been published. The Shell-D'Arcy estimate of 5600 ft, quoted in Simpson (1955) is probably a minimum thickness and the group may well exceed 10 000 ft (Geological Survey of Nigeria, in *International Stratigraphic Lexicon*, p. 36). Farrington (1952) suggested that the Albian rocks are 4000–6000 ft thick. According to Simpson (1954) a Shell-D'Arcy borehole sunk close to Ishiago railway station began in the Asu River Group and continued until the Ishiago dolerite was encountered at 2147 ft. (This well is labelled SU-7 on the Shell-BP and Geological Survey of Nigeria, Sheet 79, Umuahia.)

Age: According to the Geological Survey of Nigeria (1957) the age of the Asu River Group is "Albian and pre-Albian". Reyment (1965) assigned the Asu River Group to the Middle and Upper Albian. Adeleye (1975), without presenting the evidence, plotted "Transition Beds" with evaporites beneath the Asu River Group and thinks that Mamfé Formation underlies both the Asu River and Transition Beds. He assigned an Aptian and Albian age to the Mamfé Formation and Transition Beds (Figure 27). Bedded evaporites are not known at surface or in wells in these areas. Salt springs occur but most geologists think that the brines have been derived from Pb-Zn mineralizing solutions.

Conditions of Deposition and General Palaeogeography: Murat (1972) presented the best account of Albian palaeogeography published to date and his analysis is followed here for Southern Nigeria. A map showing the structural framework which prevailed in Nigeria from Albian to Santonian times is presented in Figures 22 and 153. The sequence of transgressions and regressions which affected Southern Nigeria are listed in Figure 21.

The Benin and Calabar hinge lines (Figure 22) roughly mark the landward position of the continental margin for this portion of the North Atlantic–West African plate and for the South Atlantic–Central African plate. Seawards of these lines, transitional and "new" oceanic crust occurs, having been generated in Early Cretaceous times when separation took place. The sea eventually flooded the area and affected a connection between the North and South Atlantic in ? Early Turonian times but separation of the continental crust must have taken place earlier. Burke (1976) placed the date of separation as pre-Valangian (120 mya).

Onshore in the Benue area trough systems probably developed over *thinned crust* and probably a trough developed over a very narrow zone of ? *oceanic crust* located beneath the Abakaliki trough. The troughs were probably initiated in Early Cretaceous time and evolved through part of Late Cretaceous time, until in Santonian time the sediments deposited in the troughs were folded, faulted and intruded. As we have said the oldest dated Cretaceous sediments exposed are Albian; older Cretaceous rocks may exist in depth within the Gulf of Guinea miogeocline and in the troughs (Reyment 1965; Murat 1972).

The Abakaliki and Benue Troughs probably

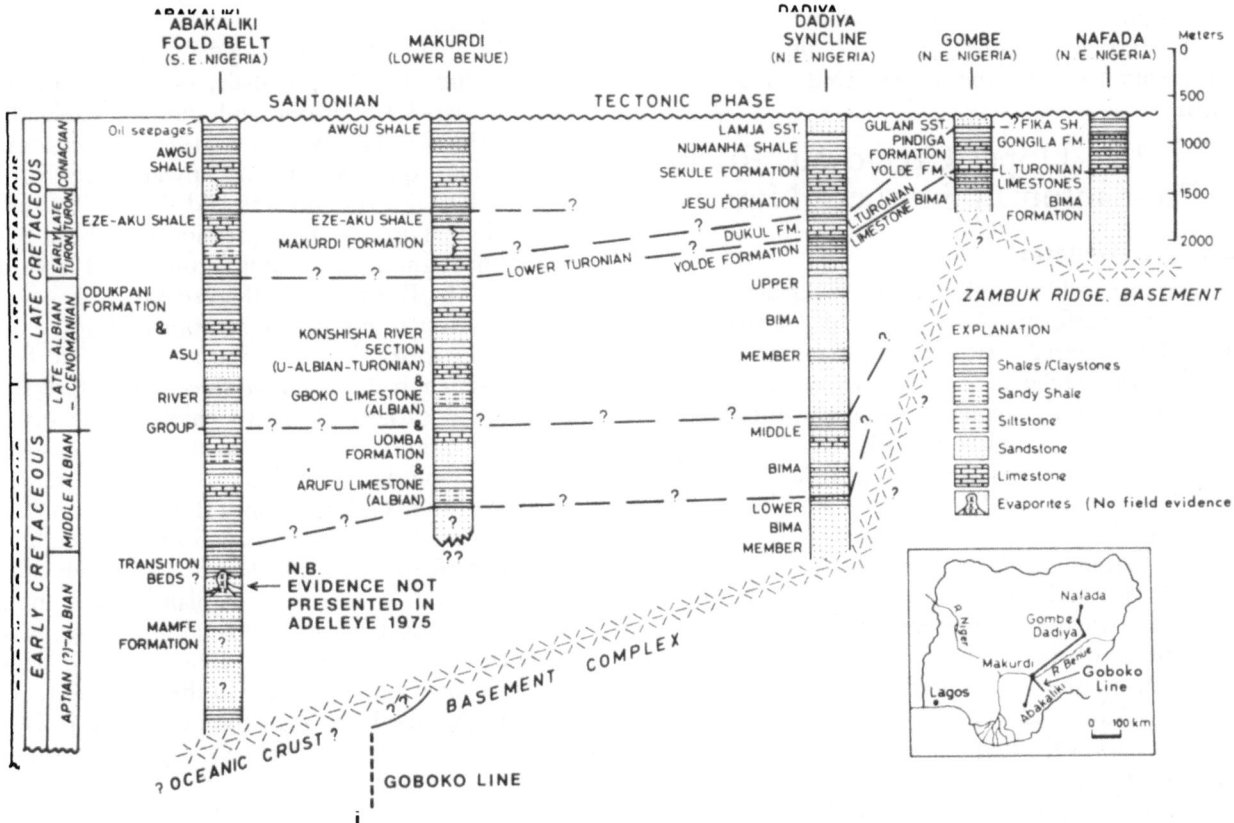

Figure 27. Stylized geological columns for Abakaliki Fold Belt, Makurdi, Dadiya Syncline, Gombe and Nafada areas. Adapted from Adeleye (1975) who based part of his original on sections compiled by Whiteman (1971 MS). Data from Simpson (1964); Carter *et al.* (1963); Reyment (1965). N.B. It is not certain whether the transition beds and Mamfé Formation plotted by Adeleye (1975) underlie Asu River Group as shown.

formed part of a triple junction which underlies the present day Niger Delta (Figure 7). On the continental side of the Benin and Calabar flexures, which in part bounded the triple junction area, a mainly gneissose and granitic Basement Complex acted as a major sediment source. Arguing from first principles these areas probably then stood considerably higher both structurally and topographically than they do now simply because the heat flux patterns involved in the development of expanding plate boundaries would produce flanking uplifts (cf. Present Day Afar area, Figures 189, 190 and the Gregory Rift).

The Anambra Platform and Ikpe Platform flanked the Abakaliki Trough but platform areas did not exist on the southern part of the Benue Trough according to Murat (1972, Figure 3), (Figures 22 and 25). With time these flank areas and the triple junction cooled and subsided and the sea transgressed into these areas.

The Asu River Group was deposited in the Abakaliki Trough as deep water marine clastics, under conditions highly favourable for the accumulation of organic debris. Unfortunately no source rock and maturation data have been published for these rocks but in places Asu River Group shales on petrographic grounds are clearly overmatured. Shallow marine clastics accumulated on the Anambra and Ikpe Plat-

forms, which were stable areas at the time. During the Medial Albian transgressive phase, limestones were deposited at the platform edge. Murat (1972) thinks that limestones were probably laid down also along part of the margin of the Anambra Platform (Figure 25). The Gboko Limestone was deposited on the southeastern Benue hinge line (Figure 22) and these, and the "Lower Limestones" were all laid down probably about the same time during the first transgressive phase of the first cycle (Figure 21). Shallow marine carbonates, the "Lower Limestones", were deposited on the Calabar flank where the pattern of sedimentation was complicated because of the existence of the Ikang Trough and Ituk High. Limestones were deposited both inshore and on the seaward side of Ikang High. The Uomba Formation of Reyment (1965 and 1967) is thought to have been laid down about the same time as the Gboko Limestone under paralic and marine conditions in the Lower Benue Trough (Figure 25). Pre-Albian–post Aptian pyroclastics have been recorded from the Abakaliki area (Uzuakpunwa 1974). These consist of agglomerates and tuffs. The agglomerate clasts consist of moderately altered basalt set in an andesitic matrix.

Wells have not penetrated deep enough beneath the Niger Delta Complex to encounter Cretaceous rocks above what was the active triple junction area and geophysical studies made to date have not been able to unravel the complexities of sedimentation and volcanicity beneath the Niger Delta sedimentary prism.

2.4.2.2. Upper Cretaceous Rocks Pre-Santonian Southern Nigeria

2.4.2.2.1. General

Pre-Santonian Upper Cretaceous rocks occur on the Anambra Platform, Abakaliki and Calabar Flank areas of Southern Nigeria, but the main outcrop areas and thickest sequences are in the Benue and Abakaliki Troughs. Thinner sequences occur on the flanks of the Abakaliki Trough especially on the Anambra Platform and on the Calabar Flank. Palaeogeographic sketch maps for the Early Turonian Transgressive Phase, Early Turonian Regressive Phase, Late Turonian Regressive Phase, Late Turonian Transgressive Phase and the Coniacian–Early Santonian Regressive Phase are presented in Figure 22. The distribution of the pre-Santonian Upper Cretaceous formations is shown in Figures 9, 24, 25, 27 and 28.

In Late Albian (Early Cretaceous) and Cenomanian (Late Cretaceous) times a major regressive phase developed and extensive deltaic deposits spread down the Benue Trough (Figure 28). The Keana sandstones of the Shell-BP geologists (Murat 1972) and part of the Keana Sandstone of Offodile (1976) were laid down during this phase. They are thought to be the down valley equivalents of the Muri sandstones of Falconer (1911) and the Bima Sandstones of Carter *et al.* (1963) (Figure 37). Braided river systems, deltas etc. are thought to have occupied most of the Benue Troughs in Late Albian times. Whether the deposits laid down at this time were continuous sand bodies, or whether the shoulders of the trough locally supplied coarse sediments in quantity, which in places coalesced to form large individual fans or deltas, separated by areas of finer grained sediments, is not clear from published data (Figure 28).

This regressive phase was followed in the Abakaliki Trough and flank areas by an Early Turonian Transgressive Phase and parts of the Odukpani Formation, the Eze Aku Group and perhaps the upper part of the Asu River Group were laid down (Figure 29). On the Anambra Platform and on the flanks of the Abakaliki Trough and the Calabar Flank shale–limestone sequences were deposited. It has been suggested that it was about this time that the waters of the proto-North and South Atlantic Oceans met (Figure 29). An Early Turonian Regressive Phase is thought to have succeeded in the Abakaliki Trough and flanking areas and the Eze Aku Shale Group was deposited in the Abakaliki Trough (Figure 32). Interdigitating sand facies were deposited on the flanks and in the Lower Benue Trough. These formations include the Agala Sandstone; Ameseri Sandstone; the Konsisha River Group and the Makurdi Sandstone.

In Late Turonian times in Southern Nigeria in the Abakaliki Trough and adjacent flanking areas another major transgression affected the area and united South and North Atlantic waters penetrated deeply into the Benue Trough. Deposits laid down at this time include the Awgu Shale; Nkalagu Limestone; the Calabar–3 Limestone and similar "flank" limestones (Figure 33). At the northeastern end of the Abakaliki Trough and in the Lower Benue Trough the Wadatta Limestone was laid down. The effects of the Late Turonian transgression have been recognized throughout the Benue Trough and have been noted in the Bornu Basin (Figure 33).

In Coniacian and Early Santonian times a major regression took place and the megatectonic framework which had prevailed since Early Cretaceous times (and earlier) began to change probably as a result of modifications of the stress field associated with the evolution of the long ridge–ridge transform complex of the Gulf of Guinea (Figures 34 and 224a). The Agbani Sandstone was laid down and the Anambra Platform, which had acted as a slowly subsiding platform area since Albian times, became markedly subsident and the Anambra Basin was initiated. The Abakaliki Fold Belt began to develop and volcanics were extruded in the Umuna–1, Ikono – 1 and Annua–1 areas on the northeastern flank of the Niger Delta area which was slowly subsiding as the newly generated continental margin began to evolve, passing through leptogeoclinal, protomiogeoclinal and miogeoclinal phases as the South American and African plates parted. Basic and intermediate intrusives were emplaced in the Abakaliki Fold Belt (Figure 238). Upper Cretaceous Pre-Santonian formations are described systematically below.

2.4.2.2.2. Odukpani Formation

General Description: The Odukpani Formation type locality is near Odukpani, Calabar Province on what is now the northeastern flank of the Niger Delta Complex (Figures 30 and 31). Reyment (1965) described the Odukpani Formation as consisting of:

TOP
5. Flaggy shales and calcareous sandstones with Lower Turonian ammonites.
4. Sandy shales with sandstone bands and calcareous sandstones with Cenomanian ammonities.
3. Alternating limestones and shale with Cenomanian ammonites.
2. Limestone and calcareous sandstones with fragmentary crinoids and algae.
1. Basal sandstones and conglomerates.
UNCONFORMITY
PRECAMBRIAN BASEMENT COMPLEX

Apart from work done by Reyment (1955 and 1956) and Reyment and Barber (1956), little has been done on the Odukpani Formation, except by Dessauvagie (1972), who gave a detailed account of the microfauna of the type section. A lithological section and a fossil distribution diagram drawn up by Dessauvagie (1972) are presented (Figure 31).

Thickness Variations: Dessauvagie (1972) estimated the thickness of the Odukpani Formation as 2460 ft. Reyment (1965) gave the thickness as 2000 ft. ˙

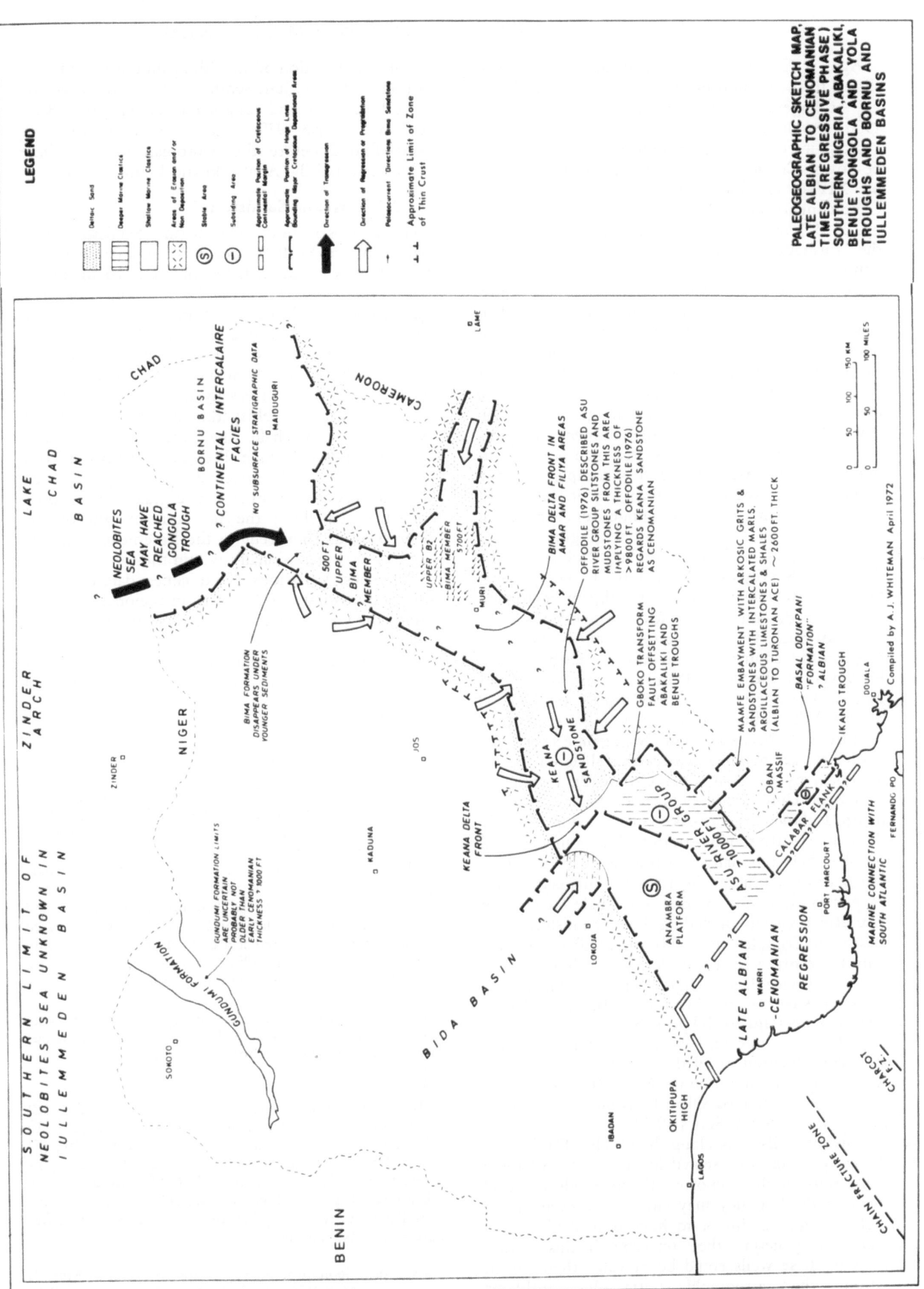

Figure 28. Palaeogeographic sketch map, Late Albian to Cenomanian Regressive Phase for Southern Nigeria; Abakaliki, Benue, Gongola and Yola Troughs and Bornu and Iullemmeden Basins. Based on various sources (see Figure 22) mainly Murat (1972) and Whiteman (1973).

Age: A Cenomanian-Turonian age has been assigned to the Odukpani Formation (Geological Survey of Nigeria, in *International Stratigraphic Lexicon* 1956 (Dessauvagie 1968 and 1972), but as Dessauvagie pointed out the basal unfossiliferous beds may be Lower Cretaceous (Albian). Reyment (1965; Table XIV–6) classified the Odukpani Formation as Cenomanian, but both in the text (Reyment 1965, pp. 34–36) and in Figure 6 (op. cit.) he stated that the formation also contains Lower Turonian ammonites.

Conditions of Deposition and Palaeogeography: The Odukpani Formation was deposited under shallow water nearshore conditions (Reyment 1965; Dessauvagie 1972). Murat (1972), although not mentioning the Odukpani Formation by name in his paper, mapped deltaic sand, shallow marine and deep marine clastic facies in this area belonging to the Asu River Group. The pattern of the sedimentation plotted by Murat (1970) for this area incorporates unpublished sub-surface data and is complex because sedimentation around the Oban Massif, Ikpe Platform, the Calabar Flank and Hinge Zone, the Ikang Trough and Ituk High are involved (Figure 28).

Cartographic and Stratigraphic Problems: The Geological Survey of Nigeria (1964, *Geological Map*, Scale 1 : 2 000 000) show the Odukpani Formation resting on Basement Complex and overlain by the Eze-Aku Shale Group. The Odukpani is shown as being both Cenomanian and Turonian in the legend. However, on the Shell-BP and Geological Survey of Nigeria Oban Hills Map (1957, Scale 1 : 250 000 Sheet 80) the Odukpani area (including the area covered by both Reyment's and Dessauvagie's sections, Figure 30) are mapped as Eze-Aku Shale Group (3) underlain by the Asu River Group (2), resting on Basement Complex. the Nkporo Shale Group is mapped above the Eze-Aku Shale Group. Clearly there is a cartographic-stratigraphic nomenclatural problem here especially as Murat (Shell-BP Geological Map of Southern Nigeria 1969) mapped Asata-Nkporo Group (5), on Eze-Aku Shale Group, on Asu River Group (including Basal Sandstone), on Basement Complex. Probably these differences will only be resolved by a consideration of both surface and sub-surface data. In figure 9 the Odukpani Formation has been mapped resting on Basement Complex and overlain by Nkporo Shale. This must be regarded as an interim solution and it may be that after consideration of all the data the best solution will be the one shown on the Shell-BP and Geological Survey of Nigeria Oban Hills Sheet 80 (Scale 1 : 250 000 Geological Series 1957).

A list of wells sunk along the Calabar Flank since the early 1950s is presented in Figure 35. All these wells start in the Cenozoic Benin Sands (Coastal Plain Sands) but they may either have been placed high enough up dip or to have been drilled deep enough to penetrate the Cretaceous in this region. Logs of these wells could be of value therefore in settling the cartographic-stratigraphic problems mentioned above.

The basal section of the Odukpani Formation consisting of fluviodeltaic sediments has been called the Awe Formation and assigned an Albian Age (Adeleye and Fayose 1979). They suggest that the beds were folded before the Santonian folding phase which affected the Abakaliki and Benue Troughs.

2.4.2.2.3. Keana Sandstone

General Description: The Keana Sandstone was named by the Shell-BP geologists in a company report. It was mentioned, but not referred to, in the bibliography of Murat (1972). It is the down-dip equivalent of the Muri Sandstone (Falconer 1911) and the Bima Sandstone (Carter *et al.* 1963). Offodile (1976) described the Keana Sandstone at its type locality in the Keana Anticline in the Lower Benue Trough (Figure 37). Dessauvagie (1974) described it as a coarse grained, feldspathic, micaceous sandstone with siltstone and shale. The formation occurs mainly in the Lower Benue Trough as defined in this account. Very little systematic mapping has been done in this area. Similar coarse grained sandstones which occur along the northern margin of the depression and adjacent to the Basement Complex are included provisionally within the formation.

Thickness Variations: No thickness data were presented by Murat (1972) but from regional considerations the sandstone must be many hundreds of feet thick.

Age: Murat (1972, Figure 2) classed these sandstones as Cenomanian, but on his Figure 6 (op. cit.)

Figure 22 A – J. Cartoon maps showing megatectonic frameworks for Medial Albian to Santonian times (A) (Figure 153) and for Campanian to Eocene times (H) (Figure 200) for Southern Nigeria, Abakaliki, Benue, Gongola and Yola Troughs and Bornu, Chad and Iullemmeden Basins. Positions of hinge lines, structural lineaments etc. bounding Cretaceous depositional features are schematic and approximate. Isopach data for Northern Nigeria is based on interpretation of French gravity data and shows only the broadest structural outline; fault troughs exist in these areas but have yet to be delimited publicly.

B – G are palaeogeographic summary maps for the same areas shown on 22A and 22H and show broad distributions, facies relationships for transgressive and regresssive episodes:

(B) Medial Albian Transgressive Phase (Figure 25); (C) Late Albian – Cenomanian Regressive Phase (Figure 28); (D) Early Turonian Transgressive Phase (Figure 29); (E) Early Turonian Regressive Phase (Figure 32); (F) Late Turonian Regressive Phase (Figure 33); (G) Coniacian to Early Santonian Regressive Phase (Figure 34); (I) Late Campanian Transgressive Phase (Figure 57); (J) Maestrichtian Regressive Phase (Figure 59).

Based on: Geological Survey Maps of Nigeria; Murat's Map (1969) Scale 1 : 1 000 000; Murat (1972); Whiteman (1973).

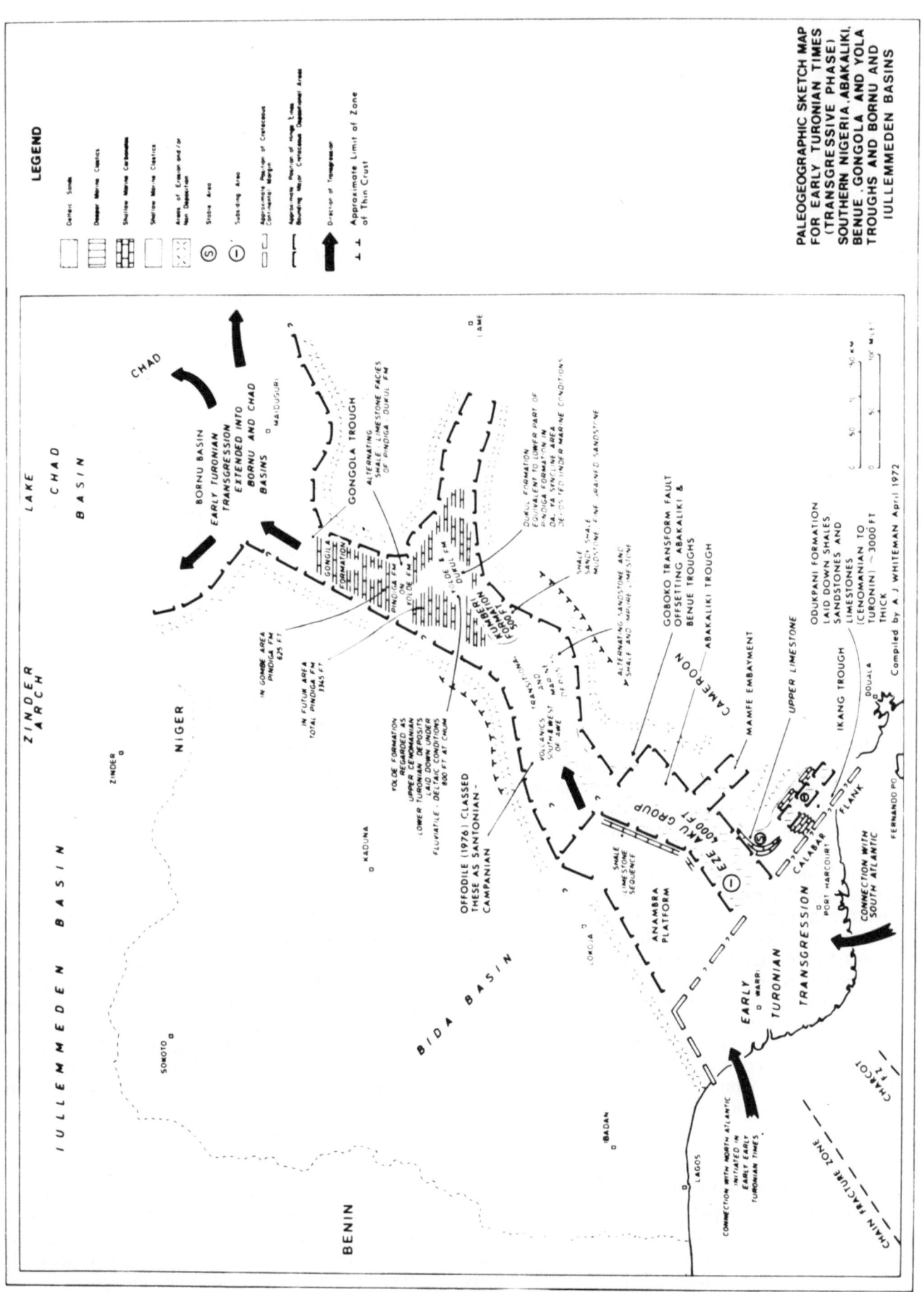

Figure 29. Palaeogeographic sketch map, Early Turonian Transgressive Phase, Southern Nigeria, Abakaliki, Benue, Gongola and Yola Troughs and Bornu and Iullemmeden Basins. Based on various sources (see Figure 22) mainly Murat (1972) and Whiteman (1973).

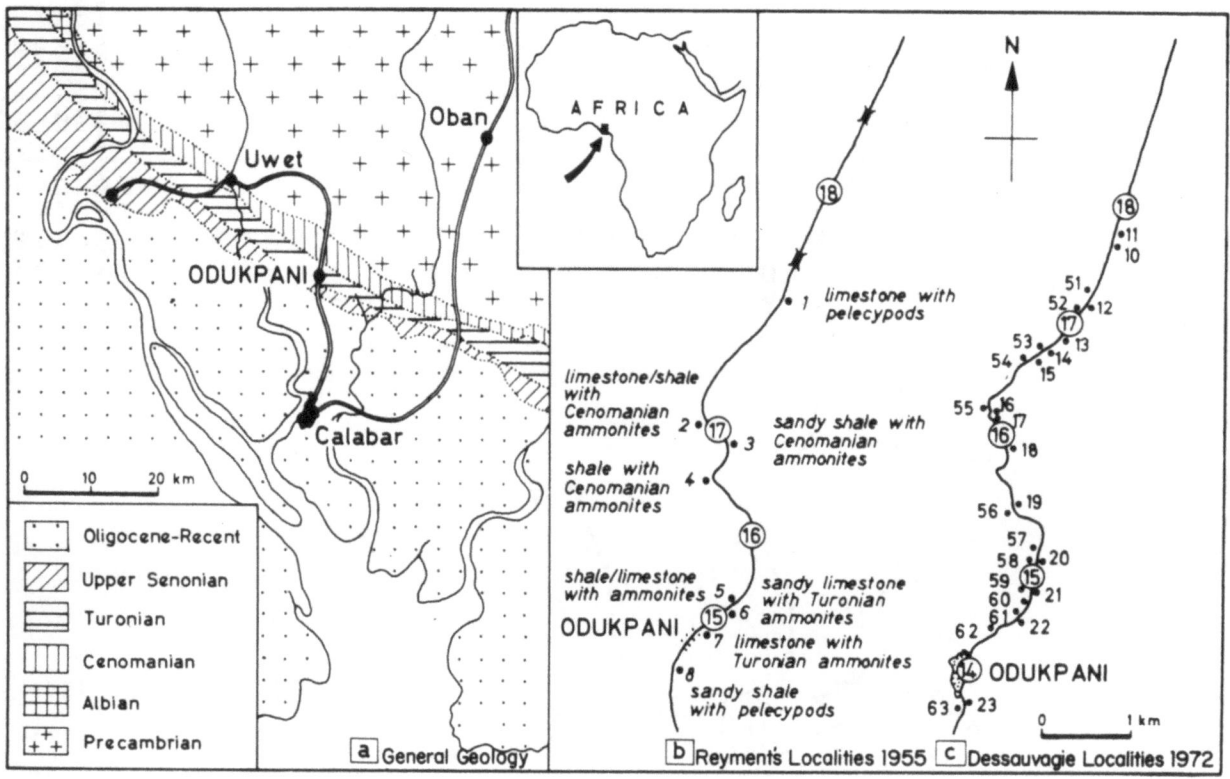

Figure 30. Locality maps, Odukpani Formation, Calabar area. Based on Reyment (1965), Dessauvagie (1972).

the Keana Sandstone is considered to be part Cenomanian and part Late Albian. Dessauvagie (1974) tentatively dated the Keana Sandstone as Albian to Cenomanian (Figure 8). Unnamed marine strata with Lower Turonian ammonites overlie the Keana at Kumberi. The formation is correlated with the Asu River Group and Bima Sandstone (Dessauvagie 1974).

Conditions of Deposition and General Palaeography: According to Murat (1972) the Keana Sandstone overlies the paralic and marine facies of the "Uomba Formation" described by Reyment (1965). The Keana Sandstones are in part the equivalent of the marine shales of the Asu River Group and the shallow marine clastic facies which developed on the Anambra Platform (Figure 28). A Keana delta front must have been situated in the area immediately northwest of Gboko and Keana deposits may have been laid down along the Calabar Flank. The formation is described in more detail in Section 2.4.3.3.3.

2.4.2.2.4. Mamfé Formation

General Description: Jaekel (1909, p. 392) referred to rocks included in the Mamfé Formation as Mamfé-Schiefer and Reyment (1955a, p. 678) used the name Mamfé Formation. The formation consists of massive arkosic sandstones and grits with intercalations of marls, arenaceous limestones and shales. Locally the beds may be carbonaceous and contain thin beds of lignite. The coarse sandstones are commonly current bedded and pebbly and plant remains are common. According to Reyment (1965) the rocks have been moderately to strongly folded along a roughly east–west axis with dips of up to 50°. Salt springs are common in the Mamfé Sandstone and periodic discharges of gas (non-inflammable) were reported by Wilson (1928). The sediments have been intruded by igneous rocks (Figure 9).

Thickness Variations: The Mamfé Formation is said to be 2620 ft thick.

Age: The beds are probably not older than Albian and not younger than Turonian. The Mamfé Formation has yielded *Proportheus kameruni* Jaekel and fossil wood. Murat (1969 and 1972) mapped the Mamfé Formation as a sandy facies of the Asu River Group of Medial and Late Albian to Cenomanian age (Figures 25 and 28). Adeleye (1975) (Figure 27) classed the Mamfé Formation as Aptian-Albian placing it beneath the Asu River Group and Transition Beds. This is not supported by published field evidence nor by palaeontology.

Conditions of Deposition and General Palaeography: Sandy deltaic facies in Late Albian–Cenomanian Times extended southwards from Gboko and merged with the coarse sandy clastics of the Mamfé Embayment. Sandy deltaic facies also extended around the Oban Massif and along the Calabar Flank (Figure 22). Shallow marine fine grained clastics formed a discontinuous belt seaward of the sandy deltaic facies from Gboko to Cameroun where the sandy phase disappears under the volcanics of Cameroun Mountain. These sands are presumably the

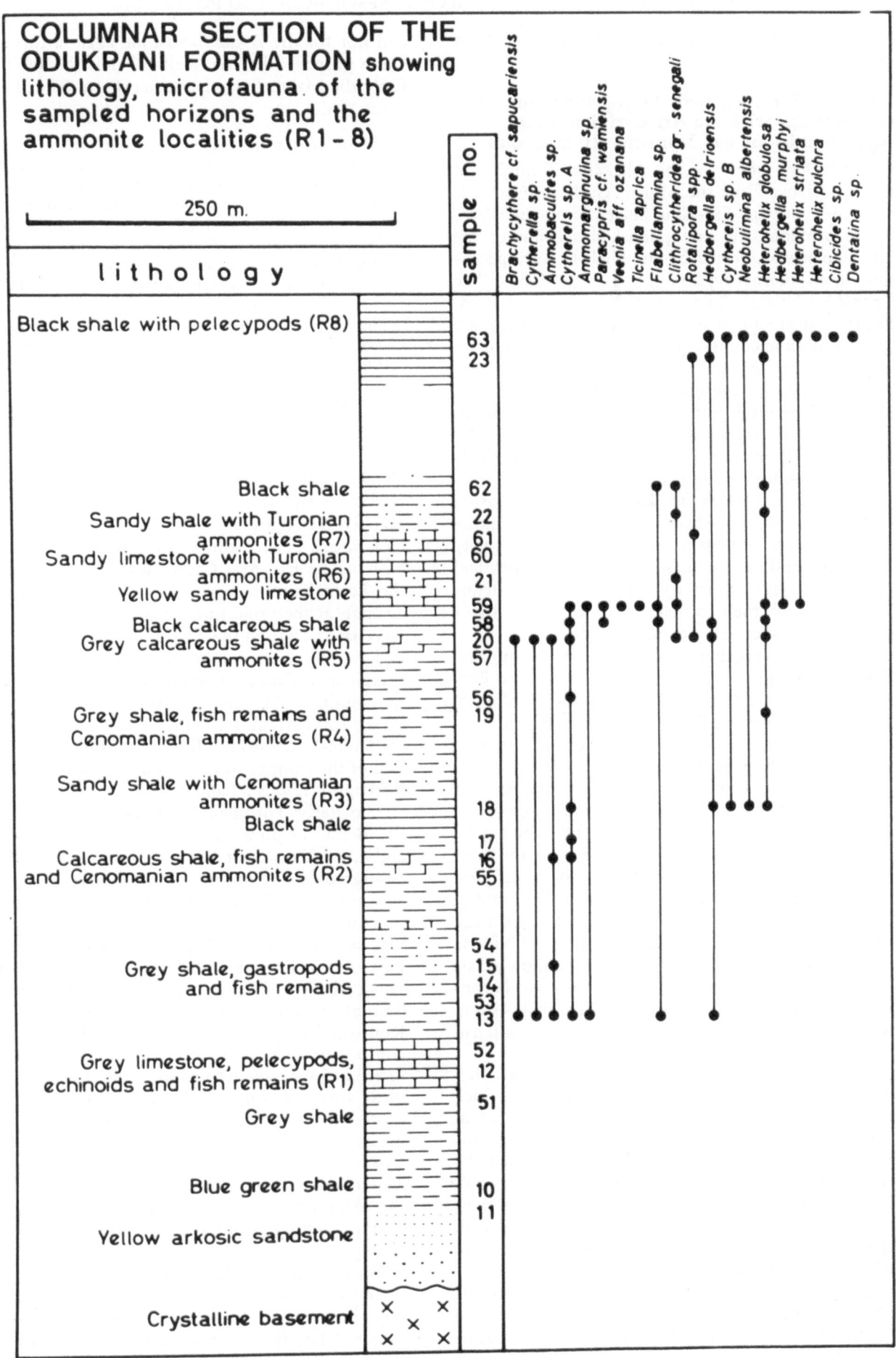

Figure 31. Columnar section of the Odukpani Formation showing lithology and microfauna of the sampled horizons and ammonite localities (R1–8). Based on Dessauvagie (1972).

equivalent of the Grès de Base of the Douala Basin of Cameroun.

2.4.2.2.5. Eze-Aku Shale Group including its relationships with the Ameseri Sandstone, Agala Sandstone and Konsisha River Group

General Description: In the Abakaliki Trough and on the Calabar Flank the Albian Asu River Group is succeeded by the Eze-Aku Shale. Simpson (1955, p. 10) provided the first published account of the unit which was erected by the Shell-D'Arcy geologists in an internal report written in the 1950s. The "Eze-Aku Shales" (Simpson 1955) consist of hard, flaggy, calcareous shales and siltstones which are grey or black in colour and which have yielded many impressions of *Inoceramus*. Minor sandstones are also present. The type locality is at Eze-Aku, Awgu Division (Geological Survey of Nigeria and Shell-D'Arcy, in *International Stratigraphic Lexicon* 1956). Murat (1972) introduced the term Eze-Aku Shale Group without defining it publicly.

In the Benue Trough the "Eze-Aku Shale" consists of flaggy, calcareous shales with bluish non-calcareous shales and siltstones with thin sandy or shelly limestones and calcareous sandstones. Cratchley and Jones (1965) in their sequence for the Lower Benue Trough show their contact with the Asu River Group as questionably unconformable.

Thickness Variations: Simpson (1955) quoting Shell-D'Arcy figures gave the thickness of the Eze-Aku Shale as 1300 to 2000 ft and the Geological Survey of Nigeria (in *International Stratigraphic Lexicon* 1956) gave a maximum thickness of 4000 ft. Reyment (1965) gave its thickness as 3280 ft.

In the Lower Benue Trough the Eze-Aku Shale is said to be 2000 ft thick (Cratchley and Jones 1965). Clearly, because of the facies changes which take place northwards and eastwards between the Eze-Aku Shale and the Ameseri and Makurdi sandstones, large variations in thickness must occur (Figure 32).

Age: The Eze-Aku Shale is of Early Turonian age (Reyment 1965) but the Geological Survey of Nigeria (1964) classed the Eze-Aku Shale including the Agala and Makurdi sandstones as Cenomanian to Turonian. Murat (1972) assigned the Eze-Aku Shale Group to the Lower Turonian remarking that the group had yielded a rich pelagic microfauna.

Conditions of Deposition and Palaeogeography: The Eze-Aku Shale is a shallow water deposit which locally passes into the Ameseri Sandstone (Reyment 1965). However, Murat (1972) reported a more complicated depositional picture, pointing out that the regressive Late Albian–Cenomanian phase was followed by a markedly transgressive phase in the Benue Abakaliki Troughs. Oil shows are associated with the Eze-Aku Shale. These are discussed in the section on petroleum prospects.

Type descriptions have not been published for the above-mentioned Ameseri Sandstone which is a Shell-BP unit (Murat 1972); nor for the Konsisha River Group which is also a Shell-BP unit mentioned first in Reyment (1965) and then by Murat (1972);

and nor for the Agala Sandstone again a formation name proposed by the Shell-BP unit (Murat 1972). Only the Makurdi Sandstone has been defined (Reyment 1965):

> the southern bank of the Benue River at Makurdi from about two kilometres west of the town to about half a kilometre beyond Wadatta. At its type locality this formation consists of massive sandstones with thin beds of arenaceous shale and calcareous shelly sandstones. At Wadatta, a suburb of Makurdi, a richly fossiliferous limestone occurs. Laterally it passes into shale–limestone sequences. The age is Lower Turonian. The name was originally proposed in an unpublished report by the Shell-BP Petroleum Development Company of Nigeria Ltd as Makurdi Sandstone, lateral equivalent of the Eze-Aku Shale, and used on maps published by Geological Survey of Nigeria in 1957. Fossils may be locally abundant. There are hoplitids and mammitids as well as other molluscs, fish remains and plant fragments.

Thickness data were not given either by Reyment (1965) or by Murat (1972) and the units are not mapped on Figure 9.

Conditions of Deposition and General Palaeogeography: The Eze-Aku Group was laid down during the Early Turonian transgression in the Abakaliki Trough at the same time as shallow marine clastics were laid down on the Anambra and Ikpe Platforms and on the eastern margin of the trough. A rich pelagic microfauna accumulated in these deposits. To the southeast "Upper Limestones" were deposited on the southern margin of the Abakaliki Trough east of Bende-1 and around Ikpe-1 and on the northwest side of the trough, south–southwest and northeast of Nkalagu (Figures 32 and 33).

The Early Turonian transgression extended into the Benue Depression and beyond into the Bornu and Chad Basins but because of the lack of detailed stratigraphic knowledge for the Lower Benue areas we cannot sketch the palaeogeography with any degree of confidence. The Early Turonian *transgression* was succeeded by an Early Turonian *regression* and part of the Eze-Aku Shale Group was laid down under shallow marine conditions in the Abakaliki Trough (Figure 32). The Ameseri Sandstone and the Konsisha River Group were deposited as continental and paralic facies of local delta complexes; the sediments having been derived from the Basement Complex areas of Ogoja. The Eze-Aku Shale may represent the pro-delta clays and deeper water deposits associated with these complexes. The Lower Benue Trough at this time was in part infilled with Makurdi Sandstone (described below), which, like the Ameseri Sandstone and the Konsisha River Group, interdigitated with the marine Eze-Aku Shale Group.

On the Anambra Platform a broad deltaic mass – the Agala Sandstone – developed, the sediments having been derived from the area which is now the Niger Valley (Figure 32). This facies occurs in: Anambra River–1, Amansiodo–1, Ihandiagu–1, Aiddo–1, Ikem–1, Okpaya–1, Bopo–1, and Adoka–1 (all of

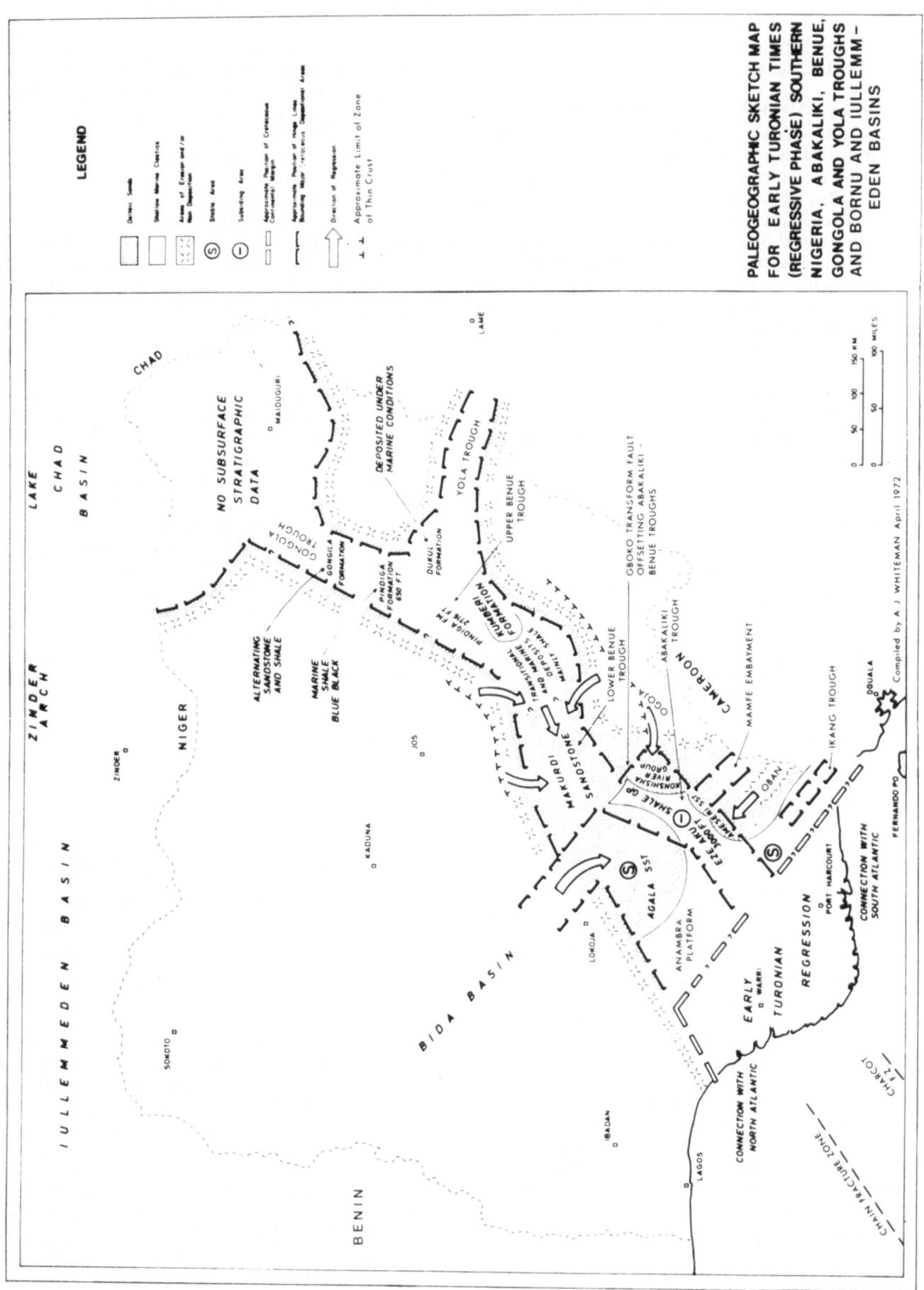

Figure 32. Palaeogeographic sketch map for Early Turonian Regressive Phase, Southern Nigeria, Abakaliki, Benue, Gongola and Yola Troughs and Bornu and Iullemmeden Basins. Based on various sources (see Figure 22) mainly Murat (1972) and Whiteman (1973).

43

which are Safrap wells except Amansiodo–1 which is a Shell–BP well).

2.4.2.2.6. Awgu Shale and Agbani Sandstone

General Description: The Awgu Shale was deposited in the Abakaliki Trough during the Late Turonian transgression (Figure 33). Originally the Awgu Shale was called the Awgu–Ndeaboh Shales (Simpson 1955; Geological Survey of Nigeria and Shell–D'Arcy in *International Stratigraphic Lexicon* 1956). Reyment (1965) shortened the name to Awgu Shale. The Awgu Shales are bluish grey, well bedded, with occasional intercalations of pale, yellow, fine grained calcareous sandstones and thin shelly limestones.

Thickness: The shales are about 3000 ft thick (Simpson 1955; Reyment 1965).

Age: Ammonites collected from the lower portion of the shales probably indicate a Turonian age and mollusca from the top of the formation indicate a Coniacian age. The Geological Survey of Nigeria (in *International Stratigraphic Lexicon* 1956) assigned a Late Turonian to Santonian age to the Awgu Shale but later Shell–BP and the Geological Survey of Nigeria (1957) gave it an Early Senonian age.

Reyment (1965) thinks that the Awgu Shale ranges in age from Late Turonian to Santonian and Cratchley and Jones (1965) dated it as Late Turonian and Coniacian. Murat (1972) assigned the formation to the upper Turonian and Lower Senonian. Reyment (1965), Cratchley and Jones (1965) and Murat (1972) drew attention to the major unconformity which cuts the Awgu Shale (Figures 9 and 23).

Conditions of Deposition and Palaeogeography: The Awgu Shales were deposited in the Abakaliki Trough as "deeper water marine clastics" and shales and other marine deposits appear to have been laid down throughout the Benue Trough during the First Transgression of the Third Cycle of Murat (1972) (Figure 23).

Shallow marine clastics were deposited on the Anambra and Ikpe platforms and inshore on the Calabar Flank and on the Ituk High (Figure 33). Limestones occur at the base of this cycle in Calabar–3 and at a similar stratigraphic position along the northeastern margin of the Abakaliki Trough in Nkalagu area and near Aiddo–1. The Wadatta Limestone which outcrops further to the north, near Makurdi is correlated with the Nkalagu Limestone (Figure 33).

Agbani Sandstone: The Agbani Sandstone was formed during the regressive phase of the Third Deposition Cycle (Figure 34). The term Agbani sandstone was used by Murat (1972) without details being given. The beds were laid down in a deltaic environment and occur in Bopo–1, Okpaya–1, Adoka–1, Aiddo–1, Ikem–1 and Amansiodo–1 and outcrop east of Enugu.

2.4.2.2.7. Umuna–1, Annua–1 and Ikono–1 Volcanics

Murat (1972) and Burke, Dessauvagie and Whiteman (1972 and 1971) mentioned that the plunge of the Abakaliki Fold Belt (which was formed probably in the Santonian times) and that part of the Calabar Flank constituting what is now the northeastern flank of the Niger Delta Complex, are overlain by "effusives" (tuffs and lava flows). These occur in Umuna–1, Anua–1 and are especially developed in Ikono–1 which penetrated an effusive sequence more than 4000 ft thick (Murat 1972) (Figure 246A).

Burke, Dessauvagie and Whiteman (1971) recorded that "more than 1300 m of andesitic and basaltic lavas and tuffs with shales were erupted at about Santonian time in the Abakaliki, Ikono–1 and Ogbabu–1 areas". No further basic information has been published about these volcanics, although their origin has become the subject of considerable discussion (Wright 1976 etc.; Burke, Dessauvagie and Whiteman 1972; Burke and Whiteman 1973; Nwachukwu 1972). The origin of these rocks and the intermediate and basic intrusives (elsewhere associated with a lead-zinc mineralization) which were intruded into largely pre-Santonian sediments is discussed in Chapter 3 Structure. "Minor and Basic Intrusives" are widespread in the Abakaliki Fold Belt especially in the Workum Hills where there is a large area of "metamorphosed shale". Little data has been published about these intrusives and extrusives or about the "metamorphosed shales".

Burke, Dessauvagie and Whiteman (1971 and 1972) suggested that the Umuna–Anua–Ikono igneous and volcanic rocks developed as a result of the formation of a *minor* "subduction zone" consequent on the development of a minor Santonian closing episode. Wright (1976) disagreed with this suggestion (see Chapter 3). The views currently held by the writer on the origin of these effusives and intrusives are presented in Section 3.3.2.3.

2.4.2.3. Hydrocarbon prospects Southern Nigeria including Anambra Platform, Abakaliki Trough and Calabar Flank Pre-Santonian Lower and Upper Cretaceous Formations

In the early days of exploration in Nigeria, Shell–D'Arcy prospected through the above-mentioned areas (Oil Concession Map 1951, Figure 296) but despite the discovery of oil shows, oil smells etc., especially in the Odukpani and Eze-Aku Formations, by 1957 Shell–BP had relinquished the Abakaliki Trough area, the Afikpo Syncline, the north and northeastern flanks of the Anambra Basin and the Mamfé Embayment. They held onto the Anambra Basin proper, the plunge of the Abakaliki Fold Belt and the Calabar Flank, relinquishing these areas some years later when they turned their main interests to the Niger Delta Province.

2.4.2.3.1. Pre-Santonian prospects, Anambra platform

Because of the great thickness of the Late Cretaceous

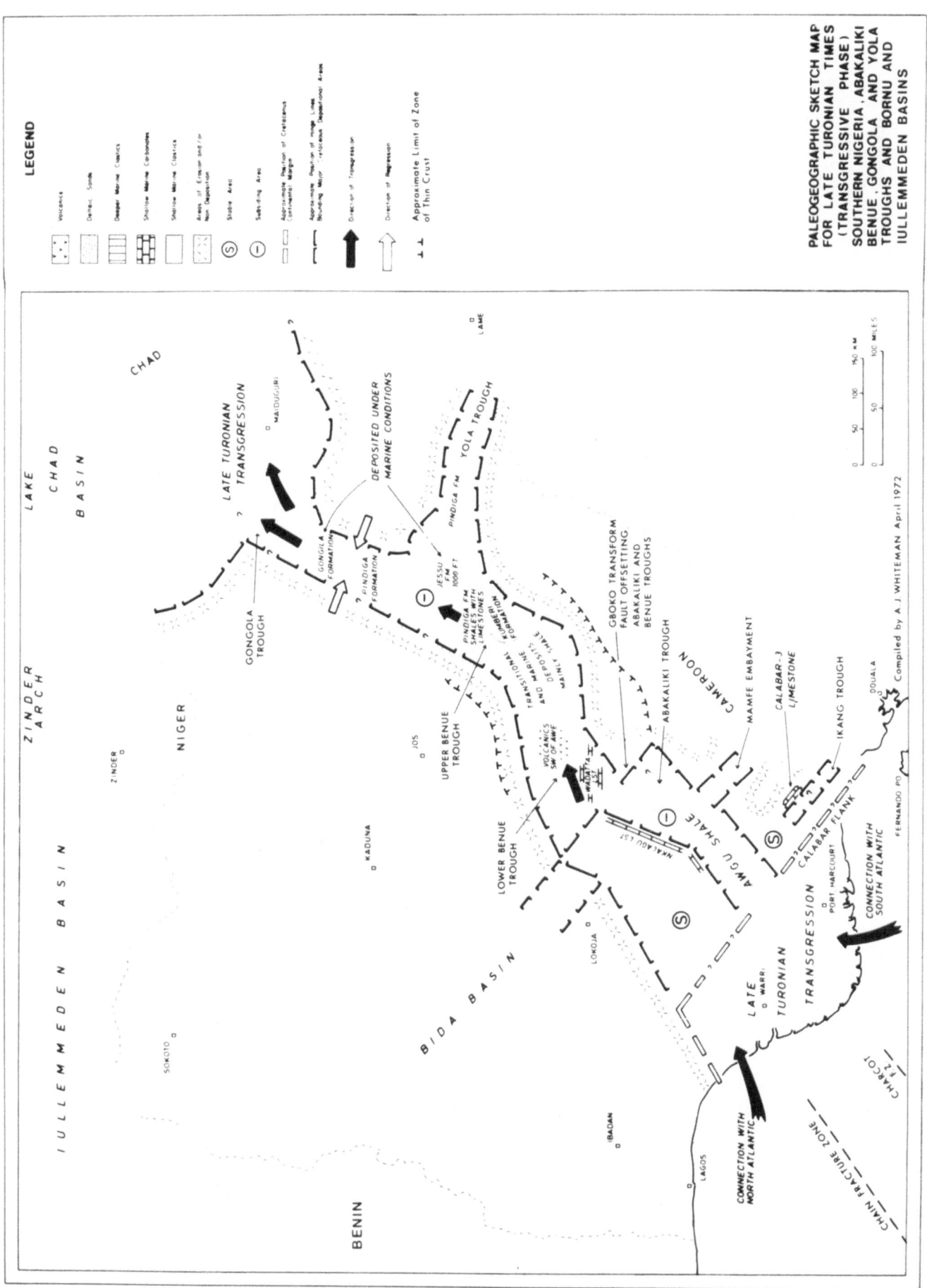

Figure 33. Palaeogeographic sketch map, Late Turonian Transgressive Phase, Southern Nigeria, Abakaliki, Benue, Gongola and Yola Troughs and Bornu and Iullemmeden Basins. Based on various sources (see Figure 22) mainly Murat (1972) and Whiteman (1973).

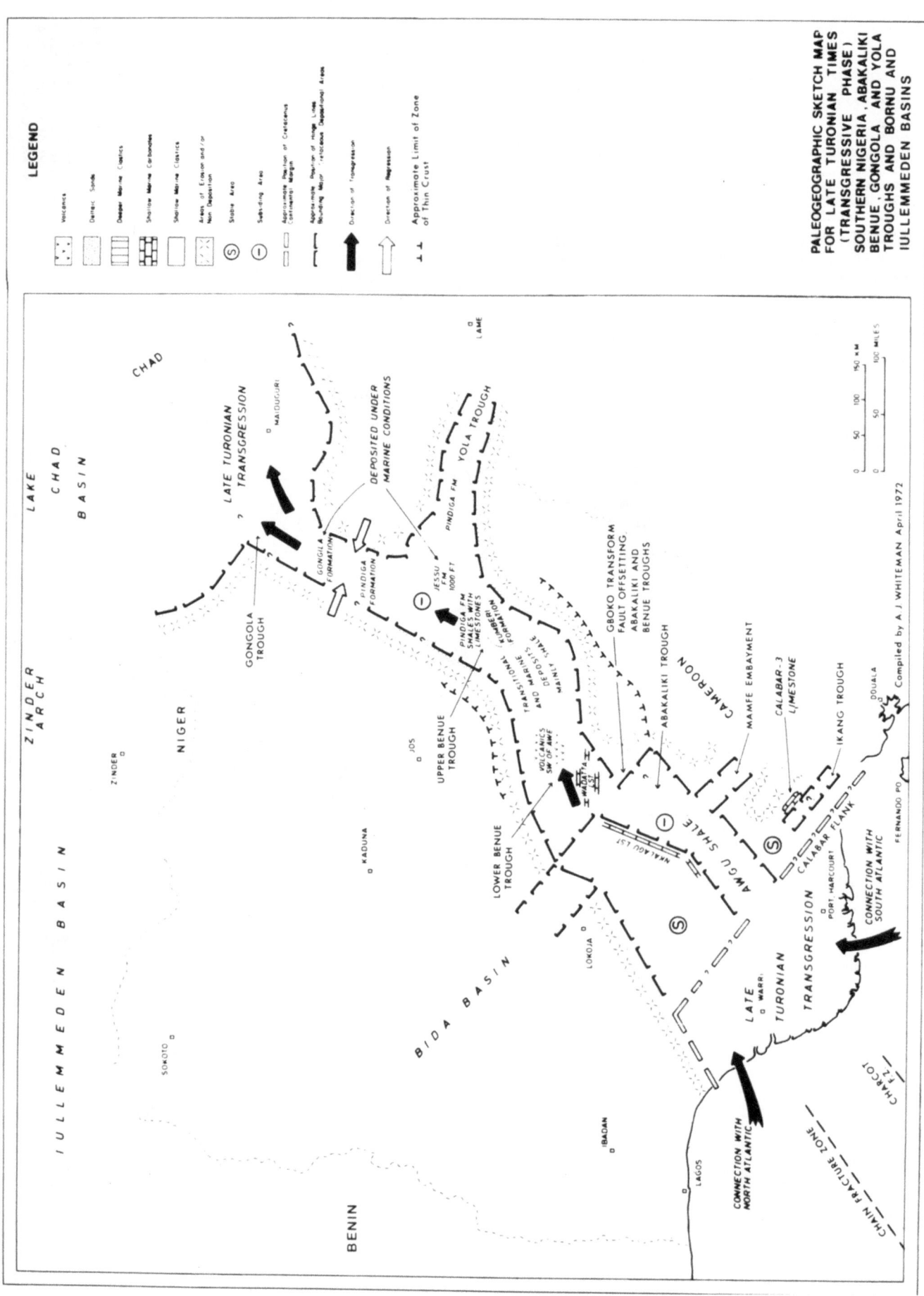

Figure 34. Palaeogeographic sketch map, Coniacian to Early Santonian Regressive Phase, Southern Nigeria, Abakaliki, Benue, Gongola and Yola Troughs and Bornu and Iullemmeden Basins. Based on various sources (see Figure 22) mainly Murat (1972) and Whiteman (1973).

46

Post-Santonian sedimentary cover which occupies much of the Anambra basin and dearth of published information about the subsurface, we can but speculate about possible Lower Cretaceous and Pre-Santonian (Upper Cretaceous) plays. From Medial Albian to Coniacian-Early Santonian times shallow water marine clastics were deposited on the Anambra Platform and marginal limestones such as the Nkalagu Limestone were deposited on the flanks adjoining the Abakaliki Trough (Figure 33). In Coniacian and Early Santonian times (Figure 34) the deposits laid down in the Abakaliki Trough were folded and raised into the Abakaliki Fold Belt or "Anticlinorium" (Figures 22 and 34). The fold belt and adjacent areas then became a source area for sediments deposited in the Anambra Basin in Campanian and Maestrichtian times.

Three possible prospects may be differentiated in the Abakaliki and Benue areas using data publicly available.

PROSPECT 1 Sub-Unconformity formations in the prolongations of the Benue and Abakaliki Fold Belts.

PROSPECT 2 Delta complexes of the Keana and Agbani, Agala, Ameseri Sandstones and marginal sandy facies of the northwest flank of the Anambra Platform.

PROSPECT 3 Southeastern flank of the Anambra Platform zone if interdigitating shallow marine platform clastics and deep water facies of the Asu River Group.

Prospect 1: Sub-unconformity formations in the Prolongation of the Benue and Abakaliki Fold Belts: On the southeastern side of the basin the highly folded Asu River Group of the Abakaliki Fold Belt disappears under the Nkporo Shale which rests upon it with strong angular unconformity at the southwestern end of the fold belt. The Eze-Aku Shale and the Awgu Shale are involved in the folding in the Abakaliki structure but whether all these rocks are present and are folded beneath the pre-Maestrichtian (Santonian) unconformity under the Anambra Basin is not publicly known.

Oil seeps are associated with the Santonian unconformity exposed on the southeastern side of the Anambra Basin and these may well have been fed by accumulations down dip to the northeast. Most of these seeps are in the younger rocks. Dependent on the depth at which the Santonian unconformity occurs in the Anambra Basin and on whether the Lower Cretaceous and pre-Maestrichtian rocks were folded in this area, there may be potential in this region. At the northeastern end of the Anambra Basin the Basement Complex is at shallow depths and was encountered at a total depth of 6018 ft in Ocheku River-1. It is not possible to say exactly how deep the Santonian unconformity may lie beneath the central part of the Anambra Basin with the data available publicly but if we take into account minimum and maximum thickness for the Nkporo, Mamu, Ajali and Nsukka formations (i.e. the post unconformity Upper Cretaceous Formations) it could lie between 3100 and 11 900 ft. On the flanks, because of the sub-concentric nature of outcrops the pre-unconformity, Lower and Upper Cretaceous rocks could well be accessible for investigation. The Basement Complex could lie below 30 000 ft at the south end of the Anambra Basin.

High geothermal gradients prevail in parts of the Anambra Basin and gas rather than oil might be expected in the deeper parts of the basin (Figure 200b).

Prospect 2: Delta complexes of the Keana, Agala, Ameseri and Agbani Sandstones and the marginal sandy facies of the northwestern flank of the Anambra Platform: On the northwestern side of the Anambra Basin, even for deposits laid down during transgressive phases, sequences increase in sandiness and grain size towards the shoulders of the Benue Trough, so reservoir potential may well increase in this direction.

During regressive phases the Keana, Agala and Agbani sandstones were probably laid down as individual deltaic sands, ideal environments for the generation and accumulation of petroleum. These delta complexes were several sizes smaller than the present day Niger Complex but nevertheless considerable quantities of hydrocarbons could have been generated and could have accumulated in them. Growth fault and rollover anticlines have not been described publicly from these delta complexes. However, Anambra River-1 is rumoured to have been located on a growth fault structure.

The Keana, Agala and Agbani delta complexes may have been folded during the Santonian period of folding. They may have been intruded by igneous rocks (granodiorites, calcalkaline volcanics), as in the Abakaliki Fold Belt which would downgrade the area for liquid hydrocarbons. Gas has been found in Anambra River-1 and in Ihandiagu-1 in this area.

Prospect 3: Southeastern flank of the Anambra Platform zone where shallow marine platform clastics and deep water facies of the Asu River Group inter-digitate: The Anambra Platform shallow marine clastics interdigitate with the 10 000 ft thick Asu River Group on the southeastern flank of the Platform. Part of this group which has superficial source rock characteristics (although no analyses are available publicly), was deposited in a trough where conditions must have been favourable for the generation of hydrocarbons.

During transgressive periods shallow marine limestones were deposited along the southeast margin of the shelf (Figures 29 and 33) and there may well be combination traps in both the zones of interdigitation with the shallow and deep water clastics and the shallow clastics, and with the limestones and the deeper water clastics. Again geothermal gradients may have been too high in these areas for liquid hydrocarbons to have been preserved.

2.4.2.3.2. Pre-Santonian Prospects, Abakaliki Trough

The petroleum prospects of the Lower Cretaceous

and the pre-Santonian Upper Cretaceous rocks in the Abakaliki Trough are not rated highly because, although the shales and other fine grained sediments accumulated to more than 10 000 ft thickness (Asu River Group etc.) and interdigitate marginally with coarser clastics, these rocks have been intruded extensively by "Minor Basic and Intermediate Intrusives" (Figure 9). Mineralization is extensive in the Abakaliki Fold Belt and Nwachukwu (1975) has suggested that inclusions found in minerals were formed in the temperature range 100 – 160°C, well into the oil generation and into the gas generating zone.

In the Workum Hills and elsewhere in the fold belts the Asu River Group shows the effects of contact metamorphism and in the Abakaliki area itself an extensive lead and zinc mineralization has been located. Also in Ikoni-1 an effusive sequence of andesitic tuffs and lavas, more than 4000 ft thick has been reported (Murat 1972 and Burke, Dessauvagie and Whiteman 1972) (Figure 34). These volcanics are classed as Coniacian-Lower Santonian (?) and the feeders of the volcanoes that produced these volcanics must cut the Lower and pre-Coniacian rocks at the south end of the Abakaliki Trough. Burke, Dessauvagie and Whiteman (1971 and 1970) identified the rocks in the area as having been formed in a minor subduction zone (Chapter 30).

The Safrap wells Ihandiagu-1 (8280 ft TD), Aiddo-1 (10 554 ft TD) and Amansiodu-1 (7516 ft TD) located on the northwestern limits of the Abakaliki "arch", started in Maestrichtian rocks and may have penetrated pre-Maestrichtian Upper and Lower Cretaceous rocks. Other wells listed in Figure 35 on the northeast and east flank of the Anambra Basin may have penetrated pre-Maestrichtian rocks also. Ihandiagu-1 is listed as a suspended gas well, having been completed just before the outbreak of the Civil War. Little information is available publicly about this well and it is not possible to say which formation yielded 4.6 MMCFD of gas and condensate on test.

The shallow Shell-BP coreholes SU 1, SU 2, SU 4, SU 6 and SU 7 located near the Ishiagu Railway Station; Afikpo 1–4 and 8, located near Afikpo; and the deep Shell-BP wells Ogabu-1 (7502 ft TD), Umuna-1 (7525 ft TD), and Dimneze-1 (7525 ft TD) may all have penetrated as deep as Lower Cretaceous but none of these wells has produced oil and gas, and all were abandoned.

Quite clearly prospecting in the difficult terrain of the Abakaliki Fold Belt would be costly and would involve detailed seismic and stratigraphic work but arguing from first principles some interesting prospects can be built up for natural gas, especially on the flanks of the fold belt where trough sediments merge with other facies. Igneous activity in combination with optimum reservoir and structural conditions might well enhance the possibilities for gaseous hydrocarbons. Another possible play in the pre-Maestrichtian Cretaceous could exist in those areas where predominantly coarse grained clastics interdigitate with the finer grained deeper water clastics of the Asu River Group; the Eze-Aku Group and the Awgu Shale.

The Mamfé Formation which was laid down in the Mamfé Embayment (Figure 9) may interdigitate with the Asu River Group and delta complexes and might well be involved in the close folds which occur southeast of Abakaliki.

Delta complexes developed on the margin of the Abakaliki Trough when the Eze-Aku Shales were being deposited. The coarse grained deltaic sandstones in this case include the Ameseri Sandstone, the Konsisha River Group and the Makurdi Sandstone (Figure 32). Again these complexes were involved in the close folding which occurs on the southeastern limits of the Abakaliki Fold Belt. Bearing these points in mind then there might well be prospects in combination stratigraphic and structural traps along the southeast margin of the Abakaliki Trough. However, Tertiary to Recent volcanics have been intruded into the sediments in parts of this area and may downgrade the prospects for liquid hydrocarbons.

The Cenomanian-Lower Santonian rocks of the Abakaliki Trough and adjacent areas have yielded a few positive signs of the presence of petroleum in the form of "oil smells". Oil smell symbols are marked along the outcrop of the Eze-Aku Shale from near Nkalagu to near Lekwesi on the north limb of the Abakaliki Fold Belt and again on the south side of this structure from almost where the Eze-Aku Shale appears from beneath the unconformity, east of Okigwe in the Afikpo area (Figure 9). Oil smell symbols are shown on the Ameseri Sandstone outcrop north east of Afikpo and where the Cross River cuts across the Ameseri outcrops which interfinger with the Eze-Aku Shale. Oil smell symbols are also shown on the Eze-Aku Shale outcrop in the Konsisha River area. Oil shows were not plotted along the outcrops of the Eze-Aku Shale nor along their sandy equivalents. Very few oil smell symbols or oil shows have been marked on the Shell-BP and Geological Survey of Nigeria 1 : 250 000 Geological Series on the outcrops of the Awgu Shale and the Asu River Group situated above and below the Eze-Aku Shale in the Abakaliki Fold Belt. This may simply reflect the density of observation, but as in South Africa, where there is a clear connection between the frequency of oil shows and dolerite intrusions in the Karroo sediments, there may be a connection between the frequency of oil shows and the occurrence of Minor Basic and Intermediate intrusives of ? Santonian age and Tertiary to Recent basalts which affect this region.

2.4.2.3.3. Pre-Santonian prospects Calabar Flank and Hinge Line

Little has been published about the petroleum prospects of the Lower and Upper Cretaceous rocks of the Calabar Flank and Hinge Line and of the Ikpe Platform (Figures 9 and 22). The Oban Massif acted as a source area and deltaic sands, shallow marine clastics and carbonates accumulated on the Ikpe Platform, Calabar Flank and on the margins of Ikang

Trough and on the Ituk High. Deeper marine clastics accumulated in the Ikang Trough and down the Calabar Hinge Line (Figures 25, 28, 29 and 34). Again arguing from first principles there could be interesting stratigraphic-structural plays in these areas because of the complex facies variations and the "down to ocean" continental margin structures which developed consequent on continental separation.

Very little direct evidence that petroleum has been generated from these sediments is available in publications, although three oil symbols and six oil smell symbols are shown on the outcrops of the Nkporo-Eze-Aku-Asu River sediments on the Shell-BP and Geological Survey of Nigeria Sheet 80, Oban Hills. Five of these occur in pre-Nkporo shale sediments classified by Reyment (1965) and Dessauvagie (1972) as Odukpani Formation (see below).

A list of wells drilled along the Calabar Flank and adjacent areas is presented in Figure 35. Although all these wells were abandoned this need not to be taken as a measure of the poverty of the petroleum prospects of the region. The discovery of oil in commercial quantities in Younger Cenozoic sediments at Oloibiri in 1956 and in Akata-1 caused interest to swing to Niger Delta Complex characterised by its growth faults and rollover structures and since then little active prospecting has taken place over the Cretaceous areas. High taxation, "buy back" prices, Nigeria's actions concerning BP etc. all mitigate against companies prospecting high risk areas, and Shell-BP's more recent relinquishments may well reflect company attitudes to holding and exploring high risk Cretaceous acreage. There may well be interesting stratigraphic and structural plays still to be found along the Hinge Line beneath the Cenozoic sediments and in the Abakaliki Trough and adjacent areas. "Lower Limestones" were laid down along the Calabar Hinge Line and Flank during the Medial Albian transgression and "Upper Limestones" were laid down during the Early Turonian transgression (Figures 25 and 29). At Calabar Cement Works partly dolomitic limestones occur in lenses up to $2 \times \frac{1}{2}$ miles in extent and 250 ft thick.

Oil show and six oil smell symbols are shown on the outcrops of the Nkporo-Eze/Aku-Asu River Sediments as mapped by Shell-BP and the Geological Survey of Nigeria on Sheet 80, Oban Hills, 1 : 250 000 series. Few of these occur in pre-Nkporo Shale sediments (see above in discussion Odukpani Formation prospects). Descriptions of these plays and prospects, which must be considered as high risk ventures, are presented in summary form in Figure 36.

2.4.3. Lower and Upper Benue Trough Pre-Santonian Lower and Upper Cretaceous Formations

2.4.3.1. General, Definition and Subdivision of the Benue Trough

The stratigraphy of Lower Benue Trough is poorly known and our knowledge of thickness variations, palaeogeography etc. is very rudimentary except for the small area described by Offodile (1976). The stratigraphy of the Upper Benue Trough and the adjoining Gongola and Yola Troughs is better known because of studies made by the Geological Survey of Nigeria (Carter, Barber and Tait 1963; Dessauvagie 1969; the University of Ibadan Benue Valley Project 1968–1972; Burke, Dessauvagie and Whiteman 1971 and 1972; Wright 1976).

A considerable mix up exists in the literature concerning what constitutes the Lower, Middle and Upper Benue areas. This has arisen mainly because of the confusion of geographical and geological names

Figure 35
List of Wells Sunk Along the Calabar Flank and Adjacent Areas Since Late 1950s to Early 1970s

Name	Owner	Lease	Date		Coordinates		TD (ft)	Comments
			Spud	Comp.	North	East		
Calabar-1	Shell-BP	open	pre-1957	—	—	—	2370	Shell-BP core holes all
Calabar-2	Shell-BP	open	pre-1957	—	—	—	2703	start in coastal plain
Calabar-3	Shell-BP	open	pre-1957	—	—	—	2046	(Benin) sands
Calabar-4	Shell-BP	open	pre-1957	—	—	—	1048	
Ikono-1	Shell-BP	OML 12	16.12.58	18.04.59	593013	123481	11050	abandoned exploration
Annua-1	Shell-BP	OML 12	16.11.58	06.11.59	610976	114050	11555	abandoned exploration
Uruan-1	Shell-BP	OML 12	10.08.59	14.10.59	617455	114597	11538	abandoned exploration
Ituk-1	Shell-BP	OML 12	06.06.54	07.09.54	619388	102317	7837	abandoned exploration
Ituk-2	Shell-BP	OML 12	24.04.56	02.11.57	622246	106200	10538	abandoned exploration
Ikang-1	Shell-BP	open	22.01.59	23.03.59	673355	184995	11189	abandoned exploration

Figure 36

Table showing summary of Petroleum Prospects and Plays, Anambra Platform, Abakaliki Trough Lower and Upper Cretaceous Series
(Albian to Lower Santonian)

Province	Region	Play	General Description	Trap Type	Shows	Prospect Class
SOUTHERN NIGERIA	ANAMBRA PLATFORM	1. Sub unconformity prolongation of Benue and Abakaliki Fold Belts	The Lower and Upper Cretaceous pre-Maestrichtian rocks may be folded beneath the Santonian Unconformity in the Anambra Basin. Basement Complex may be at 30 000 ft depth at south end of Anambra Basin. The Santonian unconformity could lie between 3100 and 12 000 ft and the folded rocks could be easily accessible on flanks of basins	Anticlines, faults, unconformities combination	Positive surface and sub-surface	GOOD TO VERY GOOD FOR OIL AND GAS. *LIMITATION* STRUCTURES IN PLACES COULD BE FLUSHED. HIGH GEOTHERMAL GRADIENTS.
		2. Delta complexes of Keana, Agbani and Agala Ameseri Sandstones and marginal sandy facies northwest flank	The Keana, Agbada and Agbani sandstones were laid down as part of individual delta complexes. Additionally these sand bodies may have been folded during the Santonian orogeny	Primarily stratigraphic with Anticlines, faults and growth faults	Positive surface and sub-surface	GOOD FOR OIL AND GAS *LIMITATION* ROCKS COULD HAVE BEEN INTRODUCED BY ABAKALIKI CALC-ALKALINE IGNEOUS SUITE. HIGH GEOTHERMAL GRADIENTS.
		3. Zone of interdigitation of shallow platform marine clastics and deep water facies of Asu River Group situated on southeastern flank of the Anambra Platform	The Asu River Group which has superficial rock characteristics and is more than 10 000 ft thick interdigitates with coarser grained shallow marine clastics on the southeast flank of the Anambra Platform. Limestones are also present at the edge of the platform and the whole sequence was folded during the Santonian orogeny	Primarily stratigraphic with Anticlines and Faults	Positive surface and sub-surface	GOOD *LIMITATION* ROCKS COULD HAVE BEEN INTRODUCED BY ABAKALIKI CALCALKALINE INGENEOUS SUITE. HIGH GEOTHERMAL GRADIENTS.
	ABAKALIKI TROUGH	Folded Delta Complex plays	The Asu River Group, Eze-Aku Shales and Awgu Shales pass laterally on the margins of the Abakaliki Trough into coarse grained paralic delta complexes. In places these complexes have been involved in strong and close folding. The shales have superficial source rock characteristics. Away from the areas of strong igneous activity the correct combination of geothermal gradient, stratigraphic and structural factors might add up to successful gas prospects	Combination structural and stratigraphic	Positive surface shows peripheral to the area	SPECULATIVE FOR GAS AND ? OIL *LIMITATION* ROCKS HAVE BEEN INTRUDED BY ABAKALIKI ALKALINE IGNEOUS SUITE. HIGH GEOTHERMAL GRADIENTS.

Based on Whiteman (1973).

and concepts. The Benue River first occupies the structural pre-Santonian Yola Trough, then the Benue Trough and eventually swings around the post-Santonian Anambra Basin to join the River Niger. For much of its course the river is called the Benue and the valley is referred to as the Benue Valley. Cratchley and Jones (1965) divided their Benue Valley into "SouthWest" including the Oturkpo-Makurdi area; "Middle Benue" including the Keana-Awe Arufu area and the Bashi-Muri-Amar area; and the "North East" including the Gombe-Lau areas.

Murat (1972) referred to the overall structural depression as the Abakaliki-Benue Trough in an attempt to sort out the confusion in terminology. Burke, Dessauvagie and Whiteman (1972) used the term Benue Depression to include the overall structural feature i.e. Abakaliki Trough plus the Benue Depression up to Yola Trough and Burke et al. (1971) used Lower Benue Depression as a synonym for the Abakaliki Trough plus the Makurdi area. Burke and Whiteman (1973) used the term Benue Trough to include the whole structure extending from the Niger Delta Triple Junction to the Chum Trilete Junction at the northern end of the Benue Trough so including the Abakaliki and Benue Troughs.

To avoid further confusion it is suggested here that the Abakaliki Trough should be clearly labelled as a separate structural and depositional unit. The depositional limits are shown in Figure 22 etc. The southwestern limit of the Lower Benue Trough is taken at a line drawn along the northwest trending sediment/Basement Complex contact south of Gboko running through Gboko to Makurdi on the Benue River to Lafia. The northeastern limit is drawn arbitrarily through Wase, Amar and Mutum Biyu, separating an area of very poorly known geology from better known geology to the northeast in the Lau-Lamurde-Yola region (Figure 9). The Upper Benue Trough merges with the Yola and the Gongola Troughs forming the Chum Trilete Junction (Burke and Whiteman 1973). So defined the Benue Trough is around 300 miles long and the sedimentary belt is around 90 miles wide.

The Abakaliki and Benue Troughs which extend from the Niger Delta Triple Junction to the Chum Trilete Junction are together over 450 miles long.

Much less is known about the Lower Cretaceous stratigraphy of the Lower Benue Trough than about the Lower Cretaceous of the Abakaliki Trough. This is because of the early concentration of petroleum exploration in the Abakaliki and Anambra areas and on the Niger Delta rim. The Lower Benue is still very much a stratigraphic "No Man's Land" and we are dependent on descriptions by Falconer (1911), Farrington (1952), Cratchley and Jones (1965), Whiteman (1973), Dessauvagie (1974) and Offodile (1976) for information about this area. Various reports belonging to oil companies, consultants etc. are in existence but have not been released for general circulation; and unfortunately the Geological Survey of Nigeria has published relatively little about the Lower Benue Trough, although internal reports on some areas exist. Vegetation and superficial cover are thick in the Lower Benue and between Makurdi and Zurak there are few rock outcrops. Except for Offodile's work (1976) systematic mapping has not been undertaken in the region and age designations, correlations etc. must be regarded as tentative. The classification of Cretaceous formations of the Lower Benue Trough recognized by Offodile (1976) and correlations with those sequences occurring in the Upper Benue Trough, Abakaliki and Anambra areas are presented in Figure 37.

As in the Abakaliki Trough the Medial Albian Sea transgressed into the Lower Benue area and Asu River Group micaceous shales, mudstones and limestones were laid down (Offodile 1976 etc.). These rocks exceed 9850 ft in thickness. The Asu River Formation is followed by the Awe Formation which is unconformably overlain by the Keana Formation. The Eze-Aku Formation is shown as a facies of the Keana Formation and is separated by an unconformity from the overlying Awgu Formation. The Santonian Unconformity overlies the Awgu Formation and the Maestrichtian Lafia Formation succeeds (Figure 37).

Farrington (1952) investigating the lead-zinc deposits of the Benue Trough presented a correlation of the Cretaceous rocks of the Lower and Upper parts of the trough. It was compiled from Geological Survey of Nigeria sources. He described Albian shales and limestones (the Arufu and Gboko limestones) followed by shales, siltstones and sandstones of the Gboko-Egedda region and then by the Turonian limestones of Makurdi and Awe. Farrington (1952) believed that these rocks are Turonian and that post-Gboko (Albian) limestones pass laterally into the micaceous sandstones of the Makurdi and Gboko regions. Falconer (1911) classified the Cretaceous beds of the Lower and Upper Benue into:

Upper grits and sandstones (Top)
Turonian limestones and shales
Lower grits and sandstones called the Muri sandstones.

According to Farrington (1952) the Muri Sandstones of Falconer (1911) are well represented in the Gwono district and in the Lamurde Hills, where they are exposed in a major anticline (Figures 9 and 170). The Geological Survey of Nigeria (Carter et al. 1963) named these sandstones the Bima Sandstone.

Cratchley and Jones (1965) proposed the correlation framework for the Lower Benue area shown in Figure 38. They reaffirmed Reyment's (1955) statement that the oldest sedimentary rocks in the Benue Depression occur in the Keana-Awe-Gboko area and are of Middle Albian age. Dark shales, siltstones and fine grained sandstones are said to pass upwards into shales and limestones of Late Albian age. These deposits overlap the older sediments and rest unconformably on the crystalline Basement Complex as at Arufu. South of Arufu, the Cretaceous sediments are thin and inliers of granitic and gneissose rocks occur. Clearly the contact is unconformable in this area and

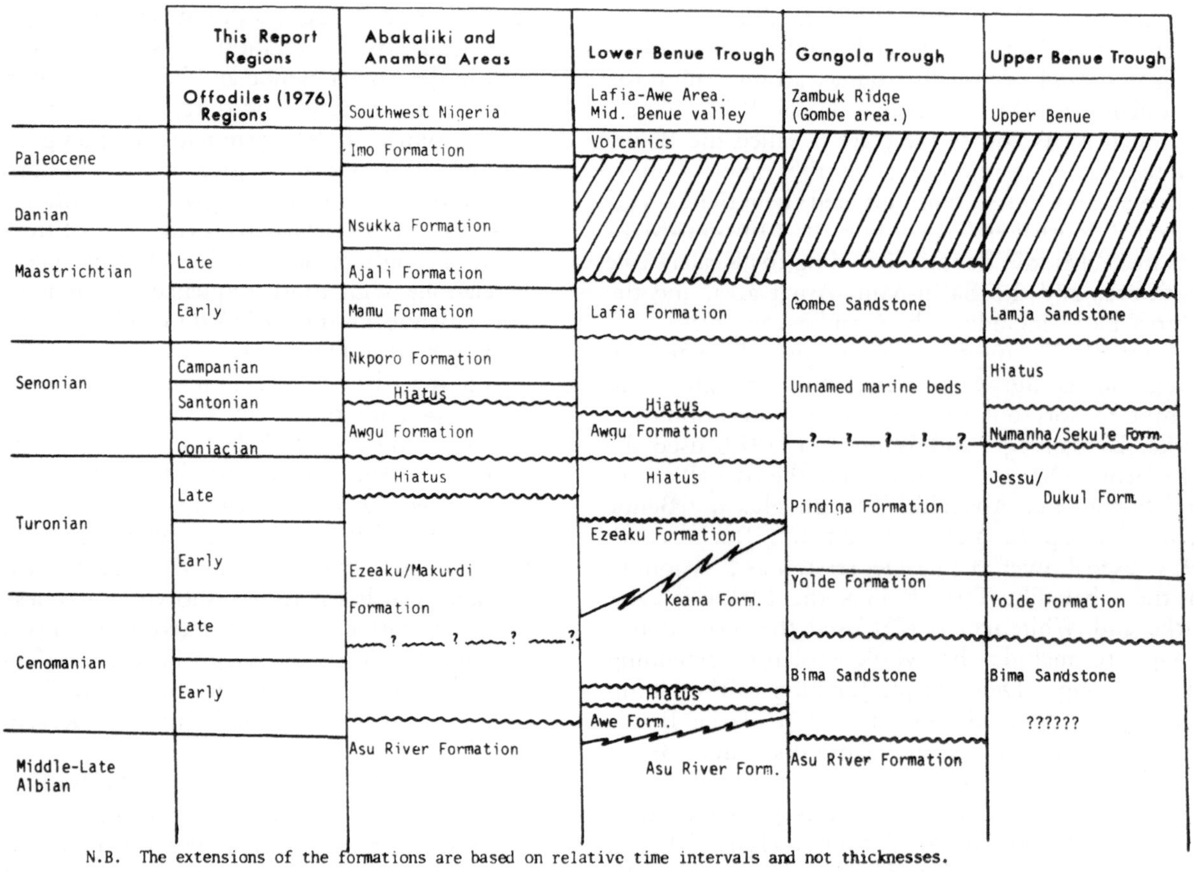

	This Report Regions	Abakaliki and Anambra Areas	Lower Benue Trough	Gongola Trough	Upper Benue Trough
	Offodiles (1976) Regions	Southwest Nigeria	Lafia-Awe Area. Mid. Benue valley	Zambuk Ridge (Gombe area.)	Upper Benue
Paleocene		Imo Formation	Volcanics	////////	////////
Danian		Nsukka Formation	////////	////////	////////
Maastrichtian	Late	Ajali Formation	////////	////////	////////
	Early	Mamu Formation	Lafia Formation	Gombe Sandstone	Lamja Sandstone
Senonian	Campanian	Nkporo Formation		Unnamed marine beds	Hiatus
	Santonian	Hiatus	Hiatus		
	Coniacian	Awgu Formation	Awgu Formation	?—?—?—?—?	Numanha/Sekule Form.
Turonian	Late	Hiatus	Hiatus	Pindiga Formation	Jessu/ Dukul Form.
	Early	Ezeaku/Makurdi	Ezeaku Formation		
Cenomanian	Late	Formation ?—?—?—?	Keana Form.	Yolde Formation	Yolde Formation
	Early		Hiatus	Bima Sandstone	Bima Sandstone
			Awe Form.		??????
Middle-Late Albian		Asu River Formation	Asu River Form.	Asu River Formation	

N.B.   The extensions of the formations are based on relative time intervals and not thicknesses.

Figure 37. Stratigraphic correlations. Cretaceous formations, Abakaliki, Benue and Gongola Troughs. Based on Offodile (1976).

Figure 38

Correlation of Cretaceous Rocks of the Lower Benue Trough in the Keana – Awe – Arufu and Bashar – Muri-Amar Areas

Keana – Awe – Arufu	Bashar – Muri – Amar
"Lafia Sandstone"	Gombe Sand
– – – – UNCONFORMITY – – ? UNCONFORMITY – – – –	
Unnamed Marine Deposits	Kumberi Formation
Unnamed transition deposits	"Passage beds"
"Keana" Sandstone	Muri Sandstone
Asu River Group	Unnamed marine deposits
– – – – – – – – – – UNCONFORMITY– – – – – – – – – –	
Crystalline Basement	

Based on Cratchley and Jones (1965)

is not a faulted one. All the sediments in the area east of Arufu have been included in the Asu River Group by Cratchley and Jones (1965) who pointed out that further mapping is needed to sub-divide the great thickness of strata present.

The Geological Survey of Nigeria (1965) mapped these deposits as Asu River Group and "Wukari Group" (a poorly differentiated unit occurring in the Lower Benue Valley). There is little that can be said about the Albian rocks, except that 20 miles ESE of Wase (Figure 9) black and brown shales and mudstones of Albian age occur. The overlying Keana and Muri Sandstones of the Lower Benue rest conformably on Albian deposits and are classed as Cenomanian.

The coarse to very coarse grained feldspathic sandstones and grits which constitute Falconer's (1911) Muri Sandstone are lithological equivalents of the Keana Sandstone of the Shell-BP geologists. Murat (1969) however recognized an Intermediate Shale Group with the Keana Sandstone overlying the Muri Sandstone. An Intermediate Shale Group is classed as Albian-Cenomanian, as is the Asu River Group, and the two units are shown as lateral equivalents (Murat 1969 and 1972).

Beneath the Intermediate Group of the Lower Benue, Murat (1972, Figure 2) plotted a Basal Sandstone of Neocomian and Barremian age, equivalent to the Basal Grit of the same age in the Anambra and

Calabar areas. No stratigraphic or palaeontological details were given in the text about the Basal Sandstone of the Lower Benue. Dessauvagie (1974) realizing the problem of mapping all these ill-defined units with information at hand decided to group the units and map only the Keana Sandstone and Wukari Group (Kwu) in the Lower Benue Trough (Figure 9).

Fortunately the stratigraphy of the pre-Santonian Lower and Upper Cretaceous sequences of the Upper Benue, Yola and Gongola Troughs is less confused. The B1 and B3 Bima Formation forms part of the Bima Delta Complex and the B2 Bima of the Lau and Futuk areas is probably the lateral equivalent of the Asu River Group of the Lower Benue and Abakaliki areas (Carter *et al.* 1963, Burke, Dessauvagie and Whiteman 1971, 1972; Whiteman 1973), although Reyment thinks that the Asu River Group may be older (Reyment in Burke, in Dessauvagie and Whiteman 1972).

Turonian marine Pindiga Shales overlie the deltaic to marine Yolde Formation in the Upper Benue Trough and adjacent areas. The Pindiga was laid down during a transgression which overtopped the Zambuk Ridge in the Gongola Trough and which extended into the Chad and Bornu Basins (Figure 22). In general terms the sea regressed in Senonian times and a major folding episode affected the whole of the Benue Trough in Santonian times. As we have already noted this folding episode is less intense in the Upper and Lower Benue areas than in the Abakaliki Fold Belt and the two belts are best considered separately. A list of formations mapped in the Upper Benue Trough is presented in Figure 40 based on Carter *et al.* (1963).

In this account Lamja and Numanha units are classed as pre-Santonian because they are folded and indeed all the litho-units mentioned above occur beneath the Santonian Unconformity and were either tilted or folded during the Santonian phase. This clearly divides these formations as is shown in Figures 21 and 23 from the post unconformity Upper Cretaceous units and provides a useful basis for separation and description. The distributions of most of these pre-lithological units are shown in Figures 9, 22 etc.

2.4.3.2. Lower Cretaceous Formations Lower and Upper Benue Trough

2.4.3.2.1. Asu River Group

General Description: The Asu River Group is the oldest formation mapped in the Lower Benue Trough and consists of micaceous siltstones, olive green to grey dark and pinkish, mudstones and shales. The Asu River Group is exposed in the Keana Anticline and elsewhere (Offodile 1976). In the core of the anticline the Asu River Group is baked where it is in contact with intrusive rocks.

Thickness: Offodile (1976) implies that the Asu River Group in the Lafia-Keana area exceeds 9600 ft (3000 m).

Age: Abundant ammonites found in the Keana area indicate a Medial Albian age.

Conditions of Deposition and Palaeogeography: The Asu River Group in the Lower Benue Trough was deposited under marine conditions as the Medial Albian seas transgressed into the region (Figure 25).

2.4.3.2.2. Awe Formation

General Description: The Awe Formation (new, Offodile and Reyment 1976) succeeds the Asu River Group in the Lower Benue Trough and its age ranges from Late Albian to earliest Cenomanian (Offodile 1976). The formation consists of flaggy medium to coarse grained sandstones interbedded with carbonaceous shales from which brines issue copiously (Offodile 1976).

Age: The formation has been assigned a Late Albian to Early Cenomanian age and is separated from the Keana Formation by a hiatus (Figure 37).

Thickness Variations: Offodile (1976) stated that the thickness is about 3200 ft (1000m).

Conditions of Deposition and Palaeogeography: The Awe Formation was deposited under transitional marine to fluviatile conditions (Offodile 1976) during the Late Albian-Cenomanian Regressive Phase of Murat (1972). To explain the copious brines which issue from the Awe Formation, Offofile (1976) postulated that hypersaline lakes and swamps developed during the regressive phase and that evaporites were deposited, although evaporites have not been recognized in either surface or subsurface sections. Another possible explanation is that the Awe brines are derived from mineralizing solutions associated with the Benue lead-zinc mineralization (Chapter 3). Physiographic diagrams portraying Albian conditions of deposition according to Offodile (1976) are presented in Figure 39. Relief was probably less "severe" and fjord-like than is shown in this diagram. Offodile while writing his Ph.D. thesis studied at the University of Uppsala, Sweden.

2.4.3.2.3 Bima Sandstone

General Description: The oldest Cretaceous rocks exposed in the Upper Benue Trough are the Bima Formation, the lower and middle parts of which are Albian in age. Reyment (1965) stated that the age of this formation is uncertain but that it is probably older than Turonian. The Bima Sandstone according to Carter *et al*, (1963) ranges in age from Late Albian to Turonian. Beds containing a Lower Turonian fauna overlie the Bima Sandstone and therefore, in the absence of unconformity the formation must be of Albian and Cenomanian age. Because of the difficulty of separating the various units within the Bima Sandstone outside the Lau area, the unit is described here as Lower Cretaceous. The formation is thought to have been laid down during two progradation phases separated by deposits laid down in a Medial Albian transgressive phase (Figures 21, 25 and 26).

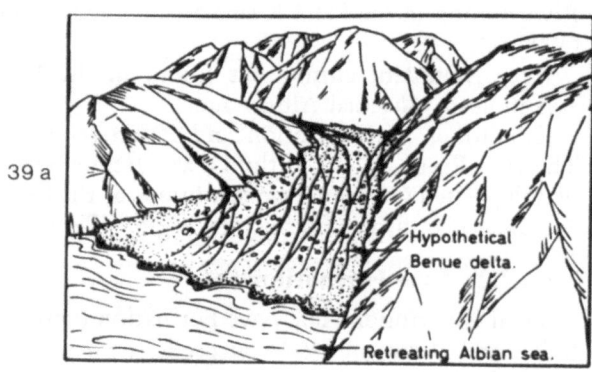

Figure 39. (A – C) Physiographic diagrams showing how the Benue Trough may have appeared in Albian times. Based on Offodile (1976). (A) Hypothetical Benue or Bima Delta situated at the northern end of the Benue Trough. (Offodile 1976). (B) Benue Trough flooded by the Medial Albian Sea (Offodile 1976). (C) Saline lakes formed as Albian Sea retreated: possible environment of deposition of Awe Formation (Offodile 1976).

The Bima Sandstone was first named by Falconer (1911, p. 159). Raeburn and Brynmor Jones (1935, p. 21) used the name Lower or Bima Sandstone Group. Longbottom (Falconer 1911, p. 184) assigned the Bima Sandstone of Yendam Hill and the Lamurde anticline to an Eocene division of the "Upper Benue Sandstones". This was shown later to be incorrect as the sandstones at Lamurde were found to underlie marine Cretaceous (Turonian) rocks. Barber, Tait and Thompson (1954, p. 18) called the unit the Bima Grits and applied the name Guyak Sandstone to similar rocks in the Lamurde-Guyak area. The *International Stratigraphic Lexicon* (1956) referred to the Bima Sandstone as "Bima Sandstone and Grits", citing Falconer (1911, p. 161) as the author of the term. The unit was said to consist of:

> Terrestrial deposits. Basal beds of purple and green false-bedded arkosic grits followed by rounded red and

white grits, which pass into red and white sandstones and grits

Additionally the Lexicon pointed out that the unit was:

> also termed the Bima Sandstone, the Lower or Bima Sandstone Group, the Bima Grits and the Lower Sandstones. Lower part of the Cretaceous succession in the eastern part of Bauchi Province. The formation is laterally equivalent to the Muri Sandstone and Guyak Sandstones. Thickness about 2000 ft. Age: Pre-Lower Turonian. Type locality: Bima Hill, Gombe Division. Distribution: Bauchi and Bornu Provinces.

The Geological Survey of Nigeria also described the Guyak Sandstones in the *International Stratigraphic Lexicon* (1956). The Cretaceous Guyak Sandstones were named by Barber *et al.* (1954) and are considered to be the lateral equivalent of the Bima Sandstones and the Muri Sandstones. They are said to consist of:

> Coarse, false-bedded, terrestrial, arkosic grits passing upwards into sandstones which become increasingly ferruginous towards the top. They form the local base of the Upper Cretaceous and are succeeded by the Transition Beds. Thickness over 3000 ft. Type locality: Guyak, Numan Division. Distribution: Adamawa Province.

Carter *et al* (1963) described the Bima Sandstone in detail. They retained Bima Hill as the type locality of the formation and although no section was obtained, they gave a general description of the beds there which is the best description available to date:

> In the eastern scarp of Bima Hill, over 1500 ft of arenaceous strata are present which consist of thick to massive bedded, medium- to coarse- and, occasionally, very coarse-grained, cream, whitish grey and buff, feldspathic sandstones. These sediments are often cross-bedded and include many scattered, rounded quartz pebbles which are often seen as layers on erosion surfaces. Interbedded with the coarser grained, poorly sorted sediments are layers of well sorted, less massively bedded, medium grained sandstones. Thin bands of whitish and dark red siltstone and mudstone are frequently intercalated among massive sandstones. The rocks exposed in nearby Wase Hill are similar to the Bima Hill sediments.

Carter *et al.* (1963) also provided a standard section in the Lamurde Anticline where the formation is said to exceed 10 000 ft (Figure 166). In the Lamurde Anticline the section is divided into Lower, Middle and Upper Bima Sandstone which are designated B1, B2 and B3 respectively on Sheet 47, Lau, Geological Survey of Nigeria (1961). The B1 or Lower Bima Sandstone rests unconformably on the Basement Complex and consists of at least 1300 ft of sandstones and argillaceous rocks. The Middle Bima Sandstone (B2) is composed almost entirely of coarse sandstones with clays and shales. The Upper Bima Sandstone (B3) is composed almost entirely of coarse grained sandstones and is 5700 ft thick. The Lower and Middle Bima Sandstones have not been recognized outside the Lamurde Anticline and it is the Upper Bima Sandstone which occurs elsewhere in the Lau and Gombe areas. Carter *et al.* (1963) did not discuss the manner in which the Lower and Middle Bima Sandstones disappear, but the Lower and Middle members of the Bima Formation must lap out onto the Basement Complex north and south of Lamurde Hill and the upper member laps onto the Basement Complex.

Neither porosity nor permeability measurements were given by the Survey but in general terms beds with good to average poroperm values are said to be common and extensive clean sands are known. Data on the ground water geology does not appear to have been published. Some of the structures developed in the Bima Sandstones have been flushed.

Thickness Variations: The Bima Sandstone varies considerably in thickness. Cratchley and Jones (1965) estimate its thickness as ranging from 1000 to 6000 ft but Carter *et al.* (1963) gave the variation as 300 to 10 000 ft. Around the type locality at Bima Hill, east of Gombe, it is said to be 1500 ft. In the Lamurde Anticline an incomplete thickness of 9715 ft has been measured and in the hills east of Gombe town the Bima Sandstone is about 300 ft thick. Dessauvagie (1969) gave a thickness of 10 000 ft for the thickness of the Bima Sandstone in the Futuk area, west of Lau (Figure 28).

Age: Few fossils have been found and these are restricted to the Middle Bima member. They consist of fossil wood and "non-diagnostic bivalves" and so little can be said about age. The Geological Survey of Nigeria Sheet 47 Lau (1961) classed the Bima Sandstone as Albian-Cenomanian but Carter *et al.* (1963) hold the view that the shaley beds of the Middle Albian member are deposits which were laid down during the Albian marine invasion. The Upper Bima "member" therefore could be Late Albian to Cenomanian or even early Turonian in age. According to Murat (1972) a transgressive phase took place in Medial Albian times and during Late Albian and Cenomanian times a major regression took place in Southern Nigeria and in the Lower Benue Trough (Figures 23, 25 and 28). The Bima Sandstone is the up-valley equivalent of the Muri and Keana Sandstones (described above) deposited during the same Late Albian–Cenomanian regressive phase. Cratchley and Jones (1965) assigned a Late Albian age to the Middle B2 Bima "member" on the basis of correlation with adjacent areas. Reyment (1970) thinks that the Bima Sandstone was laid down during the Lowermost Turonian transgression.

Conditions of Deposition and General Albian–Cenomanian Palaeogeography: The Bima Sandstone is thought to have been derived from a granitic Basement terrain as the sediments are frequently arkosic, especially in the lower part of the unit. Conglomerates and boulder beds occur along part of the boundary especially in the southern part of the trough. At Yendam Hill, for instance, the Bima Sandstone occurs as outliers on the Basement Complex and is coarsely conglomeratic. Most of the formation, however, consists of feldspathic sandstones and clays which pass upwards into coarse to medium grained sandstones with less feldspar, siltstones and clays. Cross-bedding directions suggest northerly and easterly sources in the Gombe area and easterly and southerly sources in the Lau area (Figure 28). The abundant cross bedding and highly variable lithologies indicate deposition under fluviatile and/or deltaic conditions. Massive white and purple clays are thought to have been laid down as lake deposits. The shaley B2 Middle Bima beds of the Lamurde Anticline may have been laid down as marginal deposits during an Albian marine transgression.

Burke, Dessauvagie and Whiteman (1971) regard the Lower B1 Bima "member" as a progradation deposit; the Middle B2 Bima "member" as having been laid down during the Middle Albian transgressive phase (Figure 23); and the Upper B3 Bima "member" as a Late Albian-Cenomanian regressive (or progradation) deposit. The front of the Bima delta during Late Albian-Cenomanian times was probably situated on the Filiya–Amar areas. On this interpretation the sediments constituting the Keana and Muri Sandstones were eroded Basement rocks exposed in the shoulders of the Benue Depression. The Bima Sandstone has been mapped as far north as Buri where it passes under the Chad Formation (Figure 28). The western margin of the Bima Sandstone lies under the Kerri-Kerri Formation north of Bashar.

2.4.3.3. Upper Cretaceous Pre-Santonian Formations Lower and Upper Benue Trough

2.4.3.3.1. General

Much less is known about the stratigraphy of the Upper Cretaceous rocks of the Lower Benue Trough compared with the Abakaliki Trough and the Anambra Basin and not all the pre-Santonian transgressions and regressions described by Murat (1972) have been identified in the Lower Benue Trough, but probably with more field work the pattern will become clearer (Figure 21).

Farrington (1952) described the rocks of the Lower Benue region in general terms but did not identify formations. He pointed out that rock exposures are meagre between Makurdi and Zurak except for arenaceous sediments and mentioned that there are shales, thin limestones and compact sandstones near Awe which are lithologically similar to the "shale formation south of the Benue". In present day terminology he described beds in the Keana anticline as belonging to the Asu River Group and Keana Sandstone (Albian-Cenomanian), the Eze-Aku Shale (Turonian) and the Awgu Shale (Late Turonian). Farrington (1952) mentioned that fossils from Awe, Amar and Zurak indicate a Turonian age.

The most informative regional account of the Lower Benue Upper Cretaceous rocks available to date is that given by Cratchley and Jones (1965). Offodile (1976) provided a detailed account of the pre-Santonian Upper Cretaceous rocks on the Lafia and Akiri sheets. Cratchley and Jones pointed out that, because systematic mapping had not been completed in the area, their correlations must be regarded as tentative. The correlation adopted by Cratchley and Jones (1965) for the Upper and Lower Cretaceous "Series" in the Middle Benue is presented in Figure 38.

The Keana Sandstone of the Keana anticline is classed as Cenomanian by Cratchley and Jones (1965) and, as we have already mentioned, these coarse grained deposits, named by the Shell–BP geologists, are the lateral equivalent of part of the Asu River Group of the Abakaliki Trough.

In the Lower Benue Trough the Keana Sandstone is thought to be the equivalent of the Muri Sandstone, a name first used by Falconer (1911) for a sequence of coarse to very coarse grained feldspathic sandstones with pebble and boulder conglomerates. It is now considered to be the equivalent of part of the Bima Formation. These sandstones and grits are thought to have been deposited in deltaic and fluviatile environments. Cratchley and Jones (1965) mentioned that there are similar coarse grained sandstones along the northern margin of the valley, adjacent to the crystalline Basement Complex, extending east of Kwande via Wase and Bashar (Figure 9).

The Keana–Muri Sandstones are overlain by unnamed transition deposits which pass upwards into truly marine beds in the Keana–Awe–Arufu area (Figure 38). The transitional "Passage Beds" of the Bashar–Muri–Amar region, like the "Unnamed transition beds" of Awe, consist largely of alternating sandstones and shales with thin impure limestones. The sandstones are frequently coarse grained in the lower part of the sequence and become finer and more thinly bedded higher in the sequence. The argillaceous beds of the transition group are poorly exposed and consist of calcareous shales and "clay shales" with siltstone alternations, light grey in colour.

The unnamed marine strata mentioned above are rarely exposed but consist of shales and "clay shales" with subordinate thin siltstones and shelly limestones. South and southwest of Awe these marine deposits are intermingled with volcanic material (Cratchley and Jones 1965, p. 5) (Figure 29). In the Amar–Muri district the unnamed marine strata are referred to as the Kumberi Formation (Reyment 1956). This unit consists of shale, sandy shale, mudstone and fine grained sandstone with subordinate limestone bands. The rocks are about 450 to 500 ft thick in the Kumberi syncline where Lower Turonian ammonites have been collected. The Campanian-Maestrichtian Lafia and Gombe Sandstones overlie the Kumberi Formation unconformity.

During Cenomanian times in the Lower Benue Depression Keana Sandstones were laid down in a regressive phase but it is uncertain how far up the depression these deposits extended (Figure 28). Down valley they extended almost onto the flanks of the Abakaliki Trough. Offodile's (1976) summary of the Cretaceous stratigraphy of the Keana–Awe area of the Lower Benue Trough is presented in Figure 37.

The Keana Formation which overlies the Awe Formation is thought to have been laid down during a regressive phase and may be viewed as an extension of the Bima Delta System of the Upper Benue into the Lower Benue area (Figure 28).

The Eze–Aku Formation was laid down in the Lower Benue Trough during the early Turonian Transgression which extended through into the Upper Benue Trough and beyond into the Bornu Basin (Figure 29). In the Upper Benue Trough in Turonian times thick montmorillonitic shales were deposited but because of our lack of knowledge of the detailed stratigraphy it is not possible to separate the Early and Late Turonian Transgressive Phases and the Lower Turonian Regressive Phase as Murat (1972) did for the Abakaliki Trough. These montmorillonitic shales, with limestones at their base, are called Pindiga Formation where they overlie the transitional deltaic to marine Yolde Formation (Figures 8, 22, 29, etc.). The Pindiga transgression in the Upper Benue Trough was extensive and the Turonian Sea overtopped the Zambuk Ridge and extended into the Bornu and Chad Basins. How far it extended up the Yola Depression is not clear now because of erosion but it may well have extended through into Southern Chad. The down valley equivalents of the Pindiga Formation are the Eze–Aku Group and the Awgu Shale Group. The sea regressed in Senonian times and a major folding episode affected the whole of the Benue Trough.

In the Upper Benue Trough the pre-Santonian Upper Cretaceous rocks are well known, mainly from the work of the Geological Survey of Nigeria (Carter et al. 1963). Cenomanian, Turonian and Senonian rocks have been mapped on the Futuk, Lau, Yola, Gombe, Biu and Potiskum 1:250 000 sheets covering the Upper Benue, Gongola and Yola Troughs (Figures 9 and 166). The following Upper Cretaceous, Cenomanian–Lower Senonian formations have been recognized on the sheets listed above: Bima Formation, Yolde Formation, Pindiga Formation, Gongola Formation, Jessu Formation, Dukul Formation, Numanha Shale, Gulani Sandstone and Fika Shale (Figures 9 and 40). These formations which are described below are unconformably overlain by the Campanian–Maestrichtian Gombe Sandstone and in places are overlain unconformably by the Cenozoic, Kerri–Kerri Formation.

2.4.3.3.2. Awe Formation

General Description: The Awe Formation occurs in the Lower Benue Depression and as we have already noted the Awe Formation (new Offodile and Reyment 1976) is in part Late Cretaceous in age. Descriptions of this formation are given above.

2.4.3.3.3. Keana Formation:

General Description: In the Lower Benue Trough the Keana Formation succeeds the Awe Formation

Figure 40
Correlations Upper and Lower Cretaceous Rocks in the Lau, Gombe, Gulani and Bornu Areas Benue and Gongola Troughs

				Zambuk Ridge Gulani Gombe Area Area	Chad Basin	
Pleistocene					Chad Formation	
				UNCONFORMITY		
Palaeocene					Kerri Kerri Formation	
				UNCONFORMITY		
Upper Cretaceous	Maestrichtian		Lamja Sandstone		Gombe Sandstone	
	Senonian	Campanian	Numanha Shales	Gulani Sandstone		Fika Shales
		Santonian	Sekule Formation			
		Coniacian				
	Turonian	Upper	Jessu Formation		Pindinga Formation	Gongila Formation
		Lower	Dukul Formation			
	Cenomanian		Yolde Formation			
			upper middle lower	Bima Sandstone		
				UNCONFORMITY		
Pre-Cambrian			Crystalline basement			

Based on Carter *et al.* (1963)

(Figure 37) and is separated therefrom by a hiatus (Offodile 1976). The Keana Formation is thought to be the lateral equivalent of the Makurdi Formation and Offodile (1976) postulated that the Bima, Keana and Makurdi Sandstones constitute a time-transgressive sequence younging from northeast to southwest. He assigned a Cenomanian age to the Keana Formation, and a Cenomanian–Turonian age to the Makurdi Formation. The Keana Formation consists of massive, current bedded, fine to coarse, sometimes conglomeratic, gritty and arkosic sandstones. Pebbles and boulders occur in the basal parts of the formation.

Thickness: Offodile (1976) did not provide thickness data. The Keana thickness is estimated at 4800 ft by Esso Exploration Inc. 1967, Unpublished Geological Sheets 231 and 232 Nigeria.

Age: The Keana Formation in the Lower Benue Depression has yielded few fossils and is thought to be Cenomanian in age by Offodile (1976).

Conditions of Deposition and Palaeogeography: The Keana Sandstones are thought to be delta complex deposits laid down as the Albian–Cenomanian seas regressed down the Benue Trough. Whether they form the youngest deltaic facies of the Bima Delta suite cannot be decided on the field evidence available.

2.4.3.3.4. Eze–Aku Formation

General Description: The Eze–Aku Formation has been mapped in the Lower Benue Trough and con-

sists of calcareous shales, micaceous, fine to medium grained sandstones and occasional limestones. The formation interfingers with and overlies the Keana Formation. The basal beds consist of shelly limestones composed mainly of oyster brash and marly beds.

Thickness: Offodile (1976) gave the thickness as around 300 ft for a typical locality on the bank of the River Tokura, northeast of Keana. The formation maintains this thickness in a northwest–southeast section extending from Lafia through Keana and beyond.

Age: The Eze–Aku Formation is thought to range in age from Cenomanian to Early Turonian and Offodile (1976) plotted a hiatus above the formation separating it from the overlying Coniacian Awgu Formation.

Conditions of Deposition and Palaeogeography: As in the Abakaliki Trough the Eze–Aku Formation was laid down in the Lower Benue Trough during an Early Turonian Transgressive Phase (Murat 1972) but deposition according to Offodile (1976) may have started in the Cenomanian. The formation appears to have been laid down in a shallow marine coastal environment within the Benue Trough and Eze-Aku sediments interdigitated with sandy deltaic facies. The Noko black limestone, apparently part of the Eze-Aku sequence, is thought to have been laid down in a saline reducing? lagoonal environment (Offodile 1976).

2.3.3.3.5. Awgu Formation

General Description: In the Abakaliki Trough the Awgu Shales are thought to have been laid down during the Late Turonian Transgressive Phase but Offodile (1978) without citing fossil evidence assigned the Awgu Formation to the Coniacian (Figure 37).

In the Lower Benue Trough the Awgu Formation consists of black shales, sandstones and limestones and locally seams of coal occur. The coal is confined to the northern part of the Lafia area (Offodile 1976).

Thickness: In the Lower Benue Trough the formation is less than 2200 ft thick based on borehole evidence. Sections showing the thickness and lithology of boreholes in the Obi-Lafia area which penetrate the Awgu Formation are presented in Figure 41.

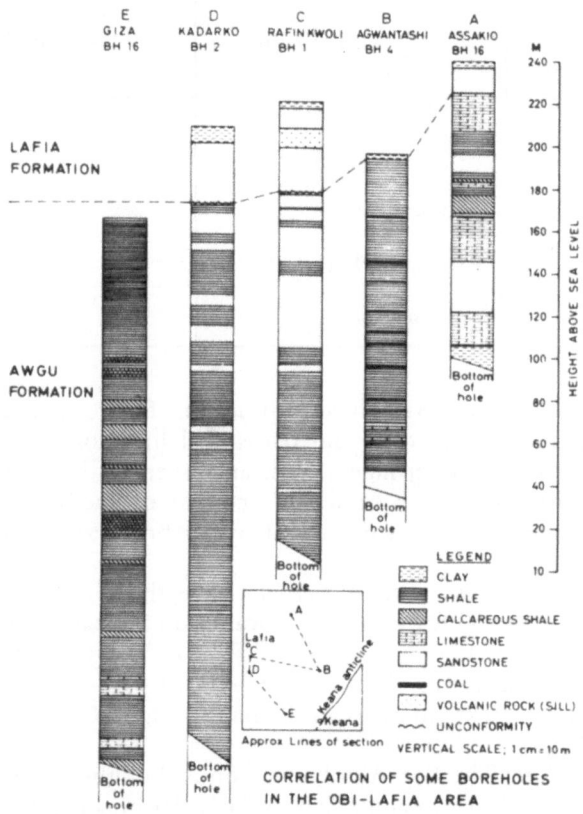

Figure 41. Sections in Lafia and Awgu Formations Obi-Lafia area, Lower Benue Trough. Based on Offodile (1976).

Age: As we have said Murat (1972) regarded the Awgu Formation as having been deposited during the Late Turonian Transgression; whereas Offodile (1976) assigned a Coniacian age to the formation.

Conditions of Deposition and Palaeogeography: Offodile (1976) thinks that the Awgu Formation in the Lower Benue Trough was deposited during "the widespread Coniacian transgression" but according to Murat (1972) Coniacian-Early Santonian times were regressive. The Awgu Shale is said to have been laid down in a shallow sea on the margins of

which swampy conditions developed in which the coals were laid down. Coal bearing conditions gave way southwards to marine conditions. Paralic conditions prevailed and at least thirty coal seams were formed. The post-Santonian Lafia Formation succeeds.

2.4.3.3.6. Bima Sandstone

The Bima Sandstone of the Upper Benue Trough has been described above in the section dealing with the Lower Cretaceous Series but because it is in part Upper Cretaceous it is mentioned here also.

The Bima Sandstone is regarded as the up-valley equivalent of the Muri and Keana Sandstones (also described above) which were deposited during the Late Albian-Cenomanian regressive phase (Figure 28).

The Lower B1 Bima "member" is probably part of a progradation system; the Middle B2 Bima "member" was laid down during the Middle Albian transgressive phase and the Upper B3 Bima "member" is a late-Cenomanian regressive progradation deposit (Figure 21). The front of the Bima Delta Complex during Late Albian-Cenomanian times was probably situated in the Amar-Filiya areas.

2.4.3.3.7. Yolde Formation

General Description: The Yolde Formation consists of a variable sequence of sandstones and shales. The sandstones are thin bedded at the base and are followed by alternations of sandy mudstones and shelly limestones.

Thickness: Thickness data for the Yolde Formation are scattered and insufficient to isopach. South of the Benue at Abare, the Yolde Formation is 540 ft thick; 800 ft near Arum; 240 ft at Lessu (Lau Sheet 47); greater than 455 ft at Gombe Town; and 700 ft at Hinna in the central part of the Gombe Sheet 36 (Geological Survey of Nigeria 1:250 000 Series) (Carter et al. 1963).

Cratchley and Jones (1965) gave a thickness of 1000 ft for the Yolde Formation in the Gombe area and 500 ft along the southern margin of the Upper Benue Trough. Reyment (1965) gave the thickness in the Dadiya syncline as 492 ft.

Age: The Palaeontological evidence is inconclusive as far as age is concerned. The Yolde is Cenomanian according to the explanation on Lau Sheet 36 (Geological Survey of Nigeria 1960). However, Carter et al. (1963) in the memoir accompanying Lau Sheet 36 gave the age as "Lower Turonian, pre-Salmurian (Upper Cenomanian)". The Geological Survey of Nigeria (1964, 1:2 000 000 Geological Map of Nigeria) however, gave the age of the "Bima Sandstone – Yolde Formation as Albian-Cenomanian". Reyment (1965) gave the age as "Lower Turonian ('Salmurian')". Cratchley and Jones (1965) assigned it an 'Upper Cenomanian-Lower Turonian' age but in their Table 1, (Figure 38 this report) they clearly assign the Yolde to the Cenomanian.

Conditions of Deposition and Palaeogeography: The Yolde and the overlying Pindiga Formation are thought to have been laid down during the Early Turonian Transgression and Regression (Figures 29 and 32). The Yolde Formation was laid down under paralic conditions transitional to the marine conditions which prevailed during the deposition of the Dukul Formation.

2.4.3.3.8. Dukul Formation

General Description: The Dukul Formation consists of shales, siltstones and limestones and rests on the Yolde Formation. Shale beds range from a few inches to 20 ft and the limestones do not exceed 2 ft. The Limestones are shaley and non-shelly.

Thickness Variations: At the type locality at Dukul in the northeast part of the Dadiya syncline the formation is 200 – 300 ft thick. The maximum thickness of the formation recorded is 500 ft.

Age: The Dukul fauna consists largely of ammonites and other molluscs and the assemblages indicate an Early Turonian (Salmurian) age.

Conditions of deposition and Palaeogeography: The Dukul Formation was laid down under marine conditions during the Early Turonian transgressive phase (Figure 29).

Lithologically the Dukul Formation is very similar to the Pindiga Formation of the Gombe Sheet 36 and the two formations are mapped on either side of the Talasse anticline and are separated by a distance of about 2–3 miles (Figure 166). Two different names may have been given to the same formation. The Dukul is best considered as a member of the Pindiga Formation, as data stands at present.

2.4.3.3.9. Jessu Formation

General Description: The Jessu Formation consists of alternating grey, white and brown shales and light brown sandy mudstones with subordinate sandstones. Sandy mudstones predominate in the lower part of the formation and shales in the higher part of the sequence. A lava flow with ash bands below occurs in the Jessu Formation near Kunini (Figure 33).

Thickness Variations: At the type locality at Jessu the total thickness is about 1000 ft. Reyment (1965) however gave the thickness as "30 metres", probably a misprint. Elsewhere it thins to 800 ft.

Age: The Jessu Formation is assigned to the Late Turonian by Carter et al. (1963) but Reyment (1965) assigned it to the Turonian ("possibly even Upper Turonian").

Conditions of Deposition and Palaeogeography: The basal Jessu Sandstones are thought to have been laid down under continental conditions (Carter et al. 1963) during the Early Turonian transgression.

2.4.3.3.10. Sekule Formation

General Description: The Sekule Formation consists of shales and limestones and is lithologically similar to the Dukul and Pindiga Formations. The limestones in places consist of a brash of oyster shells. The formation outcrops in a very limited area and again is probably best thought of as part of the Pindiga Formation.

Thickness Variations: At the type locality on the Sekule River the formation is 900 ft thick. No variations have been reported.

Age: Assemblages collected to date indicate a Late Turonian to Santonian age (Carter et al. 1963). Cratchley and Jones (1965) classed the Sekule Formation as Coniacian to Santonian.

Conditions of Deposition and Palaeogeography: The Sekule Formation was laid down under marine conditions and is abundantly fossiliferous but it is not clear, if the Coniacian and Santonian age is accepted, whether these marine deposits were laid down in a sea that connected with the proto-Atlantic via the Benue Trough or via the Chad and Iullemmeden Bassins. Coniacian and Early Santonian times, at the south end of the Benue Trough in the Abakaliki Trough and in the Anambra Basin, were marked by a strong regression. In the Abakaliki Trough and in Anambra Basin the sediments were folded and the main depocentre shifted to the Anambra Basin (Figure 34).

2.4.3.3.11. Numanha Shales

General Description: The formation is composed of shales with occasional bands of sandstones, nodular mudstone and limestone. the lower part of the section consists of grey black shales with three contemporaneous lava flows.

Thickness Variations: The Numanha Shales are between 600 – 800 ft thick.

Age: Diagnostic fossils have not been found in the Numanha Shales (Carter et al. 1963) but because they overlie beds of Coniacian-Santonian age they are tentatively assigned to the Upper Senonian (Campanian?). Cratchley and Jones (1965) assigned a Santonian age to the formation. Reyment (1965) stated that diagnostic ostracoda occur and assigned a Santonian age to the Sekule Formation. The formation occurs beneath the major unconformity produced as the result of Santonian folding and subsequent erosion and is folded with the Sekule and older formations.

Conditions of Deposition and Palaeogeography: The comments made about the Sekule Formation apply.

2.4.3.3.12. Lamja Sandstone

General Description: This formation consists of sandstones, shales and thin coals. It is the final unit on the Lau Sheet in the Upper Benue Trough and like the Numanha and Sekule Formations it is folded with the Upper and Lower Cretaceous rocks of the area. The formation therefore probably underlies the major Santonian unconformity. On the Longuda Plateau (Figure 34) the Lamja is overlain by Cenozoic basalts.

Thickness Variations: The outcrop area is small and at the type locality the Lamja Sandstone is 167 ft thick.

Age: Reyment (1965) assigned the formation to the Maestrichtian. Coal dated by the Palynological Section of Shell-BP yielded spores of ? Turonian-

Coniacian-? Santonian ages. In view of this and the fact that the Lamja sequence is folded with the rest of the pre-unconformity Upper Cretaceous rocks, the formation is best classed as Santonian-pre-Maestrichtian.

Conditions of Deposition and Palaeography: The outcrop areas are very small but we can say from the lamellibranchs which occur that the formation was deposited under marine conditions. As we have already noted, marine connections with the Middle and Lower Benue and the proto-Gulf of Guinea must have been severed in Santonian times (Figure 34) and the marine connection must then have been via the Chad and Iullemmeden Basins and the proto-Atlantic.

The Sekule, Numanha and Lamja "formations" represent the youngest Cretaceous deposits in the area. The Lamja Formation is folded and lies beneath the pre-Santonian unconformity.

2.4.3.3.13. Pindiga Formation

General Description: The Pindiga Formation consists of marine shales with limestones near the base. The lower part of the formation at the type locality consists of 250 ft of limestones and shales ("Limestone-Shale member") and is richly fossiliferous. The upper part of the Pindiga Formation consists of blue black shale ("Upper Shale Member") (Figure 17). The Pindiga Formation has not been mapped south of the Benue River, the Kaltungo Basement inlier, and the Tallasse Anticline (Figures 9 and 166). South of these areas the Dukul, Jessu and Sekule Formations have been mapped; these are thought to be the equivalents of the Pindiga Formation (Figures 8 and 9).

Thickness Variations: The total thickness of the Pindiga Formation in GSN Borehole 1504 at Gombe is 625 ft. On the Futuk Sheet 46 the Pindiga thickness is given as 3930 ft (Dessauvagie 1969). The total thickness of Dukul, Jessu and Sekule equivalents of the Pindiga Formation in the Futuk area is 2716 ft, not 3345 ft as given by Dessauvagie (1969).

Age: The Pindiga Formation is highly fossiliferous and has yielded a Lower Turonian (Salmurian) fauna except for a single specimen of *Libycoceras ismaelensis* (Zittel) which may be a mislocated specimen as this is a Maestrichtian fossil (See Reyment 1965; Carter *et al.* 1963 etc.).

Conditions of Deposition and Palaeogeography: The Pindiga Formation is clearly marine and its equivalents the Dukul, Sekule and Numanha are marine also. The Jessu and Lamja are local continental equivalents. Marine conditions then prevailed for most of Early and Late Turonian and Coniacian times in the northern Benue Trough (Figures 22 and 33).

2.4.3.3.14. Gulani Formation

General Description: The Gulani Formation consists of sandstones, massive, cross bedding and ripple marked. They are fine to medium grained. The Gulani Formation is the lateral equivalent of the Numanha Shale and Sekule Formation.

Thickness Variations: The Gulani Formation is 200 ft thick and is of limited extent. Because of erosion it is not possible to determine the original thickness.

Age: Fossils have not been collected from the Gulani Formation and as it overlies the Pindiga Formation with its rich Turonian fauna, it is dated? post-Lower Turonian to Upper Turonian. It is classed as a pre-Santonian unconformity formation also because it is folded on northerly, northnortheasterly and northeasterly axes which trend at right angles or obliquely to folds in the post-unconformity Gombe Formation.

Conditions of Deposition and Palaeogeography: (See notes for Pindiga Formation).

2.4.4. Gongola Trough, Bornu Basin and Yola Trough Upper Cretaceous Pre-Santonian Formations

2.4.4.1. General

Considerable thickness changes take place north of the Zambuk Ridge which separates the Upper Benue Trough from the Gongola Trough which in turn forms the link with the Bornu Basin. This basin is known as the Chad Basin in Nigerian geological literature but the basins are better classed as separate features. They are so labelled on Figure 2. The oldest formation exposed in the Gongola Trough is the Bima Formation (described above). Other formations exposed include the Gongola Formation, which rests conformably on the Bima Formation, and the Fika Shales.

As is pointed out in the chapter on "Structure" and in the general section on stratigraphy, the Bornu Basin merges with the Gongola Trough (Figures 2, 22 and 170) although limits are presently ill defined. The trough is bounded by the Basement Complex outcrops of the Bauchi–Kano area and of the Zinder area and by the subsurface ridge situated south of the Lake Chad Syncline. The Bornu Basin (using sediment thickness estimates based on gravity data) is "two horned" at its western end and extends eastwards to Lake Chad where Cenozoic and Mesozoic rocks taken together total more than 9840 ft. The structure may well prove to be much more complicated however and part of an extensive Chad Basin rift system.

Cretaceous rocks are not exposed in the basin, except east of Katsina, Nigeria and at the northern end of the Gongola Trough where the Fika Shales rest on the Gongola Formation (Figure 43). The validity of the Gongola Formation as a separate mappable entity is questionable but nevertheless there are over 2000 ft of Upper Cretaceous sediments which occur below the Santonian unconformity in this area (Carter *et al.* 1963).

Upper Cretaceous Pre-Santonian rocks are exposed on the south side of the Bornu Basin. The

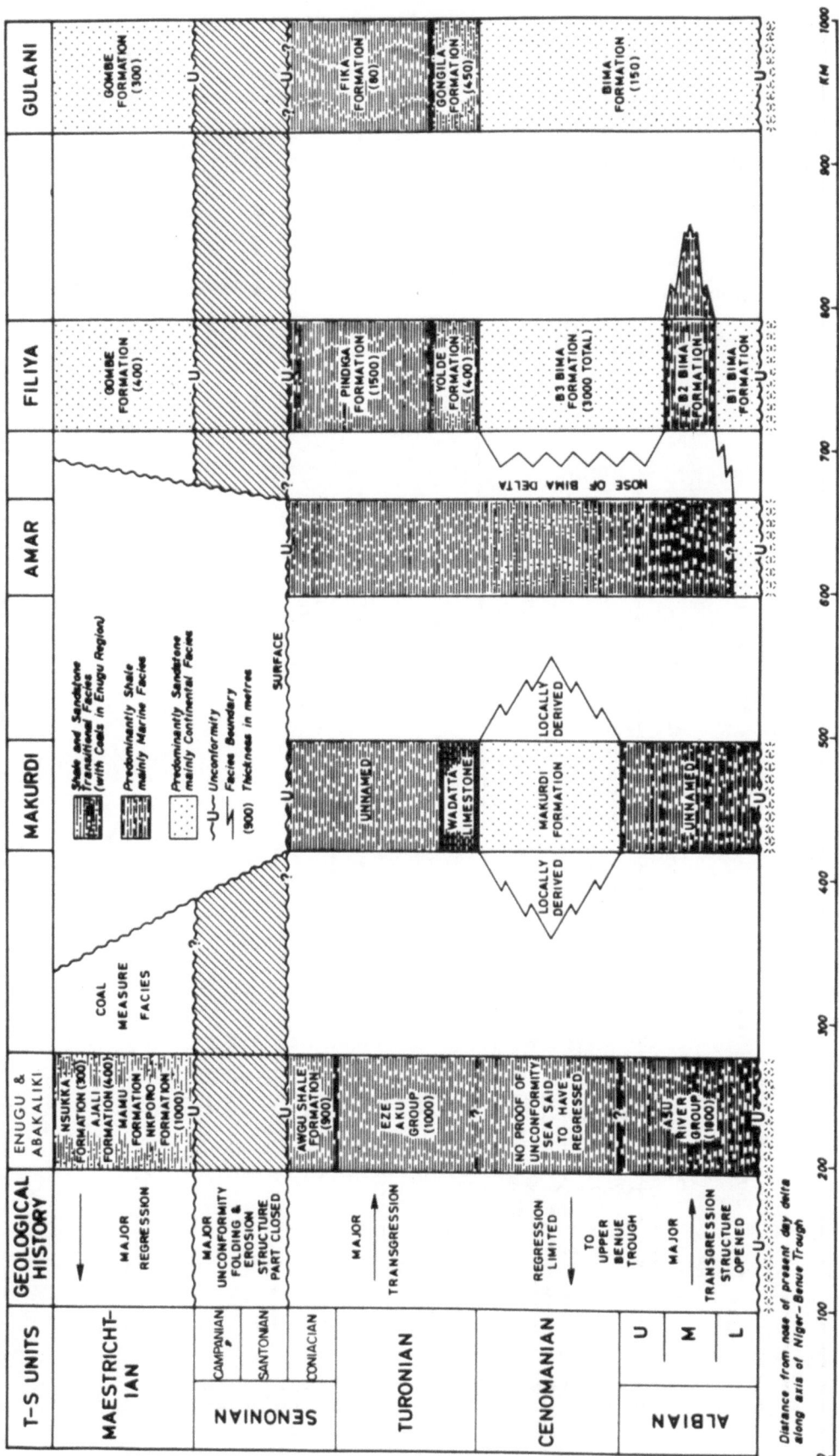

Figure 42. Correlation of Cretaceous Formations, Enugu and Abakaliki area, Makurdi, Amar, Filiya and Gulani areas. Lower and Upper Benue Trough. Based on Whiteman (1973); Carter *et al.* (1963); Farrington (1952); Cratchley and Jones (1956).

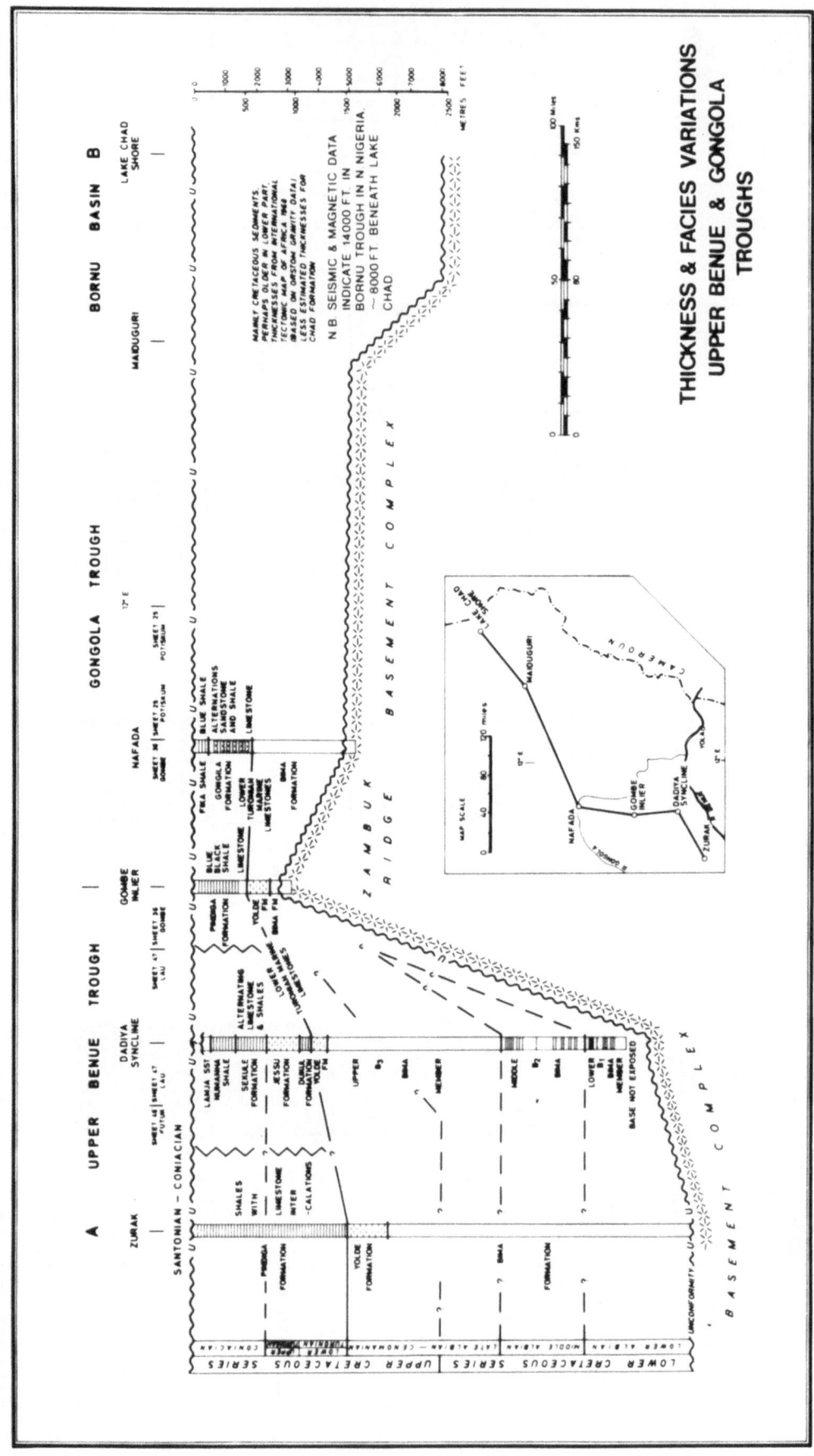

Figure 43. Sections showing thickness and facies variations Upper Benue and Gongola Troughs. Based on Whiteman (1973) (part of section used by Adeleye 1976, Figure 26 this account); and Carter *et al.* (1963).

outcrops are limited, and underlie areas along latitude 11°N. They include Fika Shale, Pindiga Formation and Bima Sandstone which rest unconformably on the Basement Complex. The Neogene formations of the Bornu and Chad Basins obscure the Cretaceous rocks all around the northern margin of the Basement Complex area centred on Kaduna. Pre-Santonian Cretaceous rocks next appear west of Katsina as the Gundumi and Illo Formations (Figure 9).

The Fika, Gongila and Bima Formations were also encountered in the Potiskum Borehole (Barber and Jones 1960) and the Fika Shales were encountered in the Maiduguri Borehole.

2.4.4.2. Upper Cretaceous Pre-Santonian Formation, Gongola Trough, Bornu Basin and Yola Trough

2.4.4.2.1. Gongila Formation

General Description: The Gongila Formation consists of basal limestones overlain by a sandstone-shale sequence. It rests conformably on the Bima Sandstone and is overlain conformably by the Fika Shale (Figures 8, 9 and 22).

Thickness Variations: In the most complete section the formation is 1369 ft thick.

Age: The limestones of the Gongola Formation, which occur near the base, are rich in fossils, mainly ammonites and molluscs, the assemblages indicate an Early Turonian (Salmurian) age.

Conditions of Deposition and Palaeogeography: The outcrop of the Gongila Formation is clearly marine in the lower part (Figure 43) and presumably the sandstone–shale sequence is marine also. The Gongila Formation passes east and south into the Pindiga Formation and a belt of Gongila and Pindiga Formation (GP) (Undifferentiated) has been mapped on Potiskum Sheet 25 and Gombe Sheet 36 (GSN 1:250 000 series). Presumably some of the Gongila Formation must be equivalent to the Yolde Formation of the Upper Benue Trough as the Gongila Formation rests conformably on the Bima Formation (Figure 43). The Gongila Formation is probably the equivalent of the Yolde and Pindiga formations and the comments made above about conditions of deposition of these formations apply.

2.4.4.2.2. Fika Shales

General Description: The Fika Shales consist of blue black shales, which are occasionally gypsiferous and which contain thin persistent limestones. Whether the gypsum is primary or secondary is not stated in the literature.

Thickness Variations: In GSN Borehole 913 at Damagum the Cenozoic Kerri–Kerri Formation overlies 320 ft of blue shales (Fika Shales). At Maiduguri 120 miles east of Damagum there are 1413 ft of "Santonian–Maestrichtian" shales which appear to be the equivalent of the Fika Shales (Carter *et al.* 1963).

Age: The fossils of the Fika Shales are mainly fish remains, chelonian fragments and reptilian remains. These indicate a Cenomanian–Maestrichtian age (Carter *et al.* 1963) but Reyment (1965) commented that several species of the Fika assemblage also occur in the Yolde Formation (Lower Turonian). The contact with the overlying Gombe Formation is not well exposed but may be unconformable and this justifies placing the Fika shales in the pre-Santonian.

Conditions of Deposition and Palaeogeography: The Fika Shales appear to have been deposited under paralic conditions but more work needs to be done on these rocks (Figure 43).

2.4.4.2.3. Bima–Yola Formation:

General Description: The sequence shown in Figure 44 was recognized in the Yola Trough (Burke and Dessauvagie 1970).

Figure 44
Cretaceous Sequence Yola Trough

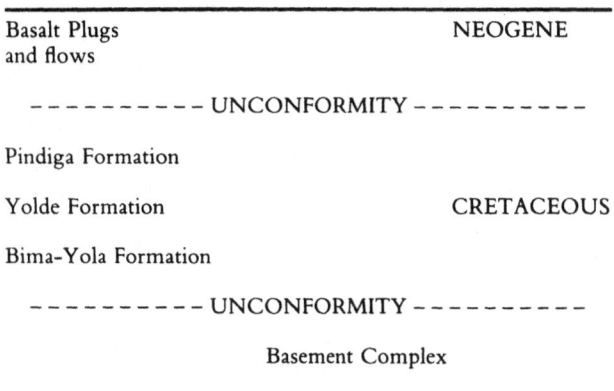

Basalt Plugs and flows	NEOGENE
– – – – – – – – – UNCONFORMITY – – – – – – – – –	
Pindiga Formation	
Yolde Formation	CRETACEOUS
Bima-Yola Formation	
– – – – – – – – – UNCONFORMITY – – – – – – – – –	
Basement Complex	

Based on Burke and Dessauvagie (1970)

2.4.4.3. Hydrocarbon Prospects of the Lower and Upper Benue, Gongola and Yola Troughs and the Nigerian Section of the Bornu Basin Pre-Santonian Lower and Upper Cretaceous Formations

2.4.4.3.1. General

The hydrocarbon prospects of the Pre-Santonian Upper and Lower Cretaceous rocks of the Upper and Lower Benue Trough can only be described as speculative and the areas classed as high risk (Figure 170). Field parties from oil companies and mining companies have reconnoitred the area but very little data has been released. No exploratory drilling has been made and there are a few subsurface details. Governments of states bordering the troughs are interested in stimulating exploration but the draw of the Niger Delta "oil patch" over the years has had the effect of detracting interest from these areas. It remains to be seen if Nigeria's financial need to increase reserves and daily production will provide a stimulus for exploration of these areas by oil companies and the Nigerian National Oil Corporation.

As is stated in Section 3.3 the Benue Trough with its side arms of Yola and Gongola constitute a series of troughs of pre-Santonian (Late Cretaceous) age formed during a rift phase connected with the development of the Niger Delta Triple and Chum Trilete Junctions. Lithospheric plate thinning caused long narrow complex depressions to develop, which in places were infilled by large thicknesses of sediments deposited in a variety of sedimentary environments. These sediments include rocks with source, reservoir and seal characteristics.

The pattern of sedimentation which developed from Albian to Santonian times was basically simple with successive delta fronts migrating down the monoclinally or fault-bounded trough during regressive phases and with marine transgressions penetrating the troughs mainly from the south via the opening proto–Atlantic.

Subsequently in Santonian times the sediments infilling the troughs were folded and andesitic volcanism, granodioritic intrusion, hydrothermal mineralization and metamorphism affected the Benue, Abakaliki, Yola and Gongola troughs. The Chum Trilete Junction (rrr) consists of the Benue, Yola and Gongola arms (Chapter 3).

Compared with folds in the Abakaliki Trough, the Benue Trough folds are much gentler and more open and may well be of the drape rather than the compressive type. However, attractive though some of these large anticlines (Figures 169 and 170) may be at first sight, uplift, mineralization, intrusion and metamorphism associated with Santonian folding downgrades some folds as far as hydrocarbon prospects are concerned, because high geothermal gradients must have prevailed regionally. No palaeotemperature or geothermal measurements are available however, and the region has only been studied reconnaissance fashion.

Gas accumulations might be targets in areas marginal to the Benue Trough. Large structures exist and striking facies changes are known. Flushing with fresh water may have adversely affected structures at the north end of the Benue Depression, where the presence of Late Cenozoic Volcanics and increased geothermal gradients may have down-graded oil prospects. Gas prospects may have been increased however in some areas because of high geothermal gradients. No maturation data have been published for the pre-Santonian Cretaceous sediments of the Benue Trough.

Isopach data are not publicly available for the Benue Trough except in a highly generalized form but in the northern Benue the pre-Santonian Cretaceous rocks may well exceed 15 000 ft. They thin down to around 3000 ft over the Zambuk Ridge which forms a threshold to the southern part of the Bornu Basin where thickness once again may well exceed 15 000 ft (Figures 2 and 43). The prospects of the post-Santonian Anambra Basin, the Gombe-Wase Trough and the Wukari-Mutum Biya Trough are discussed below.

In terms of the quality and quantity of data available for assessment of the Benue and associated basins only reconnaissance style geological and geophysical data are available publicly. Few subsurface details have been published and the little information available on thickness variations is based mainly on interpretation of gravity data obtained during reconnaissance in the Lower Benue and 1:250 000 field survey in the Upper Benue, Gongola and Yola Troughs (Figures 166 and 170). Seismic data is not available and in general terms the area is very poorly explored. However companies such as Esso, Shell-BP, Mobil, Gulf etc. and mining companies have had field parties in the area under consideration and completed many party months of survey.

Open areas in the Benue and associated basins can only be rated as a high risk and speculative. Gas rather than oil may have accumulated because of the high geothermal gradients which must have affected large parts of the folded-mineralized area. Prospects are much better in the post-Santonian basins however.

Very little information appears to be available on shows, seepages etc. Adeleye (1975) stated that there are reports of oil seepages around Muri in the Upper Benue Trough but he did not give the source of his information. Offodile (1976) referred to Adeleye's information. Cratchley and Jones (1965), Farrington (1952), Dessauvagie (1969) did not mention oil shows, seepages etc. but the reconnaissance nature of this work should be borne in mind. Offodile (1976) described the Benue Trough brine occurrences in detail. These are reviewed in Chapter 3 together with information on mineralization within the Benue Trough. Brine issues from sediments forming the flanks of major folds extending from the Gongola Trough in the north to the Abakaliki Trough in the southwest. The brines are often under pressure and attain temperatures of 40°C at surface.

2.4.4.3.2. Source Rocks

As we have said maturation studies have not been published for Benue Trough sediments, but argillaceous limestones with superficial source rock characteristics are said to occur in Asu River Group, Awe, Ese-Aku and Awgu units of the Benue Trough (Offodile 1976). The Noko fine grained, black limestone of the Eze-Aku Formation is mentioned especially in this context. Considered overall the coaly, pyritized sediments of the Awgu Formation and the bluish-green sediments of the Asu River Group might be taken to indicate deposition under anaerobic conditions. The Awgu Formation coals themselves could under suitable maturation conditions be considered as a gas source material. Offodile (1976) gave details of these coals which occur south of Lafia on the Lafia–Makurdi road and in the valley of the River Dep. Analyses of samples from the Dep outcrop are presented in Figure 45. The coals are subbituminous to bituminous. The higher grades appear

Figure 45
Proximate Analyses of Coal Awgu Formation From Sheet 231, Lower Benue Trough

Description of Samples
Two of the samples No. FF 11(C) and FF 145(VI) which are schistose in structure are dull in appearance, brownish black in colour, quite fragile with patches of limonite and greyish white earthy material. The other two samples FF 145 (VII) and FF 145 (VIII) are compact, resinous lustrous, black and also with patches of limonite.

Sample Nos	Locality	Volatiles Plus Moisture	Ash	Fixed Carbon	Carbon Ratio	Calculated to Ash free basis	
						Volatiles	Fixed Carbon
GL.60970 FF 11(C)	Rafin Kwoli	33.96	39.65	26.39	0.78	56.3	43.7
GL.60971 FF 145(VI)	Seam No. 3 R. Dep	23.83	30.40	45.77	1.92	34.2	65.8
GL.60972 FF 145(VII)	Seam No. 8 R. Dep	29.79	8.61	61.60	2.06	32.6	67.4
GL.60973 FF 145(VIII)	Seam No. 10 R. Dep	29.36	13.70	56.94	1.94	34	66

In both samples FF 145(VII) and (VIII): Colour of ash light brown;
Character of coke: dark grey and lustrous; somewhat swollen and porous; fairly strong
Carbon Ratio = Fixed Carbon/Volatiles
Analyst: 4 September 1973. J. A. Akinbehin, Geological Survey of Nigeria
Based on Offodile (1976)

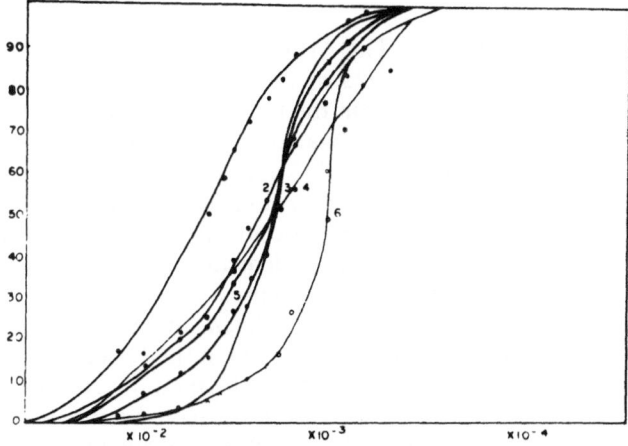

Figure 46. Graphs showing size analyses of sandstones from Sheet 231. Sample 1 = Makurdi Formation (Keana Sandstone); Samples 2–5 = Lafia Formation; Sample 6 = Ezeaku Formation. Based on Offodile (1976).

to have been produced as a result of igneous action. The coals have also been affected by "post Maestrichtian" folding although movements are small. Some 30 seams are known in the Obi anticline and the coals show ample evidence of movement in that they are folded, slicken-sided and mineralized. They could under some conditions have acted as gas source rocks.

2.4.4.3.3. Reservoir Rocks

Few direct measurements of porosity and permeability appear to be available for Benue Trough sedimentary rocks, but qualitative statements have been made by Carter *et al.* (1963); Cratchley and Jones (1965); Dessauvagie, Burke and Whiteman (1971 and 1972); Burke and Dessauvagie (1970). Generalizing about the Lower Benue Trough sandstones, we can say that permeabilities are high except in the Keana Formation which in places is cemented, especially on the crest of the Keana Anticline. Data on ground water from the Lafia and Keana area are presented in Figure 47. The Awgu Formation is the first of the pre-Santonian formations having good aquifer characteristics, although the overlying post-Santonian Lafia Formation is highly permeable and is apparently connected to the Awgu circulation. Some boreholes within the Awgu Formation are water bearing and shale beds act as seals or aquicludes. The coal bearing sequence is fractured and carries a lot of water. Awgu water is usually hard and acidic (Offodile 1976). Sand bodies have been reported from the Awe, Eze–Aku and Awgu formations in the Lower Benue Trough and from the Bima Sandstone, Yolde, Jessu, Numanha, Lamja and Gulani Formations in the Upper Benue Trough and limestones have been reported from the Pindiga and Dukul formations.

The sandstones of the Makurdi Formation are highly indurated at outcrop and almost impermeable

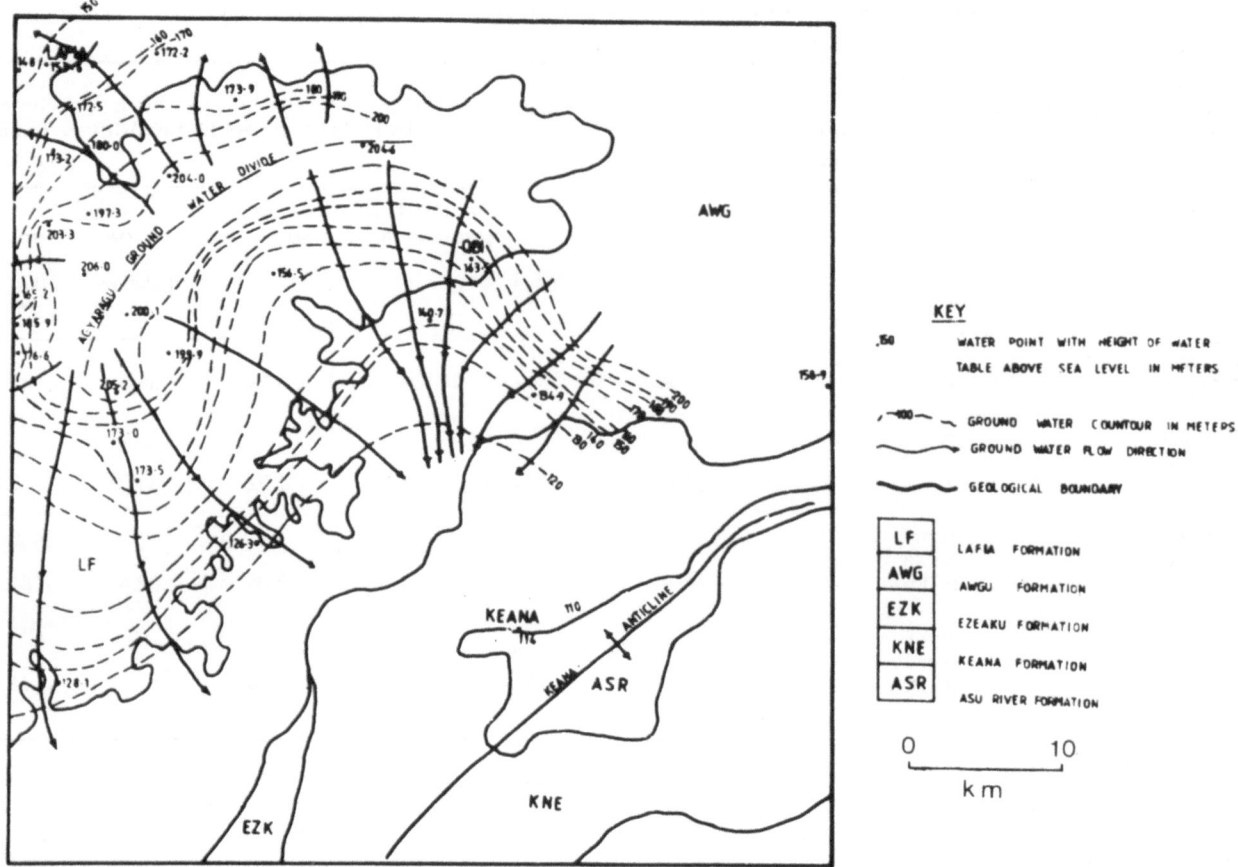

Figure 47. Map showing water table contours in the Lafia – Keana area, Lower Benue Trough. Based on Offodile (1976).

in places, whereas sand bodies in the Eze–Aku Formation, its lateral equivalent, are less indurated, more permeable and are better aquifers. Artesian conditions may occur in the Makurdi/Eze–Aku sand bodies beneath an Awgu Formation seal.

Concerning the reservoir properties of the sandstones and limestones of the Upper Benue Trough, Carter *et al.* (1963) stated that the yields of water wells drilled into the Bima Formation are often small because of low permeability due to interstitial clay and the marine shales and mudstones frequently contain salt water. Yields from open wells dug in the Bima Sandstone rarely exceed 400 gallons per hour. Sandstones, siltstones, shales and limestones occur as good aquifers in the Yolde and Gongila Formations. At Gombe the Yolde Formation yields pressure water at rates varying between 1500–4000 gallons per hour (Carter *et. al.* 1963). Pindiga Formation shales form the seal. Diagenetic studies do not appear to have been made of any of the Benue Valley sandstones.

2.4.4.3.4. Specific Prospects in Lower Benue Trough

Specific plays in the Lower Benue Trough pre-Santonian rocks may be identified. Prospects here are similar to those of the Abakaliki Trough from which the Benue Trough is offset along the Gboko line (Figure 22).

PROSPECT 1: The Arufu and Gboko Limestones occur along the flanks of the Lower Benue Trough and may well occur on the northwestern flank in the bend between the Abakaliki and Benue Troughs (Figure 9). Predominantly stratigraphic traps perhaps containing oil generated within the Uomba Formation (Middle Albian) may occur associated with such lenses if buried beneath adequate cover or within marginal deltaic sandstone lenses.

PROSPECT 2: In Late Albian and Cenomanian times a delta complex developed in the Lower Benue Trough. The Keana Sandstone formed the continental facies and the delta front must have been situated northwest of Gboko (Figure 9). Given adequate cover and well developed structures there might be considerable potential in this area. Similarly good prospects may exist in the Makurdi Sandstone deposited during the Early Turonian regression and in the Agbani Sandstone (Figures 29 and 34). However, as in the Abakaliki Trough a lead-zinc mineralization is associated with the Lower and Upper Cretaceous rocks and in Late Turonian times volcanics were formed southwest of Awe (Figure 33). Tertiary and Recent volcanics also cut the sediments. High geothermal gradients may have overmatured source materials. No maturation studies have been published for this area.

2.4.4.3.5. Specific Prospects in Upper Benue Trough

Two kinds of plays might exist in these areas in Pre-Santonian rocks:

1. Those associated with marginal deltaic sands derived from the flanks of the depression and interdigitating with shales and limestones (Figure 22).
2. Those associated with the development of the larger delta complexes of the Keana and Bima type in Late Albian to Cenomanian times (Figures 25 and 28).

A Bima Delta front probably existed in the Amar and Filiya areas and a Keana delta front existed northwest of Gboko at the southwestern end of the Benue Depression (Figure 22). Both delta complexes were folded during the Santonian orogeny and long, open folds roughly paralleling the axis of the Benue Trough were produced. In the Upper Benue Trough many of these folds e.g. the Lamurde Anticline are breached and hydrocarbons which may have accumulated have escaped. In the Middle Benue area the Keana Anticline is breached also. The Lamurde fold is over 60 miles long and dips remain mainly below 10°. Lower Cretaceous rocks exposed in this structure are intruded by Tertiary volcanics. Similar structures occur elsewhere in the Middle Benue area. However unbreached structures may exist under considerable cover down valley and might be of interest in the area of the Keana Delta Front (Figure 28).

The hydrocarbon prospects of the Upper Benue Trough are not considered to be high mainly because of igneous activity, mineralization and implied high geothermal gradients. However in the Upper Benue in the Wase–Gombe Basin the prospects are thought to be better because some 20 000 ft of Lower and Upper Cretaceous sediments probably exist in this area. A considerable part of this sequence must be post-Santonian in age. Although this trough has considerable exploration interest, and may well contain hydrocarbons, the pre-Santonian prospects must be regarded as highly speculative.

The post-Santonian Wukari–Mutum–Biyu Basin which exists in a complementary position on the south side of the Central Gravity High of the Benue Depression, is considered to be of low interest for hydrocarbon prospecting because along its southern margin the Upper Cretaceous sandstones are thought to be non-marine and the underlying Lower Cretaceous is probably thin. However very little is known about the geology of this area and it could contain surprises.

2.4.4.3.6. Prospects in the Gongola and Yola Troughs

Associated with the Upper Benue Trough, and forming a trilete junction, are the Gongola and Yola Troughs (Figure 170). The hydrocarbon prospects of all these three troughs are considered to be low for commercial gas and oil for the following reasons:

1. The sandstones of both the Lower and Upper Cretaceous in all three areas have been flushed with water.
2. Widespread interconnected poroperms exist, so water may have swept through the reservoirs on a regional scale flushing out any oil which may have accumulated.
3. Numerous igneous intrusions of post-Cretaceous age affect the region and downgrade the area.
4. Santonian hydrothermal mineralization has been recognized as far up-valley as the area between Filiya–Muri and Lau so downgrading the prospects.

The hydrocarbon prospects of the *post-Santonian* Cretaceous rocks are discussed below in Section 2.4.6.3. A short summary of petroleum prospects and plays for the pre-Santonian rocks of the Benue Depression are presented in Figure 48.

2.4.5. Sokoto "Basin" or Embayment, Nigerian Section of the Iullemmeden Basin Upper Cretaceous Pre-Santonian Formations

2.4.5.1. General

Sediments in the Sokoto part ("Sokoto Basin" of Nigerian literature) of the Iullemmeden Basin (French literature) can be divided into three main groups:

3. Cenozoic Continental Deposits
2. Maestrichtian–Palaeocene Marine Deposits
1. Pre-Maestrichtian Continental Deposits.

Taken together these sediments do not exceed 4000 ft in the Sokoto area. The overall sequence of the Iullemmeden Basin of which the Sokoto "Basin" forms part has been described by Greigert and Pougnet (1961) and by others. Exact correlations of the older Cretaceous units in Niger and Nigeria are uncertain. The stratigraphic units recognized in the Cretaceous rocks of northwestern Nigeria are shown in Figure 51. The exact ages of the Gundumi and Illo formations are uncertain but they are taken here to be pre-Maestrichtian. The Rima Group is described below. The distribution of these formations is shown in Figures 9, 49 and 53.

The Gundumi and Rima Group, together constituting the Upper Cretaceous in northwestern Nigeria, attain a maximum thickness of around 3700 ft and in the Birnin Kebi region beneath the Tertiary cover Cretaceous rocks are only about 1500 ft thick. The thinness of the Cretaceous section, lack of adequate seals and structures make the Sokoto area uninteresting to the petroleum prospector. The area was under lease to Mobil in the early days of petroleum exploration in Nigeria but was relinquished (Figure 296).

Figure 48

Summary of Petroleum Prospects and Plays, Pre-Santonian Rocks, Lower and Upper Cretaceous Series (Albian to Lower Santonian), Benue, Gongola and Yola Troughs (Figures 169 and 170).

Province	Region	Play	General Description	Trap Type	Shows	Prospect Class
BENUE TROUGH	LOWER BENUE TROUGH	Arufu and Gboko Limestones	The Arufu and Gboko Limestones occur along the flanks of the Lower Benue Trough and stratigraphic oil may have accumulated associated with these bodies having been generated within the Uomba Formation	Stratigraphic and Structural	None recorded	SPECULATIVE LIMITATION THICKNESS OF COVER. HIGH GEO-THERMAL GRADIENTS. MINERALIZATION
BENUE TROUGH		Keana Sandstone Delta Complex and Makurdi Sandstone Delta Complex	In Late Albian and Cenomanian times the Keana Delta Complex developed in the Lower Benue Valley. Likewise in Turonian times the Makurdi Sandstone Complex was deposited (Figure 36). These beds were folded	Stratigraphic and Structural	None recorded	SPECULATIVE LIMITATION THICKNESS OF COVER AND PRESENCE OF CRETACEOUS AND TERTIARY IGNEOUS ROCKS. FLUSHING, HIGH GEO-THERMAL GRADIENTS AND MINERALIZATION
	UPPER BENUE TROUGH INCLUDING YOLA AND GONGOLA TROUGHS	Flank sands interdigitating with shales and limestones.	Marginal sands derived from the flanks of the trough exist in many areas and may contain hydrocarbons. Strongly folded during the Santonian orogeny.	Anticlines, Faults, Stratigraphic	None recorded	SPECULATIVE
		Delta complexes of the Rima and Keana Formations	Delta fronts probably existed in the Amar and Filiya areas (Bima Formation) and the area northwest of Gboko (Keana Formation) (Figure 36). Strongly folded during the Santonian orogeny.			SPECULATIVE LIMITATIONS BOTH 2.2.1. AND 2.2.2. MAY BE LIMITED BY IGNEOUS ACTIVITY (CRETACEOUS AND TERTIARY AND FLUSHING. POORLY KNOWN GEOLOGY. HIGH-GEOTHERMAL GRADIENTS AND MINERALIZATION

Based on Whiteman (1973)

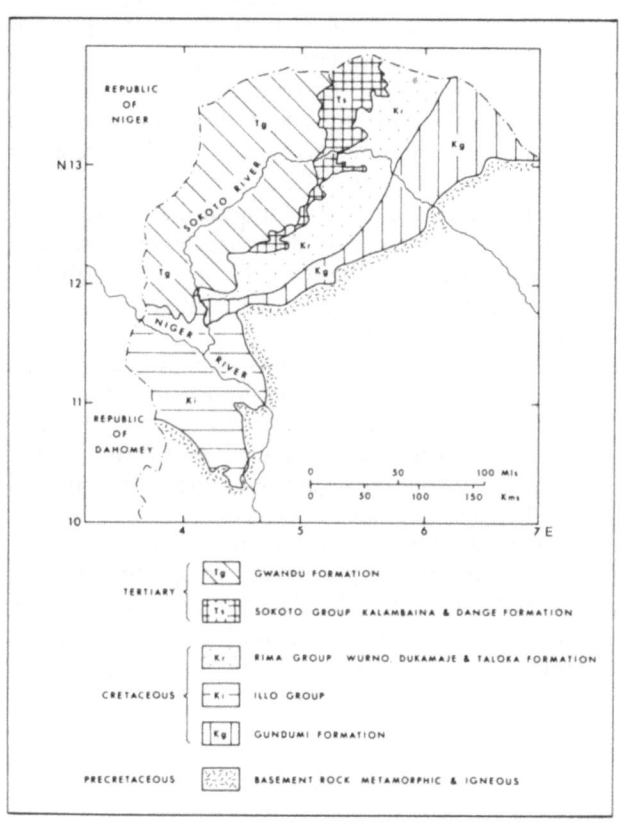

Figure 49. Generalized geological map northwestern Nigeria, so called "Sokoto Basin", southeastern margin of the Iullemmeden Basin. Based on Kogbe (1976).

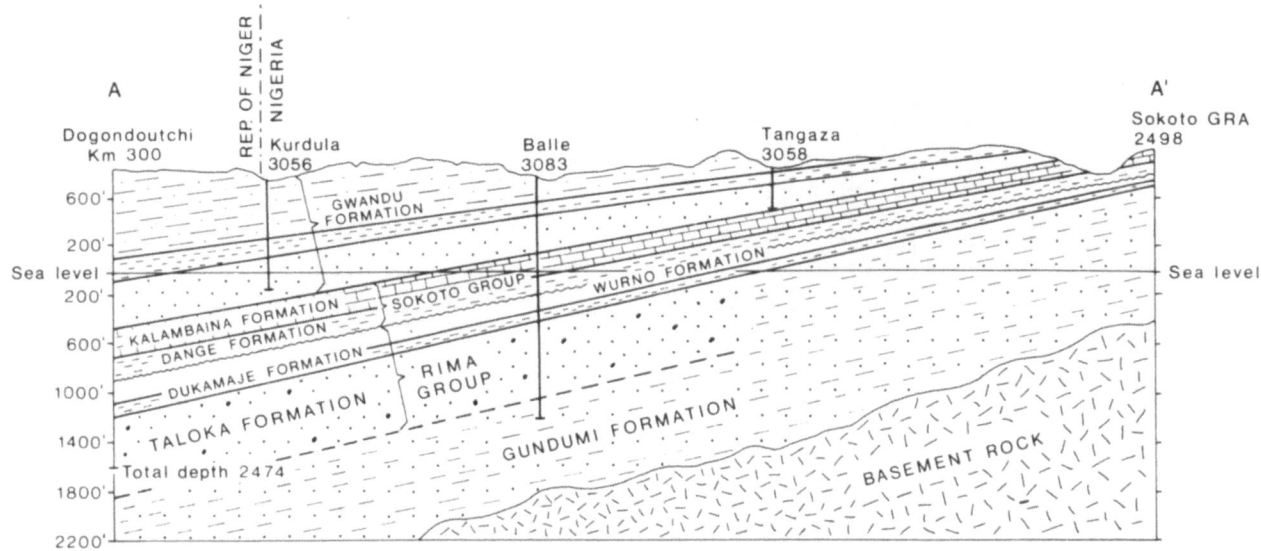

Figure 50. Generalized geological section Gwandu Formation to Basement Complex, Southwest Sokoto area. Based on Kogbe (1976); utilizes Geological Survey of Nigeria data.

Figure 51
Evolution of Nomenclature: Sokoto Section of Iullemmeden Basin

Age	Raeburn and Tattam in Swinton (1930)	Brynmor Jones (1948)	Carter (1964)	Parker (1964)	Reyment 1965	Kogbe This Report
Post Eocene	Gundumi series Gwandu series	Gwandu Clay and Grit Group	Gwandu Formation	Alluvium	Alluvium	Alluvium
Eocene	Rima series { Calcareous group, Clay-Shale group	Calcareous Group, Clay-Shale group	Sokoto group { Calcareous Formation, Dange Formation	Gwandu Formation, Kalambaina Formation	Gwandu Formation, Kalambaina Formation	Gwandu Formation, Kalambaina Formation } Sokoto group
Paleocene		–	–	Dange Formation	Dange Formation	Dange Formation
Maestrichtian	Sandstone group	Rima Group { Upper Sand-Stones and Mudstone, Mosasaurus Shales, Lower sand-stones and Mudstone	Rima Group { Wurno Formation "Mosasaurus Shales", Taloka Formation	Wurno Formation, Dukamaje Formation, Taloka Formation	Rima Formation	Rima Group { Wurno Formation, Dukamaje Formation, Taloka Formation
Pre-Maestrichtian		Gundumi grit and Clay group, Illo group	Gundumi Formation	Gundumi Formation	Illo Formation, Gundumi Formation	Illo Formation, Gundumi Formation

Based on Raeburn and Tattam in Swinton (1930);
Brynmor Jones (1948); Carter et al. (1964);
Parker (1964); Reyment (1965); and Kogbe (1976).

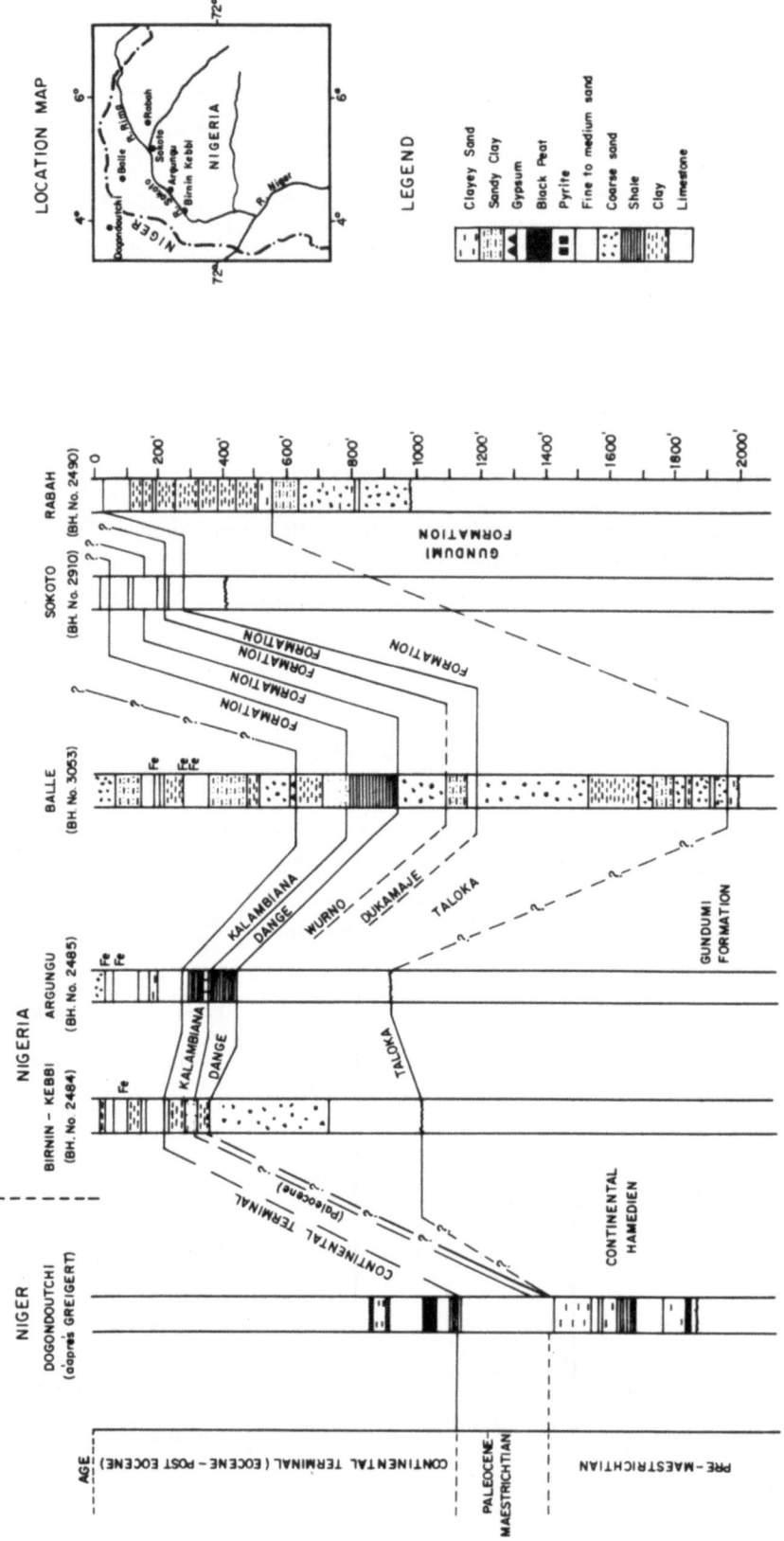

Figure 52. Correlation of Continental Terminal (Eocene – Post Eocene), Palaeocene-Maestrichtian encountered in wells drilled in Northwestern Nigeria ("Sokoto Basin") and Southeastern Niger. Based on Kogbe (1976) and Whiteman (1973).

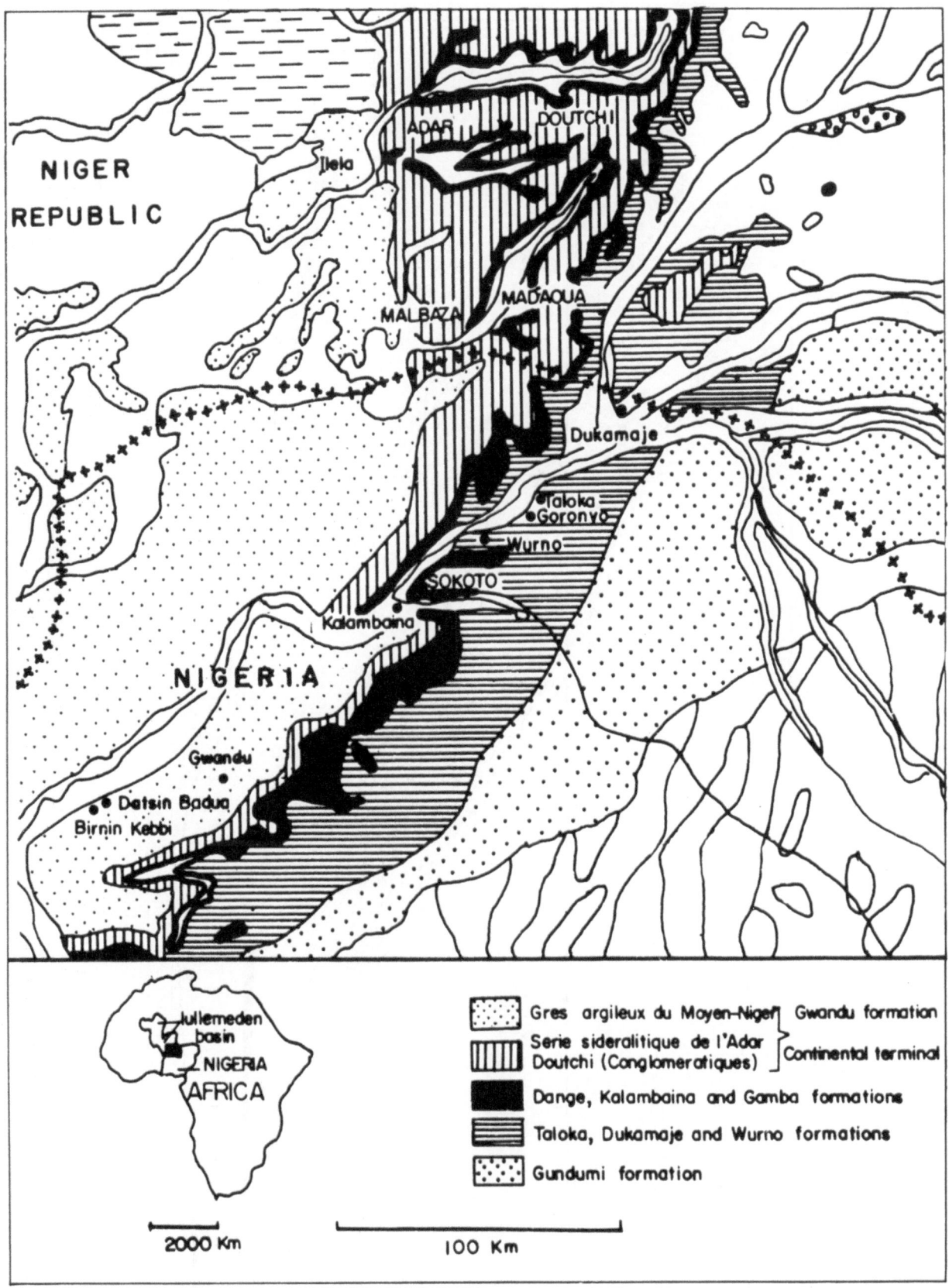

Figure 53. Geological sketch map "Sokoto Basin" Niger. Based on Kogbe (1976).

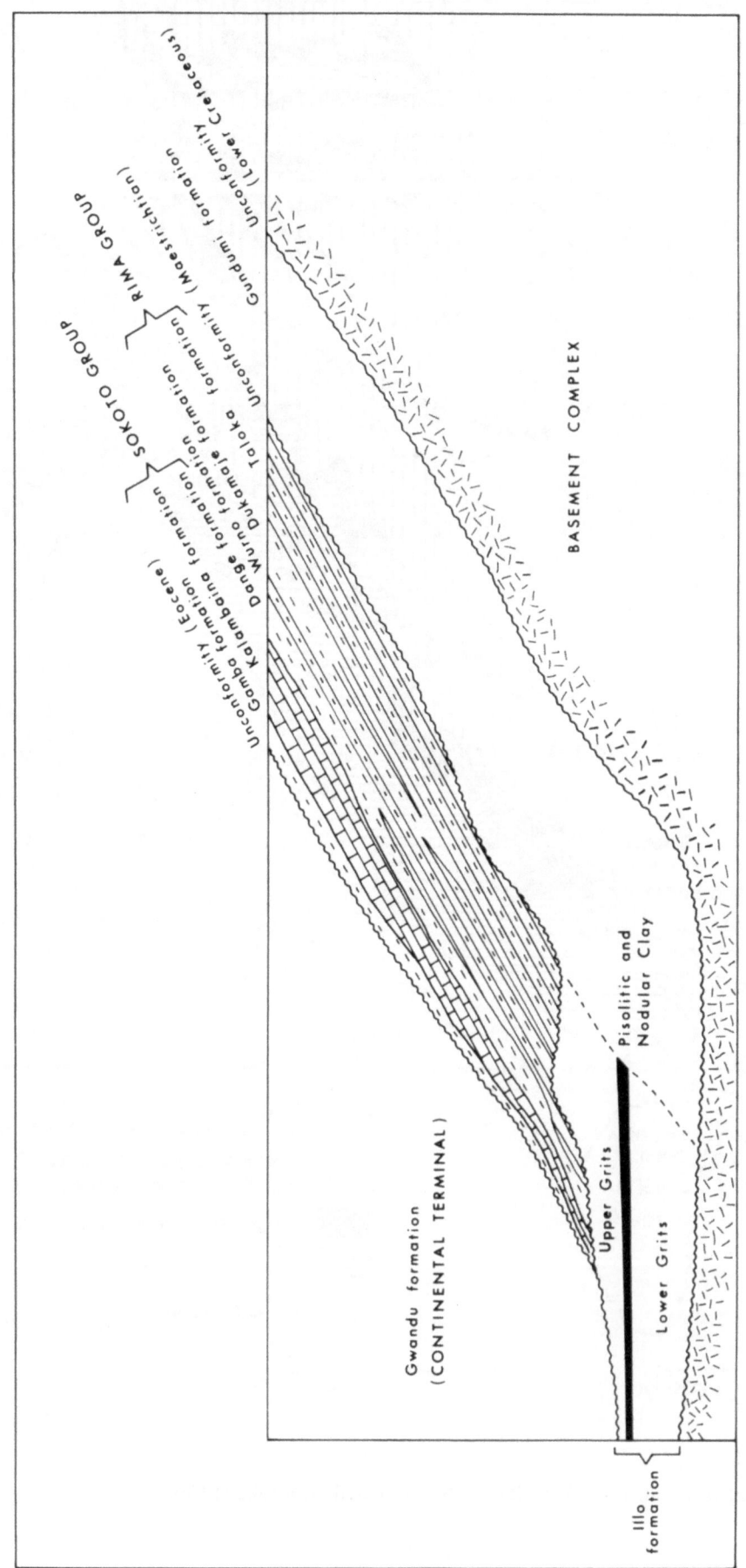

Figure 54. Diagrammatic section southeast flank of the Iullemmeden Basin Nigeria and Niger. Based on Kogbe (1976).

Figure 55
Upper Cretaceous Sequence Northwestern Nigeria

Illo Area Middle Niger Valley	Sokoto Area Iullemmeden Basin	
	Wurno Formation	
	Dukumaje Formation	Rima Group
	Taloka Formation	
Illo Formation		
	Gundumi Formation	

------- UNCONFORMITY --------

Basement Complex

Based on Geological Survey of Nigeria Maps and Reports; Reyment (1965); Kogbe (1970, 1976).

2.4.5.2. Upper Cretaceous Formations Sokoto Embayment

2.4.5.2.1. Gundumi Formation

General Description: The Gundumi Formation consists of fine to medium grained sandstone, false-bedded clayey grits, clays, basal conglomerates and arkoses. The basal conglomerate contains silicified wood. Cobbles and pebbles of well rounded quartz occur in a finer grained quartz cement. Beds attain a maximum thickness 10 ft or so and rest unconformably on the Basement Complex.

The type locality is at Dutsin Dampo, Sokoto Province (Figure 49) where the sediments consist of red to brownish red, coarse grained cross bedded grits with large well rounded quartz pebbles. The grits alternate with pinkish white silty clays and clayey sands. These beds belong to the basal portion of the Gundumi Formation, the upper beds are not exposed (Kogbe 1976). Subsurface data from Sokoto and Dange indicate that the formation becomes more clayey upwards passing into the silty Taloka Formation (Figures 50 and 52).

Thickness Variations: The Gundumi Formation is about 1000 ft thick (Geological Survey of Nigeria in *International Stratigraphic Lexicon* 1956). The Geological Survey of Nigeria 1:250 000 sheets show that it is only 300 ft thick in the Birnin Kebbi area and 1000 ft on the Tangaza Sheet. The Gundumi Formation diminishes in thickness rapidly from east to west and appears to be overstepped by the Rima Group in the neighbourhood of Gundumi.

Age: The exact age of the Gundumi Formation is unknown according to Reyment (1965), but in Table XIV–4 (*op. cit.*) it is given as pre–Maestrichtian. The Geological Survey of Nigeria (Sheets 1, 2, 3, 6, 7 and 8; Scale 1:250 000 Geological Series 1965 and 1966) gave the age of the Gundumi Formation as Turonian. Unfortunately an accompanying explanatory memoir was not published, so the evidence for this more precise age is not available. The Survey

places the formation beneath an unconformity overlain by the Coniacian–Santonian Taloka Formation (see below).

On the International Geological Map of Africa (1963) the post–Basement sediments of Katsina area and the area east and south of Sokoto are classed as K1–2 (Lower and Middle Cretaceous) and are co-extensive with similarly classified rocks in the Tegama, Niger. The Gundumi Formation of Nigeria in the area east of Sokoto Town joins with the Unit Cr^{6-5} "Cenomanian–Turonian continental" of Niger (*Carte Géologique Afrique Occidentale* 1960) but the outcrops of the Katsina area join with the unit Cr "Formation du Koutous" (Continental) of Niger. This unit extends into the Zinder area of Niger and according to the map legend is of Turonian age.

Faure (1962, *Carte Géologique de Reconnaissance du Niger Oriental*) classed the Formation du Koutous as a "Formation comprehensive" of Cenomanian to Senonian age. The formation laps onto the Basement Complex ridge of the Zinder region but the relationships of the Formation du Koutous and the Gundumi Formation of the Katsina area have not been worked out.

Bigotte and Obellianne (1968) in a geological sketch map of the Iullemmeden Basin divided the Cretaceous of Northern Nigeria into "Crétacé superieur marin (facies) and an older "Tegama et Crétacé supérieur continental". The Gundumi Formation is classed as Turonian by the Nigerian Geological Survey, it may well be older however.

To the north the Lower Cretaceous comprises the upper part of the "Continental Interclaire" sequence of continental sandstones, coarse to fine grained and cross-bedded with variegated mudstones known as the Groupe de Tegama. Fossil wood, dinosaurs, crocodiles and fish have been collected.

Conditions of Deposition and General Palaeogeography: Because of the uncertainty of the age of the Gundumi Formation we can say little about the palaeogeography and conditions of deposition except in very general terms. The sea is thought to have entered the Iullemmeden Basin in Late Cenomanian times where marine deposits contain the ammonite *Neolobites*. Southeast of Damergou the Cretaceous sediments disappear under the Cenozoic deposits of the Chad Basin (Figures 2 and 9) but it is thought that the *Neolobites* transgression reached the Gongola trough so making a through seaway which extended from west of Air via the Gao Trough into the Benue and Abakaliki Troughs and so joining with the Gulf of Guinea (Figure 22). Hitherto, this marine connection is thought to have operated during Cenomanian and Turonian times around the Tibesti area, so connecting with the Libyan shelf seas and Tethys. Klitzsch (1972) however has stated that the field evidence does not support this idea of a northeasterly connection via Libya. Marine fossils have not been found in the Gundumi Formation. The position of the shoreline of the *Neolobites* sea probably lay to the north of the Kaduna–Kano crystalline massif (Figures 22 etc.).

2.4.5.2.2. Illo Formation

General Description: The Illo Formation consists of false bedded pebbly grits, sandstones and clays, which contain silicified wood. Otherwise the formation has not yielded fossils. Jones (1948) divided the formation into three parts:

Upper Grits	7–100 m
Pisolitic Nodular Clays	3 m
Lower Grits	130 m

The rocks occur mainly along the Niger Valley and join with the sediments of the Sokoto Basin to the north and join the Nupe Group of the Bida Basin to the south (Figure 9). Field relationships are obscure at the north end of the Bida Basin. Of the three units mentioned above only the Pisolitic Nodular Clays and Lower Grits outcrop extensively.

Thickness Variations: The Illo Formation is 688 ft thick (Reyment 1965) but Adeleye (1972) gave the thickness as varying between 469–787 ft.

Age: The Illo Formation is classed simply as Upper Cretaceous by the Geological Survey of Nigeria (1956) but Reyment (1965) assigned a pre-Maestrichtian age to it. Adeleye (1972) correlated the Lower Grits of the Illo Formation with Bida Sandstones; the Pisolitic Clay with the Sakpe Ironstone; the Upper Grits with the Enagi Siltstone and the Upper Ironstone with the Batati Ironstone of the Bida area (Figure 65). There are problems however in accepting these correlations, which have not been mapped through on the ground.

Conditions of Deposition and Palaeogeography: The Illo Formation is thought to be a continental deposit. Certainly the marine horizons recognized in the upper part of the Nupe Group to the south in the Bida Basin have not been recognized in the Illo Formation. This may indicate that the degree of continentality increased northwestwards from the Bida Basin, if we accept the correlations suggested.

2.4.5.2.3. Petroleum Prospects of the Gundumi and Illo Formations

The petroleum prospects of the Illo and Gundumi Formations of Sokoto and Niger Valley areas are probably very low to nil because of the continental nature of the formation, thin cover, lack of obvious source rocks, lack of good clean reservoir sands and lack of well defined structures. The Gundumi Formation dips steadily northwestwards (Figure 9) and forms part of the southern flank of the Iullemmeden Basin. The younger (Maestrichtian) Cretaceous formations are described in Section 2.4.6.2.

2.4.6. Post Santonian Upper Cretaceous Formations

2.4.6.1. Southern Nigeria including Anambra Basin, Abakaliki Fold Belt, Afikpo Syncline and Calabar Flank

The following Post-Santonian Upper Cretaceous formations have been recognized in the above mentioned areas:

Nkporo Shale, Owelli Sandstone, Otobi Sandstone, Mamu Formation, Ajali Sandstone, Nsukka Formation, Lafia Formation.

These are described below. Their distribution and stratigraphic relationships are shown in Figures 8 and 9.

2.4.6.1.1. Nkporo Shale

General Description: The Nkporo Shale, a formation established by Shell-D'Arcy (Simpson, 1954), consists chiefly of blue or dark grey shales with occasional thin beds of sandy shale and sandstone. The formation rests unconformably on the Albian-Santonian formations of Abakaliki Fold Belt. It passes upwards into the so-called "Lower Coal Measures", now called the Mamu Formation. North of Uturu it passes laterally into the "Awgu Sandstones". On the eastern margin of the Abakaliki Fold Belt the Nkporo shale overlies the Eze-Aku Shale Group but at Uturu on the southwest end of the Abakaliki structure it oversteps onto the Albian Asu River Group.

Beds belonging to the Nkporo Shales are shown on the Shell-BP Geological Survey of Nigeria maps (1:250 000 Series) as the Asata-Nkporo Shale Group which passes laterally into the Nupe Sandstone, Otobi Sandstone, Awgu Sandstone and Afikpo Sandstone.

Thickness Variations: The Nkporo Shale is 1100 ft thick (Simpson 1954, Figure 4) but the Geological Survey of Nigeria gave the thickness of 6000 ft (in *International Stratigraphic Lexicon* 1956). Reyment (1965) gave a thickness of 3280 ft. Because the Nkporo Shale passes laterally into the Owelli (Awgu) Sandstones, thicknesses will vary dependent on location (Figure 56).

Age: Molluscs and occasional fish remains were collected from Shell D'Arcy coreholes. These were examined by Dr L. R. Cox who assigned a Maestrichtian age to this assemblage. The Geological Survey of Nigeria placed the unit within the Upper Senonian.

Reyment (1965) assigned the Nkporo Shales of the coastal areas of Western Nigeria (Araromi Shale) to the Upper and Lower Maestrichtian and the Nkporo of the inland areas of Eastern Nigeria to the Lower Maestrichtian. Reyment (1965) holds the view that the Araromi Shale is of the Nkporo type. The Araromi Shale occurs in "coastally located boreholes" (presumably Araromi-1 and 2 and Benin West-1,

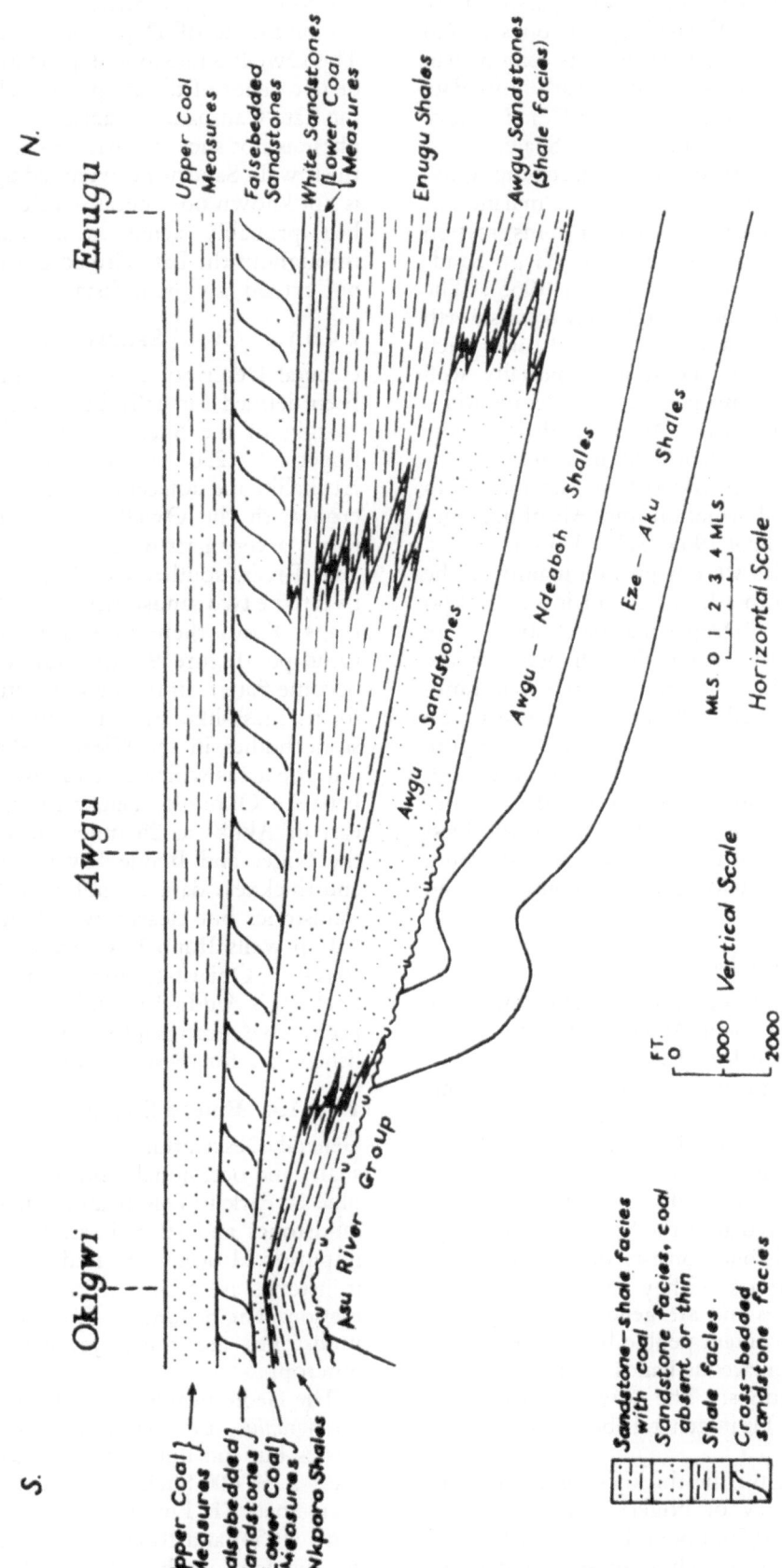

Figure 56. Section showing facies changes and structural relationships in the Okigwi – Awgu – Enugu areas. Based on Simpson (1954).

Figures 5 and 38). These beds are said to pass laterally into "coal cycle beds". Cratchley and Jones (1965) classed the "Enugu-Nkporo shales" as Campanian and they pointed out that the shales clearly post-date the Santonian unconformity. Murat (1972) also gave an Upper Senonian age to the Nkporo Shale.

Conditions of Deposition and Palaeogeography: The Nkporo Shale was laid down in a marine environment during the Late Campanian transgression (Figure 57). It passes laterally into the Awgu Sandstone (Owelli Sandstone of Reyment 1965). The Nkporo shales become increasingly arenaceous north of Enugu and eventually pass into sandstones such as the Otobi Sandstone. These are especially well developed between Oturkpo and the Benue River below Makurdi. Reyment's Araromi Shale of the Mid-West Region was deposited in a marine environment also but is considered to pass into Ajali Sandstone, Nsukka Formation and Abeokuta Formation inland (Reyment 1965, Table VII–7).

The Nkporo Shales were deposited mainly in the Anambra Basin and to a lesser extent in the Afikpo Syncline (Figure 57). "Deeper marine clastics" were laid down also in the Ikang Trough between the Calabar Flank and the Ituk High. To the west limestones were encountered in Benin West-1 resting on Basement Complex rocks and Murat (1972, Figure 11) plotted a strip of limestone along the southeastern edge of the Benin Flank (Figure 57). All these deposits listed above were laid down during the First Post-Orogenic Cycle of Deposition in the Anambra Basin and the Dahomey Miogeocline (Figures 21 and 22).

2.4.6.1.2. Owelli Sandstone

General Description: The term Owelli Sandstone was proposed by Reyment (1965) to clarify nomenclatural problems which had arisen over the use of the place name Awgu in formational names. Considerable confusion had arisen because the name Awgu which had been used for the Awgu Shales, deposited prior to the Santonian paratectonic orogeny, had also been used for the "Awgu Sandstones" (Grove 1957) deposited after the Santonian orogeny.

The Owelli Sandstone consists of massive, hard ferruginous sandstones. They are prominently crossbedded in places and are generally medium to coarse grained and contain pebble bands. Occasionally thin silty or argillaceous bands are present.

Thickness Variations: The Owelli Sandstone ("Awgu Sandstone") is in general about 1500 ft thick (Simpson 1954).

Reyment (1965) gave the thickness as 1968 ft but the Geological Survey of Nigeria (in *International Stratigraphic Lexicon* 1956) noted that the thickness is about 700 ft. Variations such as these would occur because the Nkporo Shales are developed at the expense of the Awgu Sandstone.

Age: As the Owelli Sandstone is regarded as a facies of the Nkporo Shale its age is most probably Maestrichtian. However, we should bear in mind that Murat (1972) gave a Late Senonian (Campanian)

age to the Nkporo Shale.

Conditions of Deposition and Palaeogeography: The Owelli Sandstone is probably a coarser grained deltaic facies of the Nkporo Shale laid down during the Late Campanian transgressive phase. The deltaic phase is not shown separately in Figure 57. How far the Owelli Sandstone extended up the Benue Trough is not known now because of erosion but as the fold belt probably acted as a sediment source area throughout the Late Cretaceous time it probably did not extend any great distance.

2.4.6.1.3. Otobi Sandstones

General Description: The Otobi Sandstone has not been defined formally but the name appears on the margin of the Shell-BP and Geological survey of Nigeria 1 : 250 000 Geological Sheets covering the Abakaliki and adjacent areas. It is continuous at outcrop with the Owelli Sandstone which thins and forms a very narrow outcrop north northeast of the Afar River, southeast of Enugu. The boundary between the two sandstones is not shown on the official maps. From there to Otobi, south southwest of Oturkpo (Figure 9), the outcrop widens and dips become flatter, until north of Oturkpo the sandstone forms mesa-like and meandriform outcrops. Sandstone bodies in the Nkporo Shales form extensive outcrop areas which extend towards the Benue River between Okokolo, below Makurdi and east northeast of Akukwa, 25 miles above the confluence of the Niger and Benue Rivers. Names, descriptions and thickness data are not available for any of these sandstones but clearly availability records from the Safrap wells Bopo-1, Okpaya-1, Adoka-1, Ocheku River-1, Opiaru-1, Ihandiagu-1 and Ikem-1 and Shell-BP's Aiddo-1 would help in building up a comprehensive stratigraphy for the region. Basic published data for these wells is presented in Figure 58.

2.4.6.1.4. Enugu Shales

General Description: The Enugu Shales consist of shales and occasional sandstones. The shales are grey blue or dark in colour and contain occasional white sandstones and striped sandy shale beds. Bands of impure coal occur and nodules and lenticles of clay ironstones are common towards the top of the formation. The bedding is often poorly defined and the mudstones are much jointed and fragmented at outcrop.

The Geological Survey of Nigeria (in *International Stratigraphic Lexicon* 1956) described the Enugu Shales but did not map them as a separate unit on the 1 : 250 000 Shell-BP – Geological Survey of Nigeria Sheet 71, Enugu (1957). However from Simpson's (1954) and Reyment's (1965) descriptions they are included in the "Asata-Nkporo Shale" i.e. the Nkporo Shale as it is currently called. Reyment (1965) pointed out that in the Awgu area the Enugu Shale can no longer be distinguished and the Owelli Sandstone is directly succeeded by the Mamu Formation (the old Lower Coal Measures).

Thickness: The Enugu Shales are about 3000 ft

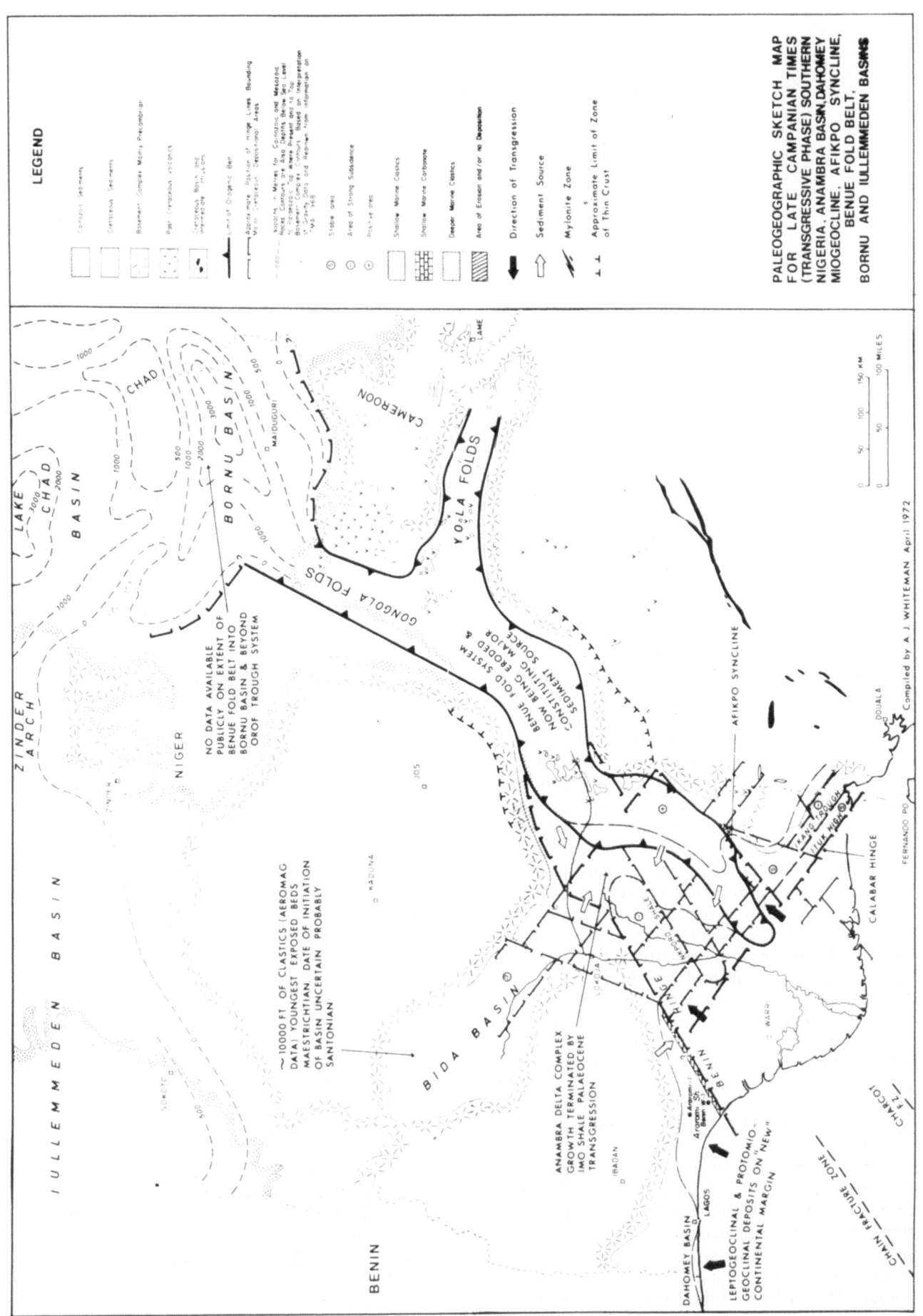

Figure 57. Palaeogeographic sketch map, Late Campanian Transgressive Phase, Southern Nigeria, Anambra Basin, Dahomey Miogeocline, Afikpo Syncline, Abakaliki – Benue Fold Belt, Bornu and Iullemmeden Basins. Based on various sources (see Figure 22) mainly Murat (1972) and Whiteman (1973).

Figure 58
List of Wells Sunk along the Northeast and East Flank of the Anambra Basin 1954–1967

Name	Owner	Lease	Date		Coordinates		TD (ft)	Comments
			Spud	Comp.	north	east		
Bopo-1	Safrap	OPL 38	30.10.54	01.03.55	601739	401243	3195	Abandoned exploration
Okpaya-1	Safrap	OPL 38	30.10.54	01.03.55	602495	390722	3243	Corehole – abandoned exploration
Adoka-1	Safrap	OPL 38	16.01.55	18.02.55	612057	382643	3200	Corehole – abandoned exploration
Ocheku River-1	Safrap	OPL 38	18.02.67	23.03.67	610911	409355	6021	Abandoned exploration corehole
Opiaru-1	Safrap	OPL 39	14.05.55	?.06.55	612633	333058	3113	Abandoned exploration corehole
Ihand-iagu-1	Safrap	Open	11.01.67	21.03.67	568635	311020	8280	Suspended gas well
Ikem-1	Safrap	OPL 40	02.03.55	28.04.55	582938	306367	3239	Abandoned exploration corehole
Aiddo-1	Shell-BP	Open	07.03.58	28.08.58	580825	314335	10544	Abandoned exploration

thick (Simpson 1954; Geological Survey of Nigeria 1956).

Age: The formation has yielded poorly preserved molluscs and plant remains. *Libycoceras angolense* was found in the lower part of the Enugu Shales (Shell-BP data) so indicating a Maestrichtian age (identification by Dr Spath, British Museum of Natural History). Reyment (1965) assigned the Enugu Shales to the Campanian and the extreme lower part of the Maestrichtian. Clearly besides the age problem a considerable cartographic problem exists in this area.

Conditions of Deposition and General Palaeogeography: Most probably the Enugu Shales were deposited in a paralic environment. A palaeogeographic "cartoon" map for Late Campanian times is shown in Figure 57. Neither on his map, nor in his stratigraphic synopsis did Murat (1972) recognize the Enugu Formation. Instead he believes the paralic Mamu Formation (Lower Coal Measures) interdigitates with the marine sequence of the Nkporo Shale. These units together with the continental Ajali Sandstones form part of a large delta complex laid down during a regressive phase (Figure 23).

2.4.6.1.5. Mamu Formation (Lower Coal Measures)

General Description: Reyment (1965) renamed the Lower Coal Measures of Simpson (1954) the Mamu Formation using the same type lithologic description. The Mamu Formation consists of sandstones, shale, and sandy shale with coal seams. The sandstones are fine to medium grained and yellow in colour. The shales and mudstones are dark blue or grey and frequently alternate with the sandstones to form a characteristically striped rock. Coal seams vary in thickness from a few inches to 12 ft.

The Mamu displays a rhythmic pattern of deposition consisting of:

5. Shale and sandy shale
4. Coal occasionally with shale at the top
3. Carbonaceous shale
2. Sandstone with occasional shale
1. Sandy shale or shale

The rhythm is repeated more than five times. The coals are black to brownish black and have an air-dried calorific value usually between 10 000 and 13 000 BTU/lb. They have a high volatile content and are classed in Groups A and B of the sub-bituminous class (American Society of Testing Materials 1951). The coals are of medium quality and have poor coking qualities. They are friable and weather rapidly in the "corrosive" Nigerian climate. The coals are rich in resins and waxes and could be used with advantage for chemical manufacture. However they have high ash content and low fixed carbon content. Reserves (indicated and inferred) are over 350 million tons (260×10^6 indicated and 90×10^6 tons inferred). Seams less than 3 ft 6 in are not included in these estimates. Reserves are estimated at around 47×10^6 tons for the Enugu, Ezimo, Orukpa, Okaba and Ogboyoga areas. The coals, dependent on their maturation chronology, depth of burial etc. could have been sources of gas in the Anambra Basin. The coals were described in detail by De Swardt and Casey (1963).

Thickness: The Mamu Formation varies from 300 ft at Okigiwi to about 1500 ft at Awgu and 2000 ft between Awgu and Enugu (Simpson 1954). Crachley and Jones (1965) gave a thickness of 1300 ft.

Age: The Mamu Formation is Maestrichtian in age (Simpson 1954, Reyment 1965 etc.).

Conditions of Deposition and Palaeogeography: Cratchley and Jones (1965) stated that the sediments of the Mamu Formation are shallow water deposits. The formation was laid down as part of the paralic facies of a large delta complex of which the Ajali Sandstone and Nsukka Formation form part (Figures 9 and 23).

2.4.6.1.6. Ajali Sandstone

General Description: The Ajali Sandstone consists of friable, white, crossbedded sandstones with thin beds of white mudstone near the base. Large scale crossbedding is common with dips as high as 20°. Irregular bands of intra-formational breccia occur. Originally the formation was called the White-Bedded False Sandstone and the White Sandstone. Reyment (1965) changed the name to Ajali Sandstone.

Thickness Variations: The thickness of the Ajali Sandstone is highly variable. Simpson (1954) quoted a thickness of 40 – 50 ft. Between Enugu and Ekana the formation is 250 ft thick and along the Enugu escarpment north of the Oji River the "False Bedded Sandstone" of Simpson (1954) is said to be 1500 ft thick. Northwest of Enugu the Ajali Sandstone could be over 1700 ft.

Age: Only fragmentary plants and "worm" tracks have been recorded from the Ajali Sandstone but on stratigraphic position the formation is classed as Maestrichtian.

Conditions of Deposition and Palaeogeography: The Ajali Sandstone is a continental sequence interdigitating with the paralic Mamu Formation, which interdigitates with the marine facies of the Nkporo Shales. The three formations form part of the Maestrichtian regressive delta complex (Figure 59). The Ajali Sandstone was deposited mainly in the Anambra Basin and in the Afikpo Syncline. Sediment was swept in from the Middle Niger Basin and deposited in the Anambra Basin and from the positive areas of the Benue Trough and its flanks.

2.4.6.1.7. Nsukka Formation (Upper Coal Measures)

General Description: the name Nsukka Formation was proposed to replace the term "Upper Coal Measures" (Reyment 1965) and consists of alternations of sandstones (sometimes cross bedded), shales and coal seams. At the top of the sequence thin limestones occur. There are few important coal seams present in the lower horizons of the Nsukka Formation.

Thickness Variations: The Nsukka Formation is classed as Maestrichtian to Palaeocene by Simpson (1954) and as Maestrichtian-Danian-Palaeocene by Reyment (1965).

Age: The Geological Survey of Nigeria (1956) assigned an age of Maestrichtian to locally Palaeocene but on the Shell-BP – Geological Survey of Nigeria 1 : 250 000 maps it is classed as Upper Senonian to Lower Eocene-Palaeocene.

Conditions of Deposition and Palaeogeography: The Nsukka Formation was deposited under paralic

conditions which prevailed during the second post-Santonian transgressive cycle (Figures 9 and 21).

2.4.6.1.8. Lafia Formation

General Description: The Lafia Formation is the youngest formation to have been deposited on the northwestern flank of the old Benue Trough. The outcrop area is now part of the Anambra Basin and is locally known as the Lafia Basin. The Lafia Sandstone or Formation has been described by several workers over the years and considerable confusion has arisen over its definition and extent. Sandstones now called the Lafia Formation were called "Upper Grits and Sandstones" by Falconer (1911) and Farrington (1952) referred to the sandstones as Lafia/Makurdi sandstone adding to the confusion because the Keana Formation can be traced southwards into the Makurdi Formation which is thought to be a lateral equivalent. The Makurdi Formation is thought to be Cenomanian to Turonian (Offodile 1976). Cratchley and Jones (1965) described the Lafia Sandstone as:

> Coarse grained sandstones outcropping in the Lafia District with black coal occurring a few miles south of that town.

They classed the Lafia Sandstone as post-Santonian and equivalent o the Gombe Sandstone of the Bashar-Muri-Amar area and to the Upper Coal Measures, Falsebedded Sandstone Lower Coal Measures and Enugu-Nkporo Shales (Figures 8, 9 and 23).

Esso West Africa (1967) regarded the Lafia Formation as a northerly extension of the "Enugu Formation" and Dessauvagie (1972 MS and 1974) regarded the Lafia Sandstone (Kla Figure 9) as equivalent to his mapped Ke or Enugu Shale and Sandstone Member. It is hoped that these nomenclatural problems will be resolved either by the Nigerian Geological Survey or in the forthcoming new edition of the Lexique Stratigraphique International.

The Lafia Formation as used by Offodile (1976) rests unconformably on the Awgu Formation. In the Lafia-Awe area the formation appears to be wedging out and is less than 160 ft thick. It consists of

> continental ferruginized sandstones, red, loose sands, flaggy mudstones and clays. The type locality is in and around the town of Lafia.

Thickness: Esso West Africa put the thickness of the Lafia Formation between 1600–4800 ft, although it is only said to be 160 ft in the type area because of pinch outs (Offodile 1976).

Age: Identifiable fossils have not been collected from the Lafia Formation although carbonized plant remains are known. Because of its post-Santonian position and because of its regional correlatives (Figure 9) the Lafia Formation is considered to be Maestrichtian.

Conditions of Deposition and Palaeogeography: The Lafia Formation consists of fluviatile sandstones and finer grained clastics considered to have been

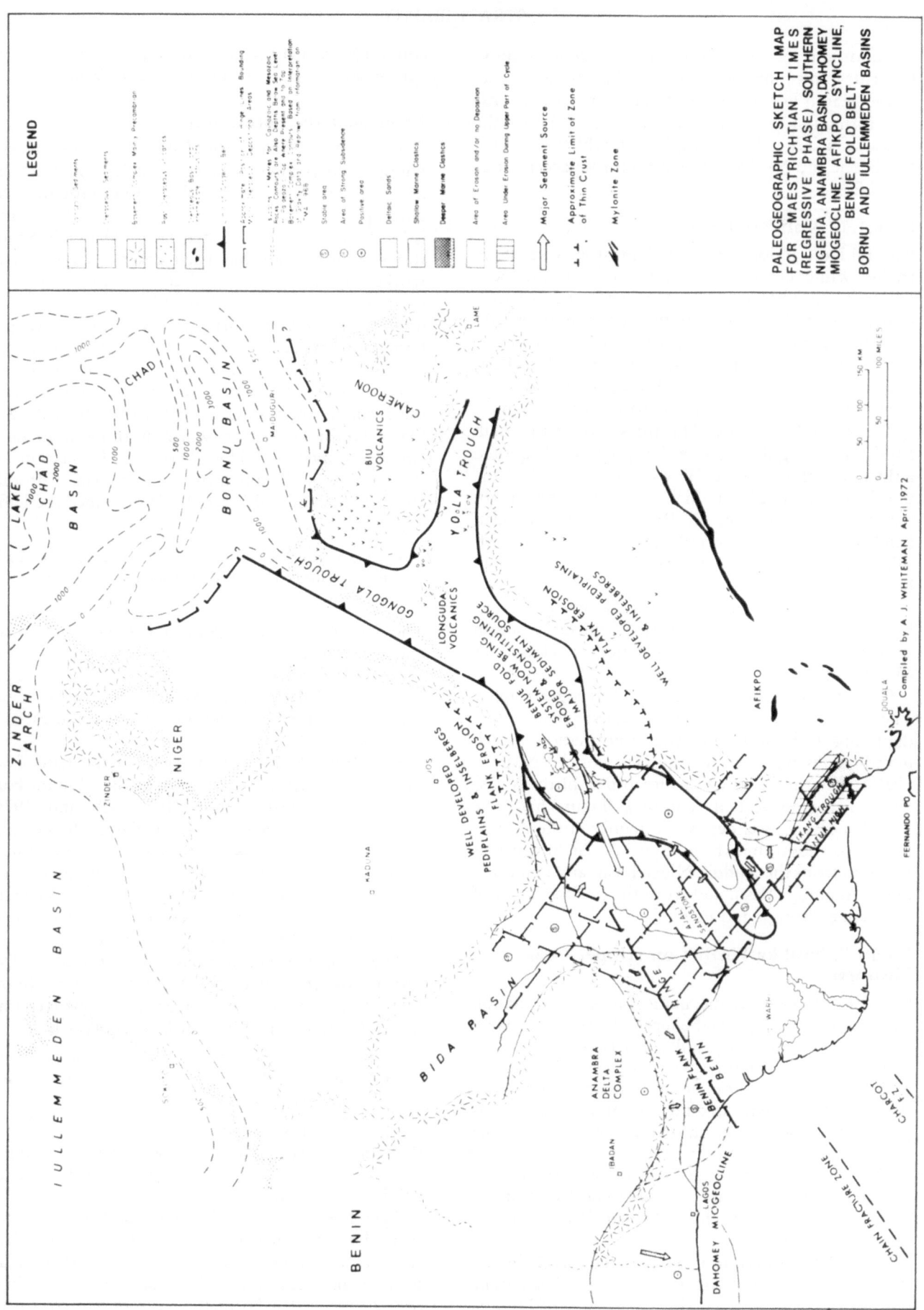

Figure 59. Palaeogeographic sketch map Maestrichtian Regressive Phase, Southern Nigeria, Anambra Basin, Dahomey Miogeocline, Afikpo Syncline, Abakaliki-Benue Fold Belt, Bornu and Iullemmeden Basins. Based on various sources (see Figure 22) mainly Murat (1972) and Whiteman (1973).

derived from the erosion of the folded rocks of the Benue Fold Belt and the Basement Complex formations of the flanks of the Benue Trough. The vast amount of erosion which took place in Late Cretaceous time is to some extent indicated by the distance of the main escarpment from the centre of the Benue Depression.

2.4.6.1.9. Gombe Sandstone

General Description: The Gombe Sandstone outcrops on the northwestern side of the Upper Benue Trough and on the western side of the Gongola Trough (Figure 9). It consists of well bedded sandstones, grits, siltstones, flaggy sandstones and clays which become increasingly ferruginous upwards. Ironstone bands are common and thin coals occur.

Siltstones and flaggy sandstones form the greater part of the Gombe Sandstone. The formation is thought to rest unconformably on older strata. Frequently a basal contact is not exposed and the existence of an unconformity has been deduced by the difference in the degree of folding and disturbance between the Gombe and the older Cretaceous rocks. The Geological Survey of Nigeria (1956) states that the Gombe Sandstones pass down into the Clay Shales (now part of the Pindiga sequence) and that the Gombe is succeeded by the Ako Sandstone (part of the Kerri-Kerri Formation Figure 9).

The Gombe coals range in thickness from a few inches to 5½ feet and are known only to shallow depths. Analyses of the Molko Coal Seam are presented in Figure 61. The Molko seam is a black lignite.

Thickness Variation: At the type section near Gombe the formation is over 1000 ft thick but in the Futuk area it is 1968 ft thick (Dessauvagie 1969). The Geological Survey of Nigeria (1956) at the type locality at Gombe gave the thickness as around 600 ft.

Age: The Geological Survey of Nigeria (1956) gave the age of the Gombe Sandstone as Turonian to Senonian.

Reyment (1965) assigned a Maestrichtian-Campanian age to the Gombe Sandstone and Cratchley and Jones (1965) equated the Gombe with the Lafia Sandstone and assigned a Maestrichtian age. Carter et al. (1963) assigned a Late Maestrichtian and pre-Palaeocene age to the Gombe Sandstone. Shell–BP geologists have dated the Gombe coals as Late Senonian and the Kerri-Kerri Formation which unconformably overlies the Gombe sandstone has been dated palynologically as Palaeocene.

Conditions of Deposition and General Palaeogeography: The Gombe Sandstone consists of sequence of estuarine and deltaic sandstones, siltstones, shales and ironstones. Originally it was thought that the Gombe Sandstone had been deposited in a small basin situated to the west of the Zambuk Ridge (Carter et al. 1964) but as the formation has been mapped further to the south this idea is no longer acceptable (Figure 9). The restriction of the outcrop of the Gombe Sandstone to the northwestern side of the Benue Trough and the western side of the Gongola

Figure 60
Proximate Analyses of Coal Exposed in the Numanha Stream

	Centre	Top
Moisture	6.53	5.00
Fixed carbon	53.56	28.15
Volatiles	18.78	29.59
Ash	21.13	37.26

Analyst: J. A. Akingbehin, Geological Survey of Nigeria.
Carter et al. (1963).

Figure 61
Proximate Analyses of the Molko Coal seam

	Middle			Bottom		
	Air dried	Ash free dried	Parr basis	Air dried	Ash free dried	Parr basis
Moisture	10.2	—	—	9.9	—	—
Ash	10.5	—	—	9.3	—	—
Volatiles	38.8	49.0	48.2	38.4	48.4	47.2
Fixed carbon	40.5	51.0	51.8	42.4	52.0	52.8

Analyst: Powell Duffryn Technical Services Ltd.
Carter et al. (1963).

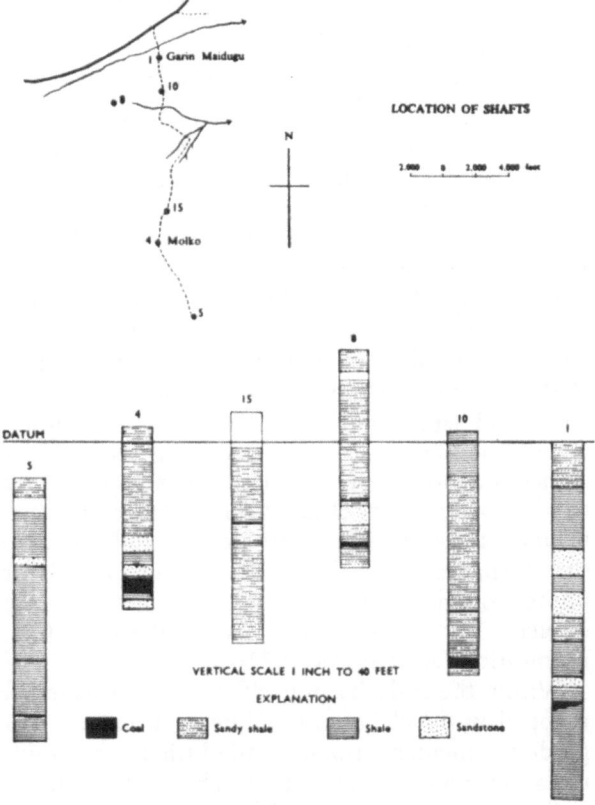

Figure 62. Sections and map showing the distribution of the Molko Coal in the Gombe Sandstone, Upper Benue Trough Nigeria. Based on Carter et al. (1963).

Figure 63
Ultimate Analyses of the Molko Coal Seam

	Middle			Bottom		
	Air dried	Ash free dried	Parr basis	Air dried	Ash free dried	Parr basis
Moisture	10.2	—	—	9.9	—	—
Ash	10.5	—	—	9.3	—	—
Carbon	57.3	72.0	73.4	58.3	72.0	73.6
Hydrogen	4.1	5.2	5.1	4.2	5.2	5.2
Nitrogen	1.4	1.8	1.8	1.4	1.7	1.8
Sulphur	0.7	0.9	—	0.8	1.0	—
Oxygen (diff.)	15.8	20.1	19.7	16.1	20.1	19.4
Calorific value (BTU/lb)	9700	12200	12000	10000	12300	12700

Analyst: Powell Duffryn Technical Services Ltd.
Carter *et al.* (1963)

Trough indicates strong tectonic control and the Gombe Sandstone may be "molasse" which was shed off the Upper Benue and Gongola Santonian fold belt. The Kerri-Kerri Formation may have been formed in a similar manner as were the post-Santonian deposits in the Anambra Basin and in the Wukare-Mutum Biyu Basin (Figure 9).

2.4.6.2. Bida Basin and Adjacent Areas Upper Cretaceous Post Santonian Formations

2.4.6.2.1. General

The oldest rocks exposed in the Bida Basin (or Middle Niger Basin or Nupe Basin) are Maestrichtian in age but sedimentation may have started earlier bearing in mind that beneath the Bida area itself there may well be as much as 10 000 ft of sediments (estimated from aeromagnetic observations). The rocks outcropping have been called variously the Nupe Series, Nupe Sandstone Series, Kontogora—Nupe Beds, part of the Lokoja Series of Falconer (1911) and Geological Survey of Nigeria (1956). Jacques (1945) divided the Bida Basin Sandstones into Nupe Sands and Sandstones and the Nupe Basal Conglomerate. The Geological Survey of Nigeria (1956) assigned a Senonian age to these rocks. Fossils found to date among the Bida Basin surface sediments (Nupe Sandstone, Figure 9) are few and the age of the formation has been determined mainly by photogeological correlation with the post-Santonian sequence of the Anambra Basin (Personal Communication Dessauvagie 1972).

Within the Bida Basin itself small areas have been mapped in detail because of the occurrence of low grade sedimentary iron ores; but little has been written about regional correlation. The stratigraphy and sedimentation of the Upper Cretaceous rocks of the Bida Basin have been studied in detail by Adeleye (1972) and Adeleye and Dessauvagie (1972) in a small area around Bida. Our knowledge of the ironstones

and associated strata of the Aghaja Plateau situated north of Lokoja, is mainly based on Jones (1958). The Upper Cretaceous strata around Bida constitute the Nupe Group (Adeleye and Dessauvagie 1972). These units are described below.

Part of the Bida Basin of this area was under lease to Great Basins Petroleum Company but only surface geological studies and aeromagnetic work were done before the area was relinquished. The acreage has never been taken up since. Prospects for this basin are considered to be high risk and poor (Figure 299).

2.4.6.2.2. Nupe Group

The name Nupe was originally used as a formation name by Russ (1931) and as a group name by Adeleye (1972) and Adeleye and Dessauvagie (1972). The type area for the Nupe Group is the country around Bida. The name Nupe is taken from the Nupe Tribe who live in the district. The Nupe Group as defined by Adeleye (1972) consists of four new formations:

TOP
4. Batati Ironstone Formation
3. Enagi Siltstone Formation
2. Sakpe Ironstone Formation
1. Bida Ironstone Formation
BASE – Not exposed

The maximum thickness of the Nupe Group is not known because the base is not exposed in the type area. Constituent formations at the surface total more than 1167 ft. Gravity measurements (Ojo and Ajakaiye 1976), indicate that the basin fill exceeds 3000 ft but is less than 6500 ft and that the margins of the Bida Basin are largely unfaulted. Aeromagnatic data (Great Basins Petroleum Company, Mr A. Hawley, Personal Communication) indicate that maximum basin fill could be around 10 000 ft. Adeleye (1972, 1976) did not discuss whether these subsurface sediments are part of his Nupe Group or not.

Adeleye (*op. cit*) suggested a Campanian–Maestrichtian age for the exposed Nupe Group but Murat (1972) and the writer hold the view that the Nupe Group was deposited during the Late Campanian transgression which affected the Anambra Basin. Dessauvagie (1974 Text) suggested a Senonian age for the Nupe Group but on his map (Dessauvagie 1972 and 1974) (Figure 9) and Nupe Group is shown as Campanian and post Maestrichtian and as equivalent to the Lafia Sandstone and the Patti and Lokoja formations. The sedimentological evidence suggests that the group was derived from the predominantly migmatic Basement Complex of southwestern Nigeria and laid down in fluviatile and marine environments.

The following description was made of "Nupe Sandstones" by the Geological Survey of Nigeria (in *International Stratigraphic Lexicon* 1956):

Nupe Sandstones   Upper Cretaceous

Russ, W. 1931. The Minna-Birnin Gwari Belt, Part 38. *Ann. Rep. Surv. Nigeria* 1930, p. 12

Estuarine series of sandstones, grits and conglomerates, gritty clays and oolitic ironstones.

Also referred to as the Nupe Series, the Nupe Sandstone Series and the Kontagora-Nupe Beds.

Forms part of the Lokoja Series (Falconer 1911). Divided by Jacques (1945) into Nupe Sands and Sandstones and the Nupe Basal Conglomerate.

Age: Senonian (?)

Distribution: Niger, Ilorin and Kabba Provinces

Comment: Unfossiliferous except for plant remains

References: Falconer, J. D. (1911); Jones, B. (1948); Jacques, E. H. (1945 and 1947); Tattam C. M. (1944).

2.4.6.2.3. Bida Sandstone Formation

General Description: The Bida Sandstone consists of two members:

TOP	Feet
Jima Sandstone Member	295
Doko Sandstone Member	600

Doko Sandstone Member (new; Adeleye 1972):
This is the basal unit exposed in the Bida area and consists of arkoses, feldspathic sandstones, and quartz sandstones. Smaller amounts of lithic-feldspathic sandstones, subgreywackes, sandy siltstones and intra-formational breccias occur.

Along the Bida-Zungeru Road, near the northern margin of the basin coarse grained sandstones and basal conglomerates crop out. Adeleye (1972) implies that these rocks are part of the Doko Sandstone Member but he had not then mapped the area between Bida and the northern margin of the basin, and consequently the position of these coarse grained clastics was uncertain. Boulders of conglomerate occur around Wushishi Station and between mile posts 8–10 on the Zungeru-Bida Road. The conglomerates are quartzitic and range in size from pebbles 2 to 8 cm across. The pebbles have a clay ironstone cement. However, again the relationships of the basal conglomerates, exposed on the Bida–Zungeru road to the Doko Sandstone Member of the Bida areas apparently were not determined by Ajibade (1972) who mapped the area.

Apparently Adeleye (1972) and Ajibade (1972) did not determine whether there is an onlap sequence towards the basin margins or whether the marginal Wushishi conglomerate facies passes laterally into sandstone facies or whether relationships of the marginal facies to the central basin facies are structural. Russ (1931) suggested that the basal conglomerate of his Nupe Sandstone is not continuous but forms lenticles of considerable size, and that an overlap occurs near Kontagora, where conglomerates and pebble beds, which occur higher up in the succession, rest directly on crystalline rocks.

Jima Sandstone Member: The Jima Sandstone Member (new, Adeleye 1972) consists of quartzose sandstones, brownish, massive beds cross-stratified with subsidiary subgreywackes, siltstones, claystones and breccias. Rapid facies changes typify the member. Leaf impressions are the only fossils which have been found and occur mainly in the argillaceous rocks. The type section is 2.5 km east northeast of

Jima village on the south facing slope of a residual conical hill.

Thickness Variations: The Doko Member is over 600 ft thick in the Bida area taking into consideration the amount exposed and the amount recognized in Geological Survey of Nigeria Borehole 1256. The Jima Sandstone Member is about 295 ft thick but thickens southwards and westwards from the type area (Adeleye 1972). Aeromagnetic data (Great Basins Petroleum Company of Nigeria Ltd., Personal Communication, Mr A. Hawley) indicate that there may be as much as 10 000 ft of sediment beneath the Bida area but how much of this is to be included in the Bida Sandstone Formation and the Nupe Group is uncertain. Considerable thickness variations are known to occur within the Bida Basin but the reasons for this are not clear.

Age: As the Bida Sandstone Formation has only yielded leaf impressions little can be said about its age. Adeleye (1972) classed it as Late Cretaceous (Campanian?–Maestrichtian) but there may well be older beds in the thick sequences beneath the Bida area. Dessauvagie (1974) classed it as Canpanian (Figure 9).

Conditions of Deposition and Palaeogeography: The sedimentological evidence indicates deposition in a braided stream environment for the Doko Sandstone Member. The source area is thought to have been a migmatic Basement Complex which became more metamorphic westward. Grain size and sedimentary structures indicate deposition under lower stream power than that which prevailed during the deposition of the Jima Member. The presence of leaves and ferruginized channel margin deposits are taken to indicate deposition under tropical to subtropical conditions (Adeleye 1972).

2.4.6.2.4. Sakpe Ironstone Formation

General Description: The Sakpe Ironstone is divided into the Baro and Wuya Ironstone Members (Adeleye 1972). The Wuya Ironstone Member consists of oolitic and pisolitic ironstones with small amounts of locally developed claystones.

Thickness Variation: The Wuya Member is of fairly constant thickness and at the type locality is about 16 ft thick. The Baro Ironstone Member consists of oolitic ironstones and ferruginous sandstones which are locally pyritic and concretionary. Rapid facies changes characterize the unit and gradation exists between ferruginous and quartzose facies (Adeleye 1972). The type section is situated at 8°36′18″N and 6°25′30″E, 1.2 km south southeast of Baro Rest House. The member caps the entire Baro Plateau and thickness varies, attaining a maximum of 10 ft. The Baro and Wuya Members are considered to be lateral equivalents.

Age: In the Doko area the Wuya Member has yielded a Turritella, several specimens of Faunus and a bivalve indet. The Baro Member has yielded fossil wood and basal burrows. These have been identified as Skolithus (Goldring in Adeleye 1972). The occurrence of Faunus is taken by Adeleye (1972) to indicate a Maestrichtian age for the Sakpe Ironstone Formation but *Turritella* is known to range in Nigeria down into rocks as old as Cenomanian (Reyment 1965).

Conditions of Deposition and Palaeogeography: Adeleye (1972) holds the view that the Sakpe Ironstone Formation was deposited under marine conditions in a sea which was not connected with Southern Nigeria but with Tethys via a northerly connection situated in the Sokoto area of the Iullemmeden Basin. Salinities are thought to have been low in places because of the influx of freshwater into the narrow arm of the sea in which Adeleye thinks the Ironstones were deposited. Adeleye's case for a northerly connection is unproven and a more substantial case can be made for a connection to the south. The Late Campanian sea in which the Nkporo and Araromi marine shales were laid down transgressed into the Bida Basin (Figure 59) and the Sakpe and Batati Ironstone formations probably were laid down during this transgression.

Adeleye (1972) has assumed that the present limits of the Bida Basin roughly approximate to the depositional limits of the basin, whereas there is evidence that the limits are erosional. Late Campanian marine deposits may well have covered areas of Basement Complex west of Lokoja so connecting with Bida area.

The northernmost occurrence of marine fossils found to date in the Bida Basin is in the Bida area. None have been found to the northwest, and therefore to postulate that a Maestrichtian sea transgressed from this direction is not acceptable. Adeleye's (1972) interpretation of the palaeogeography appears to be incorrect. The Shell-BP and Geological Survey of Nigeria 1:250 000 Geological Series shows Upper Senonian rocks (Undifferentiated — 9) on the right banks of the Benue and Niger Rivers and on the interfluves which are coextensive with the Bida Basin deposits. On the left banks of the Niger and Benue rivers the "Unit 5a" consisting of the Otobi Sandstone and equivalents has been mapped to within about 6 miles of Idah, on the Niger River below its confluence with the Benue River. These rocks are overlain by the Lower Coal Measures (Mamu Formation of Reyment 1965). Murat (1972) interpreted these distributions, together with information from Safrap and Shell-BP wells, to indicate that shallow marine clastics were laid down in the area around Bopo–1 and in the entrance to the Bida Basin. As is shown in Figure 59 a transgression probably entered the Bida Basin from the southeast, not from the northwest as Adeleye (1972) has postulated, and extended into the Bida Basin or Middle Niger Embayment. The Bida Sandstone Formation and the Sakpe Ironstone Formation were laid down in this embayment.

In the writer's view because of the few fossils found in the Bida Basin, and their limited variety the most that can be said about the age of the Sakpe–Enagi and Batati units is that they are Late Cretaceous. The marine fossils could all have been laid down during the Late Campanian transgression and the older Bida Sandstone Formation could have been laid down at the same time of Agbani Sandstone, i.e. during the Coniacian–Early Santonian regression (Figure 22). An exact age and an acceptable correlation of these deposits will only be obtained when more fossil evidence is available but in the terms of Murat's (1972) transgression-regression hypothesis (Figure 23) the correlations and datings proposed above seem to fit the facts and limitations better than Adeleye's (1972) interpretation.

2.4.6.2.5. Enagi Siltstone Formation

General Description: This formation was proposed by Adeleye (1972) and consists of siltstones with subordinate amounts of sandstones, siltstone-sandstone mixtures and claystones. The type locality lies about 6 km to the east of Enagi at 9°8′N and 5°35′E. The formation occurs just below the tops of the major mesas such as the Doko and Nupe Plateau.

Thickness Variations: The thickness is variable and in places it is as little as 29 ft but is usually between 100–200 ft.

Age: The Enagi Formation is classed as Maestrichtian because of its stratigraphic position between the Sakpe Ironstone and the Batati Ironstone (Adeleye 1972).

Conditions of Deposition and Palaeogeography: Adeleye (1972) interpreted the stratigraphy to indicate that there was a withdrawal of "Tethys" northwards and that fluviatile sedimentation ensued. Grain size, lithology, structures and flora according to Adeleye (1972) indicate deposition under flood plain conditions. The Enagi Formation is said to show evidence of penecontemporaneous ferruginization. Adeleye considers the Enagi Formation to be the lateral equivalent of the Lokoja Carbonaceous and Shaly Sandstones or Patti Formation (Jones 1955 and 1958); of the Lower Dekina Sandy Clay Shales (Jones, 1955) and the Mamu Formation of Reyment (1965).

The Enagi and the Bida Formations may have been laid down as "the truly continental facies of an extensive delta" or that the two formations "most probably represent the deposits of coastal plain alluviation" (Adeleye 1972). Murat (1972) and the writer do not hold the view that the Enagi Siltstones were laid down in a side arm of "Tethys" which regressed northwestwards as Adeleye proposed. They think that the Late Campanian deposits were laid down in the Bida Basin which was a side basin adjacent to the Anambra Basin (Figure 59).

2.4.6.2.6. Batati Ironstone Formation

General Description: The Batati Ironstone is subdivided into the Edozhigi Ironstone Member and the Kutigi Ironstone Member. Where the two members occur together the Edozhigi Member rests unconformably on the Kutigi Member (Adeleye 1972). The Edozhigi Member consists of goethitic and oolitic ironstones whereas the Kutigi Member consists of mixed goethitic and kaolinitic oolites.

Thickness: The Kutigi Member is 7 ft thick (max-

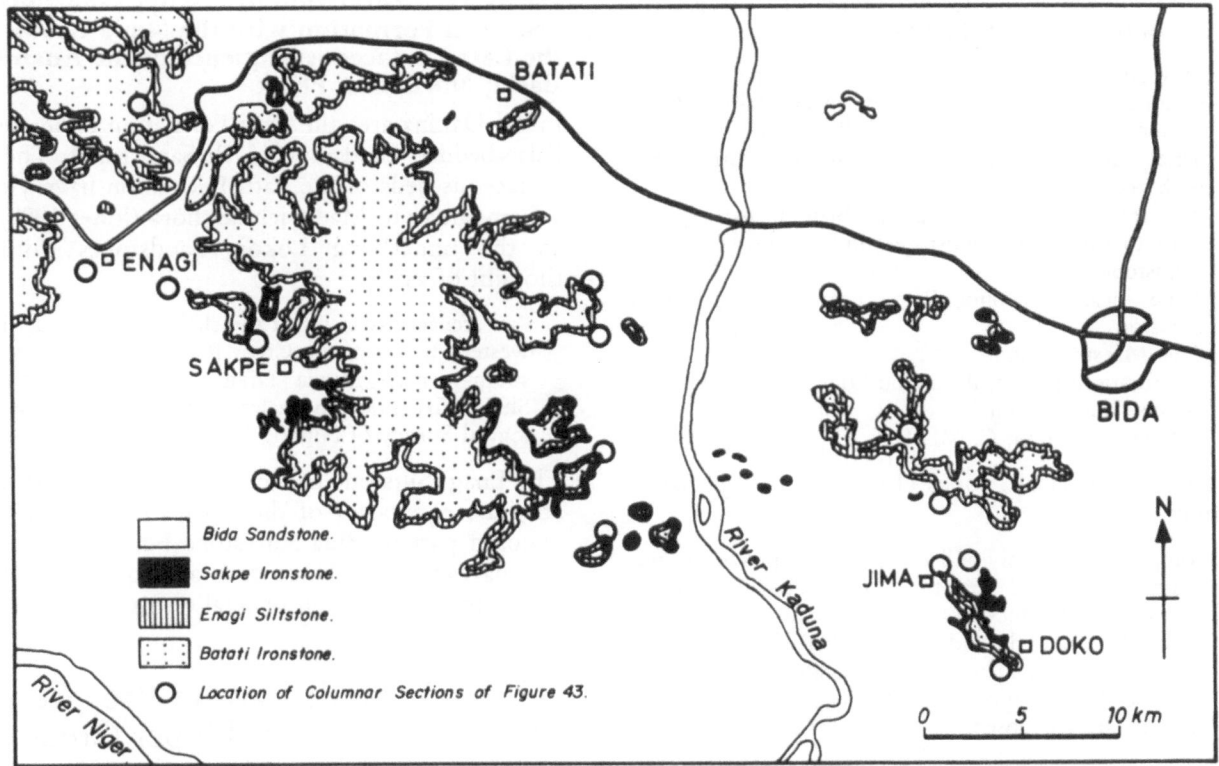

Figure 64. Map showing the distribution of the Bida Sandstone, Sakpe Ironstone, Enagi Siltstone and Batati Ironstone formations, Bida Basin, Nigeria. Based on Dessauvagie and Adeleye (1972).

imum) erosion has modified its thickness. The Edo-zhigi Member is generally between 3–15 ft thick but thins down to 1 ft and thickens to 50 ft. The variability is due to erosion (Adeleye 1972).

Age: Adeleye (1972, p. 150) has listed the following fossils from the Batati Ironstone Formation:

Ostrea, Lopha, Venericardia, Septifer?, Cardidae, Faunus beyenburgi, F. miskalensis, Turritella, and burrows of *Alpheus* sp.

This assemblage according to Adeleye (1972) indicates an age not older than Maestrichtian, although *Faunus* is said to range into the Eocene (see below).

Conditions of Deposition and Palaeogeography: The assemblage is taken to indicate deposition in near shore shallow water marine to fresh water conditions and the assemblage is taken to indicate a second marine transgression into the Central Nupe Basin (Adeleye 1972). This transgression he regards as having taken place rapidly and catastrophically and to have caused erosion. Again he thinks that it came from the northwest (Adeleye 1972, Figure 46). Additionally the marine influence was said to have been "felt further south beyond Lokoja, into southeastern Nigeria (Figure 34)". The Batati Ironstone is considered to have been laid down as a shoreline facies in this great hypothetical arm of Tethys. As mentioned above the writer and Murat (1972) hold the view that the Sakpe and Batati assemblages are Campanian in age; that they were laid down during the Late Campanian transgression of Murat (1970)

and that the area was not connected via Sokoto and Iullemmeden to "Tethys" but was connected to the Gulf of Guinea via the Anambra Basin.

The fossils listed by Adeleye (1972) are not diagnostic of the Maestrichtian. *Venericardia*, for instance, is known from Coniacian strata (Reyment 1965), *Turillites* occurs in the Odukpani Formation of Calabar Province which Reyment (1965) classed as Cenomanian but which may be as young as Turonian. The above mentioned assemblage could indicate Late Cretaceous age. Murat (1972) and the writer hold the view that the Bida Basin was in part a source area (Figure 59) in Maestrichtian times, when the Ajali regressive Sandstone was deposited.

2.4.6.2.7. Lokoja Sandstone

General Description: In the area of the Agbaja Plateau which covers an area of approximately 40 square miles stretching along the Niger from opposite Koton Karifi to within 4 miles of Lokoja, Jones (1958) defined two major units:

2. Patti Formation and
1. Lokoja Sandstone

These units occur on the left banks of the Niger and Benue Rivers and underlie the Falsebedded Sandstones (Ajali Sandstones of Reyment 1965) of Late Cretaceous (Maestrichtian age). The sedimentary rocks of the Agbaja and Lokoja regions were first described by Falconer (1911) who included them in his Lokoja Series. The Geological Survey of Nigeria (in *International Stratigraphic Lexicon* 1956) described the "Lokoja Series" as follows:

LOKOJA SERIES   Upper Cretaceous
Falconer, J. D. 1911. *The Geology and Geography of Northern Nigeria*, London, Macmillan, p. 174.
Sandstones, grits, conglomerates and oolitic ironstones with occasional clays.
Also termed the Lokoja–Passage Beds. In parts equivalent to the Nupe Sandstones and the Lower Benue Sandstones.
Type locality: Mount Patti, near Lokoja, Lokoja Division
Age: Maestrichtian
Distribution: Kabba, Ilorin and Niger Provinces
Fossils: Plant remains
Reference: Tattam, C. M. 1944.

According to H. A. Jones (1958) the Lokoja Sandstones are:

composed of pebbly and clayey grits and sandstones, coarse falsebedded sandstones and a few thin oolitic ironstones. The thickness of the formation ranges from 300 to 900 ft and depends on the relief of the Basement Complex floor. The basal conglomerate, which consists of well rounded quartz pebbles in a matrix of white clay, is rarely exposed.

The type section of the Lokoja Sandstones is exposed in the stream immediately south of the Agbaja Plateau. The sandstones extend along the western margin of the Cretaceous sediments from Koton Karifi to the Ate Hills. Dessauvagie (1974) assigned a Maestrichtian (?Senonian) age and thinks that the Lokoja is part equivalent of the Nupe Sandstone following Jones (1958).

2.4.6.2.8. Patti Formation

General Description: Jones (1958) defined the Patti Formation as consisting of:

fine to medium, grey and white sandstones, carbonaceous silts and shales and oolitic ironstones. Thin coals have been recorded and massive, white gritty clays are common. The maximum exposed thickness of the formation is 226 ft and to the north it thins out completely. The only fossils found are plant remains in carbonaceous beds. Fossil seeds and other plant fragments are preserved in detail where these rocks have been ferruginized near the surface.

2.4.6.2.9. Oolitic Ironstones, Agbaja Plateau

General Description: The plateaux around Lokoja are all capped with iron oolites and laterites. The oolite ranges in thickness from 23 to 44 ft and averages 46 ft. The laterite thickness vary. As much as 39 ft was measured near Koton Karifi but the average is about 17 ft (Jones 1958). Concerning the origin of the oolites, Jones (1958) thinks that the absence of bedding in the ironstones and their restriction to high-lying plateaux suggests that the Agbaja oolite is of lateritic origin. However the siderite and magnetite rich ironstones which occur are thought to be sedimentary oolites, despite the complete absence of fossil debris. The position of the oolitic ironstones is shown in Figure 65.

2.4.6.2.10. Correlation of Lokoja Sandstones and Patti Formation with the Nupe Group and the Late Cretaceous Sequence of the Anambra Basin, Northwest Flank

In the Dakina area the Patti Formation underlies the False-bedded Sandstone (Ajali Sandstone). The carbonaceous beds of the Patti Formation have yielded a Campanian to Maestrichtian flora (Jones 1958) and so the underlying 'Lokoja Sandstones must be thought of as:

"an arenaceous facies of the shales which underlie the western fringe of the Cross River Plain at Enugu and south-western Idoma Division" i.e. Asata–Nkporo Shales (5a) (Nkporo Shales) of the Shell-BP and Geological Survey of Nigeria 1:250 000 Geological Series.

The Enagi Siltstone of Adeleye (1972) is most probably the equivalent of the Patti Formation and the exposed part of Bida Sandstone Formation and the Lokoja Sandstone Formation are equivalent. Relationships are shown in general terms in Figure 66.

2.4.6.3. Hydrocarbon Prospects of the Post Santonian Upper Cretaceous Rocks of the Anambra Basin Abakaliki Fold Belt, Afikpo Syncline, Calabar Flank and Benue Depression and Flanking Basins

2.4.6.3.1. General

The hydrocarbon prospects of the post-Santonian Upper Cretaceous rocks of the above mentioned areas vary greatly from region to region. The following areas are considered to have prospective interest:

1. Anambra Basin and Calabar Flank
2. Dahomey Miogeocline
3. Wase–Gombe Basin flanking Benue Depression
4. Wukari–Mutum Biyu Basin flanking Benue Depression
5. Bida Basin

2.4.6.3.2. Anambra Basin Prospects*

The structure of the Anambra Basin is described in Chapter 3 Structure. More than 30 000 ft of sediments may be present in this basin and a large part of the section may be of Late Santonian and Maestrichtian age. Rocks of this age were laid down after the Santonian folding episode in the Abakaliki and Benue Troughs and consist of sediments eroded from the rising fold system and flanks. They are relatively undisturbed and do not appear to have been intruded by igneous rocks. High geothermal gradients prevail in this area. Growth faults may occur in the Anambra Basin in addition to normal tectonic structures but these have not been described publicly. Potential source rocks are said to be present throughout the Cretaceous section in the basin and adjacent areas

* Avbovbo and Ayoola (1981) of the Nigerian National Petroleum Corporation discussed the petroleum prospects of the Anambra Basin and rate the basin as gas prone and worthy of further attention.

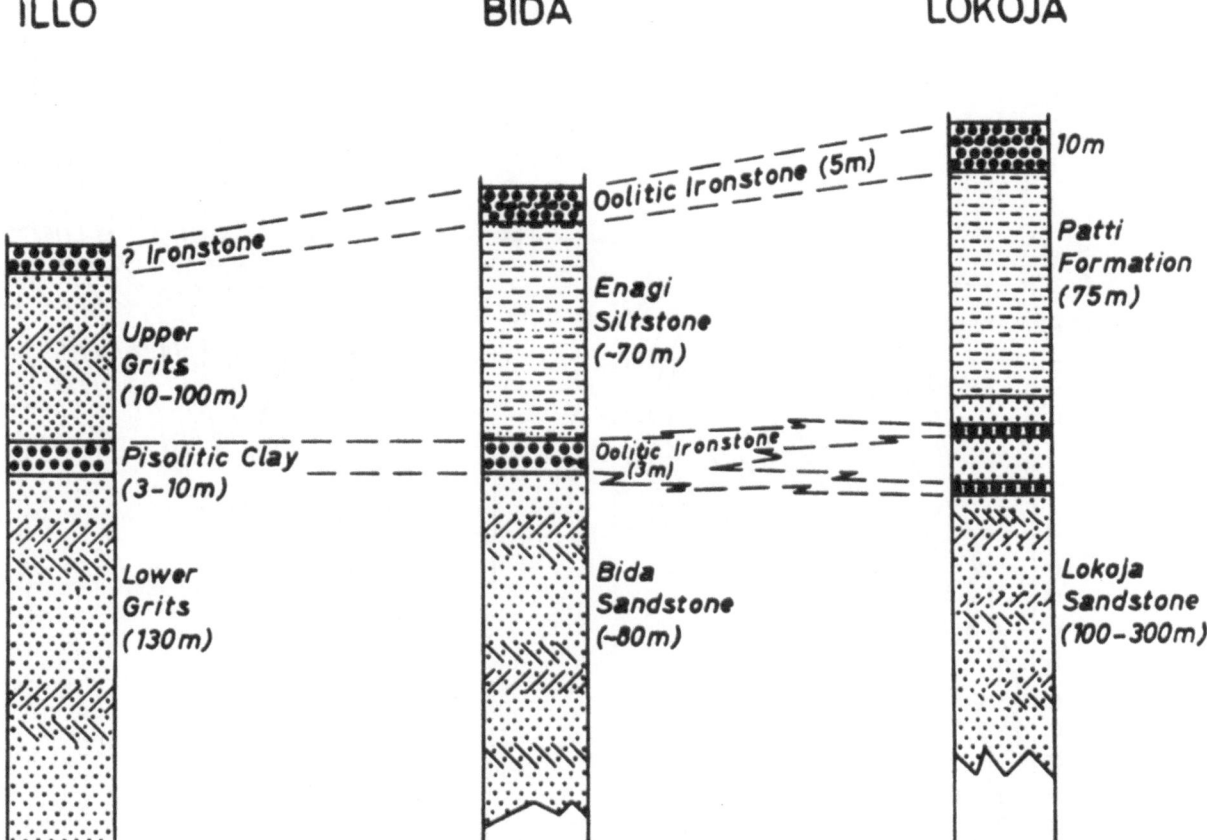

ILLO   BIDA   LOKOJA

Figure 65. Sections showing correlation of Bida Sandstone, Sakpe Ironstone, Enagi Siltstones, Patti and Lokoja Formations, Illo, Bida and Lokoja areas. Based on Adeleye and Dessauvagie (1972).

and, although analyses have not been published, oil seeps, inspissated deposits, bitumen occurrences have been reported from many areas. Seeps have been reported from around the Ishiagu High and the most active seeps occur at Ugu Eme, northwest of Ishiagu, four miles south of Awgu. The seeps occur mainly in the "Awgu Sandstone" or Owelli Sandstone which wedges out against the Ishiagu positive structure. Four exploration holes were drilled into the Awgu Sandstone the deepest of which reached 2238 ft. The sandstone was found to be impregnated with heavy black oil for over a length of one mile along a scarp face. The oil bearing sandstone attains a thickness of about 20 ft and is coarse grained and grey in colour. This unit is the basal bed and rests unconformably on the folded Eze-Aku shales. Another oil soaked sandstone outcrops at Lokpanta and oil seeps occur in the Upper Cretaceous rocks near Enugu. The oil which occurs south of Awgu may have migrated from the west and is associated with the unconformity between the Owelli Sandstone–Nkporo Shale and the underlying Eze–Aku Group, Awgu Shales, and the Asu River Group.

Much of the Anambra Basin is considered to be a good prospect for hydrocarbons, except perhaps for the northeastern part where the sandstones may have been flushed by fresh water. Beneath the central part of the Anambra Basin there may be "buried hill" structures, simple synsedimentary growth faults and rollover anticlines. The western flank of the Anambra Basin may be faulted and there may be combination structural and stratigraphic traps in this region. High geothermal gradients may limit liquid hydrocarbon occurrences to shallow depths and the basin may be gas prone.

2.4.6.3.3. Calabar Flank Prospects

Not a great deal has been published about the post-unconformity portion of the section but, because of the rapid thickening of the Tertiary section, these rocks may well be beyond economic depths limits at present. Shell-BP has recently relinquished acreage along this flank.

2.4.6.3.4. Other Basins

The prospects in the Bida Basin as we have noted are thought to be poor and high risk; and since Great Basins Petroleum relinquished the acreage no other companies appear to have been interested in it. Post-Santonian – unconformity Upper Cretaceous prospects of the Yola arm of the Chum rrr trilete junction (Burke and Whiteman 1973) are rated as low mainly because:

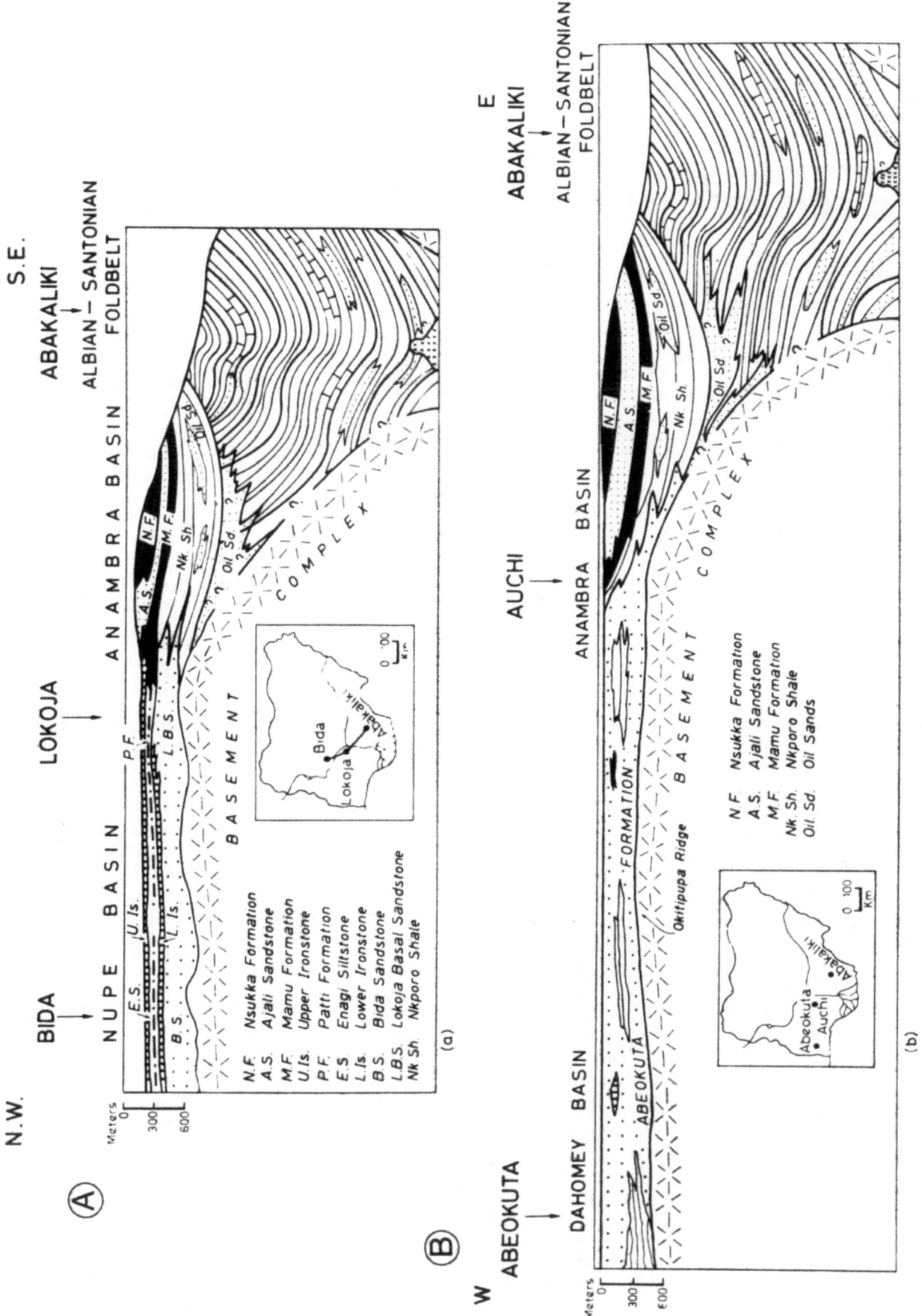

Figure 66. Sections, highly diagrammatic, of (A) Nupe Basin, Anambra Basin and Abakaliki Fold Belt; (B) Dahomey Basin, Anambra Basin and Abakaliki Fold Belt. Based on Adeleye (1975). N.B. Relationships of folded Pre–Santonian sediments to flat lying and gently folded Post-Santonian sediments are highly schematic.

88

Figure 67
List showing wells Sunk in the Anambra Basin and Adjacent
Areas, Southern Nigeria up to 1972

Well	TD (ft)	Remarks
Aiddo–1 (Safrap)	10544	Dry
Akukwa–1 (Shell–BP)	7883	High pressure gas from Cretaceous well blew out
Akukwa–2 (Shell–BP)	11992	Dry
Alada–1 (Safrap)	10222	Bottomed in Cretaceous. Dry
Amandsiodu–1 (Shell–BP)	7516	Low pressure gas, Cretaceous
Anambra River–1 (Safrap)	11263	Tested 2280 b/d from two Cretaceous sandstones
Bopo–1 (Shell–BP)	3195	Dry
Egoli–1 (Shell–BP)	3202	Dry
Idah–1 (Shell–BP)	3216	Dry
Ihandiagu–1 (Safrap)	8280	Tested 4.6 MMCFD and condensate from Cretaceous sandstones
Ikem–1 (Shell–BP)	3239	Dry
Nzam–1 (Shell–BP)	12046	Gas show ? Cretaceous
Ocheku River–1 (Safrap)	6018	Bituminous shows
Ogbabu–1 (Shell–BP)	7502	Dry
Okpaya–1 (Shell–BP)	3243	Dry
Opiaru–1 (Shell–BP)	3131	Dry
Owan–1 (Shell–BP)	2412	Dry
Ubiaja–1 (Shell–BP)	3200	Dry
Ubulu–1 (Shell–BP)	11430	Dry

Based on Whiteman (1973)

1. the sandstones of the Yola area and the adjacent parts of the Upper Benue have been widely flushed with fresh water;
2. no source rocks have been described from the sediments of the Yola arm;
3. of widespread Cretaceous and Tertiary intrusion and volcanic activity; and
4. of lack of cover rocks and adequate seals; and
5. because probably high geothermal gradients have affected these rocks.

Some of these comments apply to prospects in the sedimentary rocks of Upper Benue Trough and the Gongola Trough also. The prospects of the Wase-Gombe Basin and the Wukari–Mutum Biya Basin are classed as highly speculative (Figures 169 and 170).

2.4.6.4. Iullemmeden Basin, Sokoto Area Upper Cretaceous Post Santonian Formations

2.4.6.4.1. General and Rima Group

We have already described the pre-Santonian Upper Cretaceous sediments from the Nigerian Sokoto Section of the Iullemmeden Basin and as has been noted the pre-Maestrichtian Gundumi Formation is succeeded by the Rima Group.

Kogbe (1970 and 1976) has described the Upper Cretaceous of the Nigerian Section of the Iullemmeden Basin (Figures 49–53). The distribution of the formation is shown in Figure 49. In the Sokoto area, the Rima Group extends from Fokku on the Kia River, northeast of the Niger Valley, across the Zamfara River and Rima River into Niger. The Rima Group includes the Taloka, Dukumaje and Wurno Formations. The term Rima Series was proposed by Raeburn and Tattam (in Swinton 1930) and was changed by the Geological Survey of Nigeria (in *International Stratigraphic Lexicon* 1956) to Rima Group.

Originally the Rima Series (Raeburn and Tattam 1930) was described as consisting of:

TOP
3. Calcareous Group
2. Clay-Shale Group
1. Sandstone Group

These units were renamed by the Geological Survey of Nigeria (in *International Stratigraphic Lexicon* 1956) as Lower Sandstones and Mudstones (oldest); *Mososaurus* shales; and Upper Sandstones and Mudstones. The Rima Group was then divided by the Geological Survey of Nigeria on the 1:250 000 Geological Series (1965) into:

TOP
3. Wurno Formation
2. Dukumaje Formation
1. Taloka Formation

Reyment (1965) downgraded the Rima Group to a formation and the constituent formations to members. This was not accepted by the majority of geologists working in Nigeria. The group varies in thicknesses from 3000 ft in the Dange area to 800 ft in the Birnin Kebbi area, to about 1000 ft in the Tangaza (Figure 49). Reyment (1965) gave the thickness of the Rima Group as 492 ft.

The age of the Rima Group is taken as Maestrichtian by the Geological Survey of Nigeria (in *International Stratigraphic Lexicon* 1956) but on the margins of the Geological Survey of Nigeria 1:250 000 Geological Series maps (1965) the Taloka Formation ("Sandstone Group") is given as Coniacian to Santonian and Campanian, the Dukumaje Formation ("Clay-Shale Group") is classed as Campanian and Maestrichtian and the Wurno Formation ("Calcareous Group") is classed as Maestrichtian. The Geological Survey of Nigeria (in *International Stratigraphic Lexicon* 1956) described the Rima Group as follows:

Figure 68
Table showing Plays and Prospects of Post-Santonian (Campanian to Maestrichtian) Upper Cretaceous Rocks, Southern Nigeria,
Anambra Basin, Calabar Flank, Dahomey Miogeocline, Benue Trough and Bida Basin

Province	Region	Play	General Description	Trap type	Shows	Prospect Class
SOUTHERN NIGERIA	ANAMBRA BASIN		More than 30 000 ft of sediments may be present in this basin and a large part of the section belongs to the Late Cretaceous. The sediments are relatively undisturbed and are not intruded by igneous rocks	Anticlines Faults, Unconformities, Stratigraphic Growth Faults, Buried hills	Positive shows of oil and gas in wells and at surface	GOOD TO VERY GOOD FOR OIL AND VERY GOOD FOR GAS *LIMITATION* IN THE NORTHEAST TRAPS MAY BE FLUSHED. HIGH GEOTHERMAL GRADIENTS
	CALABAR FLANK		No basic information available about the post-Santonian strata	Anticlines, Faults, Stratigraphic Growth Faults	Not recorded	FAIR TO GOOD
	DAHOMEY MIOGEOCLINE		A thick prism of Upper Cretaceous sediments extends westwards from the Okitipupa Ridge first thickness and then thins seawards	Faults, Drape Folds Unconformities	Surface oil seeps and "tar sands". Positive shows of non-commercial accumulations of oil and gas in adjacent Dahomey	GOOD *LIMITATION* FRESH WATER FLUSHING
BENUE TROUGH	WASE-GOMBE BASIN		Thick Cretaceous sediments occur in this trough situated on the northwestern side of the Mid-Benue High. Very little is known of the stratigraphy.	Anticlines, Faults, Unconformities, Stratigraphic	Not recorded	SPECULATIVE
	WUKARI-MUTUM-BIYA BASIN		Thick Cretaceous sediments occur in this trough southeast of the Mid-Benue High	Anticlines, Faults, Unconformities, Stratigraphic	Not recorded	SPECULATIVE
BIDA BASIN			Late Cretaceous sediments may be as much as 10 000 ft thick in the deepest part of this basin. Great Basins Petroleum Limited had this area under lease but no holes were drilled.	Small Anticlines, Faults, Stratigraphic	Not recorded	FAIR TO POOR, HIGH RISK AREA

Based on Whiteman (1973)

RIMA GROUP   Upper Cretaceous
Raeburn, C. and Tattam, C. M. 1930. A preliminary note on the sedimentary rocks of Sokoto Province. *Bull. Geol. Surv. Nigeria* No. 13, p. 58.
Fine grained sandstones and mudstones, with fossiliferous shales and limestones near the top.
Originally termed Rima Series. It covered both Eocene and Cretaceous formations. As the Rima Group it is restricted to the Cretaceous.
The Rima Group is divided into the Upper Sandstones and Mudstones, the *Mosoaurus* shales and the Lower Sandstones and Mudstones. As the Rima Series it included the Calcareous Group, Clay Shale and the Sandstone Group.
Age: Maestrichtian
Distribution: Sokoto Province
Fossils: Species of *Mososaurus, Lamna, Schizorhiza, Ste-*

phenodus, Veniella, and *Libycoceras?* sp. have been recorded.
References: Tattam, C. M. (1944); Jones, B. (1948).

2.4.6.4.2. Taloka Formation

General Description: Formal type descriptions do not appear to have been published by the Geological Survey of Nigeria for the formations of the Rima Group.

The names Taloka, Dukumaje and Wurno formations appear on the map margin of Sheets 1, 2, 3, 6, 7 and 8 of the 1:250 000 Geological Series, Geological Survey of Nigeria (1965). The Taloka Formation, referred to as the Lower Sandstones and Mudstones by Jones (1948), consists of interbedded grey to dark brown mudstones and light-coloured

Figure 69
Cretaceous and Cenozoic Formations "Sokoto Basin", Nigeria

Age	Formation	Type section
Eocene and probably younger	Gwandu Formation	Birnin Kebbis – Argungu Road near Dutsin Bardua 2½ miles North of Birnin Kebbis 4° 15′E – 12°28′N
Palaeocene	Gamba Formation	Quarry site of the Cement Factory of Northern Nigeria near Kalambaina and Gamba villages
	Kalambaina Formation	5°11′E – 13°03 N
	Dange Formation	Dange village Road cutting 16 miles SE of Sokoto on Sokoto – Gusau Road 5°20 E – 12°52′N
Maestrichtian	Wurno Formation	Wurno village 5°25′E – 13°18′N
	Dukamaje Formation	Dukamaje village, 5°49′E – 13°46′N
	Taloka Formation	Taloka, village 5°42′E – 13°27′N

Kogbe (1970)

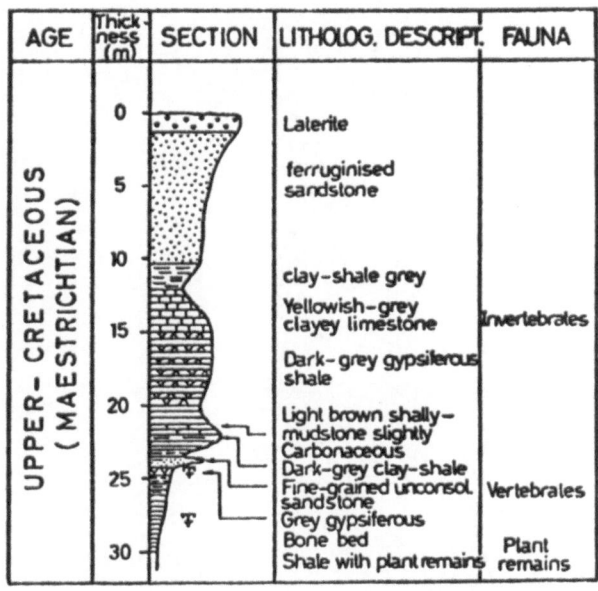

Figure 71. Type section of the Upper Cretaceous Dukamaje Formation, Dukamaje Village, "Sokoto Embayment" of the Iullemmeden Basin, Nigeria. Based on Kogbe (1976).

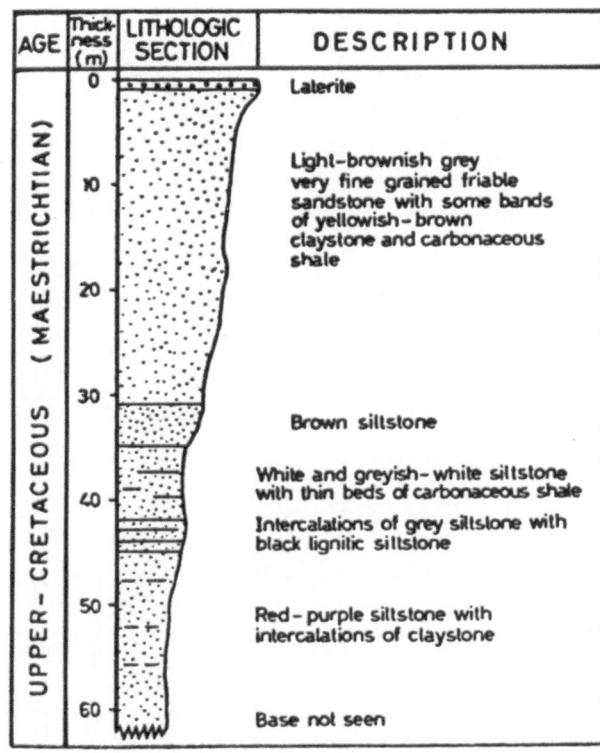

Figure 70. Type section of the Upper Cretaceous Taloka Formation, Taloka Village, "Sokoto Embayment" of the Iullemmeden Basin, Nigeria. Based on Kogbe (1976).

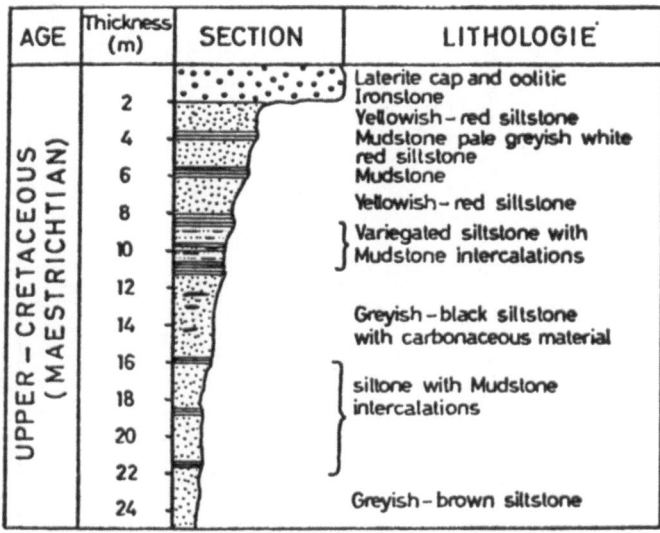

Figure 72. Type section of the Upper Cretaceous (Maestrichtian) Wurno Formation, "Sokoto Embayment" of the Iullemmeden Basin, Nigeria. Based on Kogbe (1976).

91

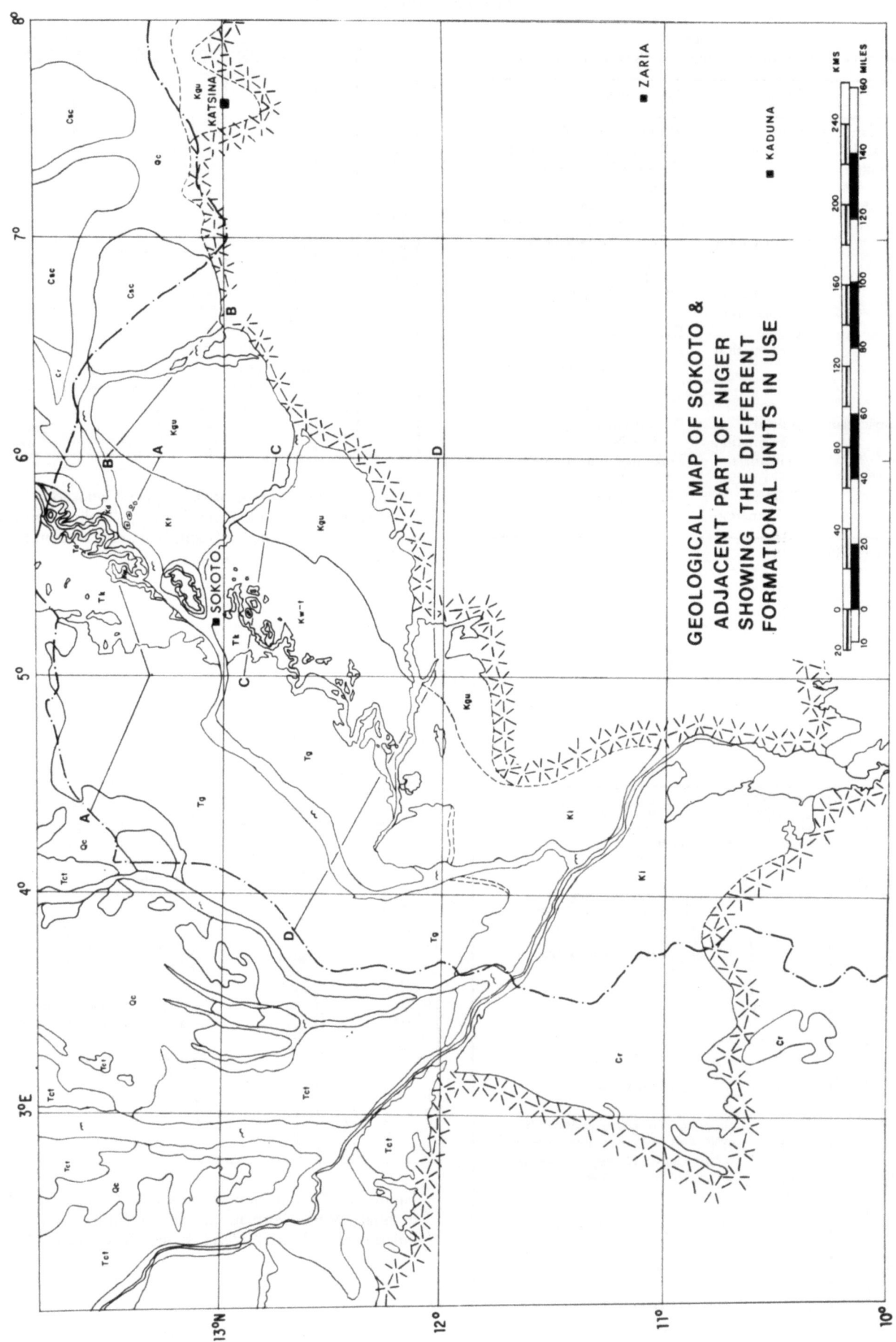

Figure 73. (A) Geological map of Sokoto and adjacent parts of Niger and (B) Key to stratigraphic units. Based on Dessauvagie (1972) and Whiteman (1973). Section lines for Figure 80 are shown on this figure.

STRATIGRAPHIC UNITS USED IN THIS ACCOUNT FOR NIGERIA AND ADJACENT AREAS

Compiled by
T.F.J. DESSAUVAGIE & A.J. WHITEMAN 1972
(Similar to upgraded data Dessauvagie 1974)

medium to fine grained sand with some thin bands of carbonaceous shale or lignite. According to Ogilbee and Anderson (1965):

> In borehole Geol. Surv. Nigeria 3053, at Balle, the Taloka attains a thickness of more than 600 ft and is predominantly fine to coarse grained, heterogeneous sand and sandstone with thin interbedded shale.

Kogbe (1972) stated that the type section of the formation is at Toloka, near Goronyo, Sokoto Province and that the formation consists of white, fine grained, friable sandstones with intercalated mudstones and carbonaceus mudstones. He gave the maximum thickness as 393 ft which is considerably less than that given by Ogilbee and Anderson (1965) at Balle some 60 miles to the east. The "Clay-Shale" contains masses of gypsum and ferruginous material (Raeburn and Tattam 1930). South of Wurno the Taloka and Wurno Formations cannot be differentiated because the Dukumaje Formation thins out.

Age: Kogbe (1972) mentions that Late Cretaceous dicotyledons occur in the Taloka Formation and Reyment (1965) mentions that poorly preserved mussels occur around Gilbidi. Mososaurian bones and fish teeth occur (Raeburn and Tattam 1930) which suggests a Maestrichtian age. Reyment (1965) assigned a Maestrichtian age to the formation but on the map margin of Sheet 2, Sokoto, 1:250 000 Geological Series, Geological Survey of Nigeria (1965) the Taloka Formation is classed as Coniacian–Santonian and Campanian.

2.4.6.4.3. Dukumaje Formation

General Description: The Dukumaje Formation was originally called the "Clay-Shale Group" by Raeburn and Tattam (1930) and *Mososaurus* shales by Brynmor Jones (1948) and Carter (1960). The Dukumaje Formation consists principally of dark fossiliferous shale with white spheroidal nodules and some thin limestone bands (Ogilbee and Anderson 1965). It is present in outcrop north of the Rima River (Figure 49). Kogbe (1970 and 1976) stated that the Dukamaje Formation contains 98 ft of limestone and is locally gypsiferous. The formation consists mainly of shales.

Thickness Variations: At Balle Borehole (GSN 3053) 88 ft of Dukumaje Formation were penetrated.

Age: Most exposures of the Dukumaje Formation include a bone bed near the base which has yielded boney plates of turtles, skull fragments, vertebrate and limb bones of *Mososaurus nigeriensis* Swinton; several species of fishes including *Lamna appendiculata* (Agassiz), *Lamna libyca* (Quass), a sawfish *Schizorhiza stroemeri* (Weiler) and *Stephanodus libycus* (Dames). Invertebrates occur mainly in the marl and Reyment (1965) recorded tentative identifications of *Daradiceras?* sp., *Veniella undata* (Conrad), *Lopha semiplana*. A Maestrichtian age was assigned to the formation (Reyment 1965) but on Sheet 2, Geological Survey of Nigeria 1:250 000 Geological Series the Dukumaje Formation is classed as Campanian–Maestrichtian. Kogbe (1976) mentions that the Maestrich-

tian ammonite *Libycoceras* has been collected from Dukumaje.

2.4.6.4.4. Wurno Formation

General Description: The Wurno Formation, originally the "Upper Sandstones of Mudstones" consists of thin friable fine grained sandstones intercalated with soft mudstone and shale (Ogilbee and Anderson 1965). Kogbe (1972) states that the formation can only be distinguished north of latitude 15°N, and south of Rabat (Figure 49) the Dukumaje Formation is absent and the Wurno and Taloka formations merge. The Wurno Formation locally contains carbonaceous matter and iron sulphide suggesting deposition under stagnant conditions (Reyment 1965). The type locality of the formation is at Wurno (Figure 49).

Thickness Variations: In surface exposures the Wurno Formation near the type locality attains a thickness of 75 ft but at Balle it consists of 150 ft of fine to coarse grained, loosely cemented sandstone with a little mudstone (Ogilbee and Anderson 1965).

Age: The Wurno Formation is classed as Maestrichtian by Reyment (1965) and by the Geological Survey of Nigeria on Sheet 2, Sokoto, 1:250 000 Geological Series (1965).

The top of the Wurno Formation is taken as the mappable Cenozoic/Cretaceous Boundary in northern Nigeria with fine grained Maestrichtian sandstones of the Wurno overlain conformably by Palaeocene limestones and shales, indicating little change of environment from Late Cretaceous to Early Tertiary times.

2.4.6.4.5. Conditions of Deposition of the Rima Group and General Palaeogeography

The Taloka, Dukumaje and Wurno formations of the Rima Group form part of an interdigitating marine and non-marine sequence extending from Nigeria into Niger. As noted above in Niger, the *Neolobites* sea transgressed into the Iullemmeden Basin in Cenomanian times (Figure 28), probably from the northwest, and a zone of interdigitating marine and continental sediments was laid down north of the Nigerian frontier. During Turonian times a similar series of interdigitating marine and non-marine deposits were laid down but it is difficult to obtain precise limits from the literature. Again it was certainly situated north of the Nigerian frontier in the Sokoto area. During Early and Medial Senonian times limestones, mudstones, and sandy mudstones were laid down in the Iullemmeden Basin where Greigert and Pougnet (1967) reported that they are little more than 164 ft thick. A marine transgression marked by the occurrence of *Libycoceras*, began during the Medial Maestrichtian and extended as far south as latitude 13°N (Greigert and Pougnet 1967) (Figure 22). This was followed by a second *Libycoceras* transgression towards the end of Maestrichtian time. The transgression extended into the southern Iullemmeden Basin. It was probably this *Libycoceras* sea that reached the Sokoto and Bornu areas of Nigeria. Until more stratigraphic information becomes available and detailed correlations of the Late Cretaceous deposits of Niger and Nigeria can be made, we can do little more than discuss the palaeogeography in the general terms presented above.

2.4.6.5. Hydrocarbon prospects of the post Santonian Upper Cretaceous Rocks on the Nigerian Section of the Iullemmeden Basin

The hydrocarbon prospects of the Senonian to Maestrichtian rocks of the Sokoto section of the Iullemmeden Basin are considered to be very low because of the thinness of the section, lack of structures and adequate cover. Those of the Bornu Basin however are rated higher and prospects improve northwards into Niger in the Iullemmeden Basin. Mobil held the whole of the old Northern Nigeria Region under lease in the 1950s as the concession map for 1957 (Figure 296) shows.

2.4.6.6. Dahomey Miogeocline Western Nigeria, and Benin Flank Upper Cretaceous Post Santonian Formations

2.4.6.6.1. General

Upper Cretaceous Post Santonian rocks extend from the River Niger and the Western Anambra Basin, over the Ilesha Spur or Okitipupa High in western Nigeria and beyond into (Dahomey) Benin and Togo. They rest disconformably on the Basement Complex and west of the Okitipupa High consist mainly of coarse grained clastics known as the Abeokuta Formation in western Nigeria and "Maestrichtian sableux" (Slansky 1962) in Benin (Dahomey).

Shell-BP and Geological Survey of Nigeria (1:250 000 Geological Maps) shows undifferentiated Upper Senonian rocks, overlain by Palaeocene – Eocene Imo Shale Group extending from the River Niger via Ifon to about 8 miles east of Ijebu Ode. West of Ijebu Ode the Abeokuta Formation is overlain by the Cenozoic Ewekoro Formation (Figure 74).

On the 1:2 000 000 scale Geological Map of Nigeria the Upper Cretaceous east of Ijebu Ode is classed as Upper Coal Measures (i.e. Nsukka Formation). Dessauvagie (1974) showed the Abeokuta Formation passing into undifferentiated Sandstones of the Lower Benue Valley succession (Figures 9 and 74). Adeleye (1975) showed a further variation suggesting that the Abeokuta Formation passes into the pre-Mamu Formation Nkporo Shale. Clearly there is divergence of opinion about details here (Figure 66).

Reyment (1965, p. 62) recorded a poor quality coal in these sandstones at Adda stream near Usen. If such coals are present in depth they could be gas source rocks under suitable conditions. No subsur-

face details are available however.

Down dip the Abeokuta Formation has been recognized in Mobil's Ofowo-1 from 3145 to 7061 ft (Fayose 1970). The Araromi Shale Member is the uppermost member of the Abeokuta Formation according to Fayose (1970) and consists of Araromi Shale characterized by alternations of marl and sandy marl passing into dark shales. Towards the top of the unit the shale becomes sandy. Fayose (1970) did not give any information about the remainder of the Abeokuta Formation in depth, nor did he make it clear whether his description applies to the Abeokuta Formation found in Ofowo-1 or whether it is a general description for Nigeria. The Imo Shale (Palaeocene) (2770 – 3145 ft) overlies the Abeokuta Formation in Ofowo-1.

Deep water facies of the Abeokuta Formation have not been described from this or any of the coastal wells drilled between the Nigerian frontier and Okitipupa Ridge (onshore) but conceivably deep water Upper Cretaceous rocks might be present in Union Oil's (Dahomey) Benin wells and are almost certainly present beneath the deeper water offshore (Figures 74). The Upper Cretaceous rocks are said to shale up east of Lagos towards the Okitipupa Ridge.

Down dip on the Benin Flank, marine Araromi Shale (Reyment 1960) was deposited and presumably this is a shaley deep water facies of the so-called Maestrichtian Sandstone etc. (Undifferentiated, Dessauvagie 1976) which rim the Basement Complex between the Okitipupa and the Anambra Basin (western flank).

Local details of the nature of the Upper Maestrichtian Basement Complex contact and of limestones which occur in the Awle River, near where it is crossed by the Agbor-Anchi Road are given in Reyment 1965:

The contact between Upper Maestrichtian and Precambrian in the Auchi area of Mid-Western Nigeria may be studied in the River Ogle, where an ill-sorted grit rests on granite rocks. The most complete section available occurs in the Owam River:

Conglomerates and pebbly sandstones	18 m
Grits and feldspathic sandstones	85 m
Carbonaceous shale	45 m
Grit	60 m
Brown sandstone with *Inoceramus coxi* Reyment	1 m
False bedded, micaceous sandstone	1.1 m

The basal sequence passes into a series of sandstones, shales and mudstones. Thin coal seams have been found in a tributary of the Owam River.

A section in the Obeze River, east of Ifon, discloses friable shales, micaceous sandstones and a brown sandstone with *Sphenodiscus studeri* Reyment, as well as casts of pelecypods and gastropods. a limestone on the banks of the Awle River, near where it is cut by the Agbar-Auchi Road (about mile 65¾) is of Maestrichtian age. It is brown and contains pseudo-oolites, the cores of which consist of ostracods and foraminifera. Minute clay pockets contain *Afrobolivina afra* Reyment. An analysis of the limestone made some years ago by the Government Chemist gave the results shown below:

CaO	47.7%
Al_2O_3	6.5%

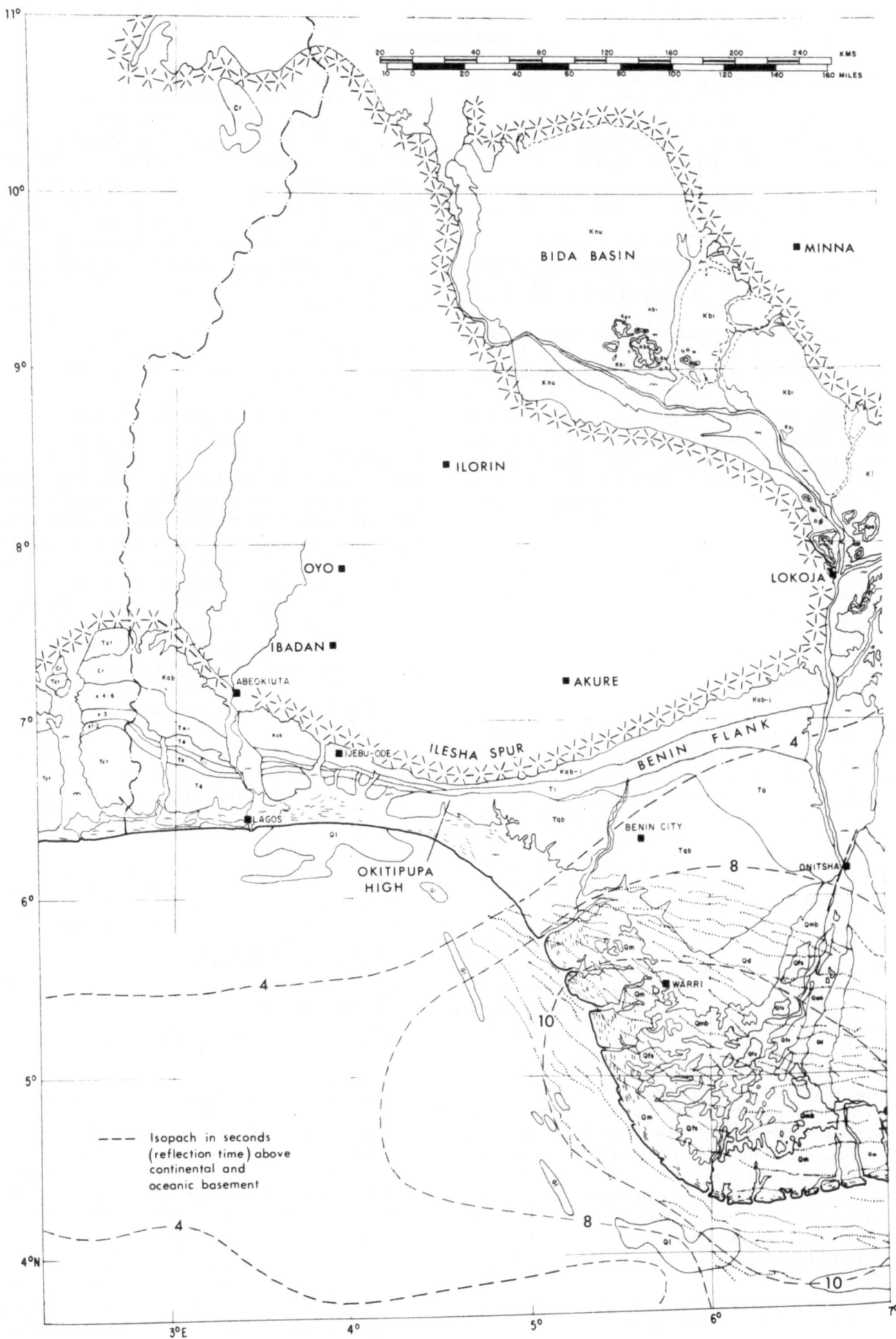

Figure 74. Geological map showing the distribution of Cretaceous and Cenozoic formations west of the River Niger and in the Anambra Basin. Based on Dessauvagie (1972) and Whiteman (1973). (Key as in Figure 73B.)

Fe₂O₃

Mg	0.7%
SiO₂	6.0%
loss on ignition	39.1%

It contains the ostracods *Bythocypris* sp., *Cytherella* sp., *Buntonia (Protobuntonia) ioruba* Reyment, *Cytherella kunradensis* van Veen, *Buntonia (Buntonia) crassicostata* Reyment *Brachycythere* sp. (Reyment 1960c).

In summary then Upper Cretaceous (largely Maestrichtian) rocks at outcrop and as far as the present day shore are predominantly of sandy facies and probably were laid down during the first post-Santonian sedimentary cycle of Murat (1972) (Figure 23).

Down dip in the subsurface along the Benin City Flank, around the Ilesha Spur on Okitipupa High and probably in the offshore these coarse grained continental clastics pass into the Araromi Shale of Reyment (1960 and 1965) and Murat (1972). Murat (1972) pointed out that a similar shale facies occur on the Calabar Flank and that on both flanks "the equivalent" of the upper part of the sequence deposited in the Anambra Basin is missing due to erosion or non-deposition. Only the Upper Cretaceous Abeokuta Formation has been described formally from the area west of the Niger.

2.4.6.6.2. Abeokuta Formation

General Description: The Abeokuta Formation consists of grits, arkosic sandstones, sandstones, siltstones and clays. A basal conglomerate is present but not at all localities. The sands overlying the basal bed are coarse grained, poorly sorted, micaceous and more or less clayey. Cross bedding is common and the rock is soft and friable. A thin lignite bed has been recorded at one locality in the basal sands but lignite beds and carbonaceous materials are more common in the overlying clays. The predominantly sandy strata near the base of the Abeokuta Formation are followed by clays and shales and show a transition from continental to truly marine conditions (Figure 75).

Thickness Variations: At the Dahomey Border the Abeokuta Formation is about 800 ft thick and in the type area southwest of Abeokuta it is about 600 ft thick (Figure 74). The Abeokuta Formation is given as 388 ft thick in Ofowo-1 with the formation extending from 3145 ft to about 7000 ft (Fayose 1970). It possibly extends to TD 7061 ft. This well penetrated Basement beneath the Abeokuta Formation which there would be about 3816 ft thick. Unfortunately tops and lithological descriptions are not available publicly for the Union Oil Company of Dahomey's wells situated offshore just west of the Benin Dahomey – Nigeria border.

Age: Because the Abeokuta Formation underlies the Palaeocene Ewekoro Formation, Jones and Hockey (1964) assigned a Late Senonian age to the formation. The formation is certainly diachronous but does not appear to be older than Maestrichtian onshore. Shell-BP palynologists identified Senonian pollen from intervals 95 – 100 ft in Wasimi Borehole

GSN No. 1585 and Senonian (probably Maestrichtian) microfossils from below 193 ft in Itori Borehole GSN No. 1583.

Conditions of Deposition and Palaeogeography: At least part of the Abeokuta Formation was deposited under fluviatile conditions and the existence of a major river draining southwards from the region of Ishaga has been postulated by Jones and Hockey (1954). Hystricosphaeridae are abundant near the top of the Itori Borehole 1583, presumably about 193 ft (Figure 75). Reyment (1965), having assessed foraminiferal ostracod and mollusc assemblages collected from 144 ft in the Wasimi Borehole, assigned a Maestrichtian age to the Abeokuta Formation. The Abeokuta Formation passes laterally southwards into the Nkporo Shale (Reyment 1965) but Murat (1972) holds the view that it passes into the Araromi Shale which Reyment regards as a facies of the Nkporo Shale).

2.4.6.7. Hydrocarbon Prospects, Dahomey Miogeocline, Nigeria Section Post Santonian Upper Cretaceous Rocks

The wedge of sediments which constitutes the Dahomey Miogeocline extends westwards from the Okitipupa Ridge, really part of the Benin Hinge zone (Figures 2, 9 and 74), towards the Volta Delta. The outcropping sediments are predominantly sandy continental clastics but these interdigitate with paralic and deeper water marine facies in the subsurface. The whole section thickens markedly seaward before thinning down and is probably thickest near the present shelf break (Figure 74). No detailed information is available publicly about the thickness variations of the post-Santonian Upper Cretaceous sediments but they may well follow a similar pattern. Details are not available either about structures, but "down to ocean" continental margin accommodation structures, such as those described from the Abidjan Basin probably exist.

Oil and gas were found in the Union Oil Company of Dahomey's wells drilled just west of the Benin-Nigeria frontier and Benin is expected to produce 15 000 b/d from its offshore Seme Field in 1981.* All other deep wells drilled between the frontier and the Benin Hinge Line, both on shore and offshore have proved to be dry (Figure 22). However along much of the outcrop of the Abeokuta Formation extending from Dahomey to the Okitupa Ridge, there are records of oil seeps and inspissated deposits and in some of the shallow wells drilled by the Nigerian Bitumen Corporation (1907–1914) heavy oil still stands to the surface today. This oil is thought by some geologists to have been derived from the Tertiary sediments which once overlay the region but from a structural point of view this is an unlikely explanation. The Abeokuta and associated seeps could well be evidence of Cretaceous oil.

* This is a highly optimistic estimate as the field has yet to be developed.

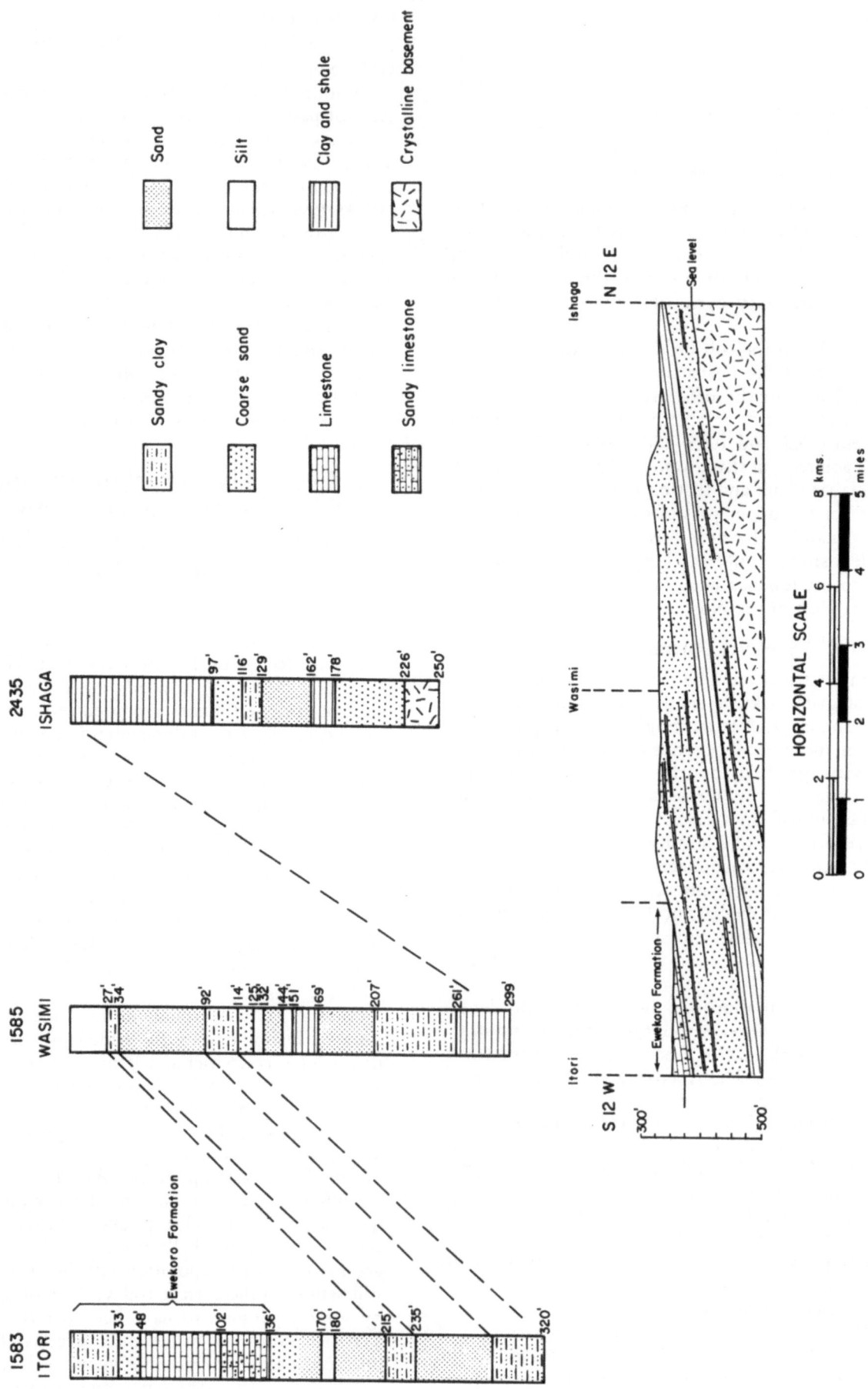

Figure 75. Sections showing water well correlations, type area of Abeokuta Formation, Dahomey Miogeocline. Based on Jones and Hockey (1964) and Whiteman (1973).

Figure 76
Dahomey Miogeocline General Prospects Upper Cretaceous Series (Upper Senonian to Maestrichtian) Post-Santonian Unconformity Sedimentary Rocks

Province	Region	Play	General description	Trap Type	Shows	Prospect Class
	DAHOMEY MIOGEOCLINE (ONSHORE)		A thick prism of Upper Cretaceous sediments extends westwards from the Okitipupa Ridge (Figures 13 and 51) first thickens and then thins seawards	Faults, Drape Folds Unconformities	Surface oil seeps and "tar sands". Positive shows of non-commercial accumulations of oil and gas in adjacent Dahomey's (now called Benin) Seme Field.	GOOD *LIMITATION* FRESH WATER FLUSHING

Based on Whiteman (1973)

line may be due to the fact that (1) too few holes have been drilled to date; (2) that the philosophy of searchers was not really geared to developing stratigraphic plays; and (3) simply that because parts of the matured sedimentary prisms may have been open-ended in places and oil that may have been generated may have escaped except in such localities as the Seme Field, Benin.

It is a peculiar fact that in West Africa only a few commercial accumulations of hydrocarbons have been found to date west of Ghana, in the Abidjan Basin, in offshore Sierra Leone nor in Senegal. Another problem in the Dahomey Miogeocline is that large areas have been flushed with fresh water, so the successful explorer has to contend with defining unflushed structures. Data is not available publicly as to which of the structures drilled have been flushed. Large over-sedimented fault structures are known from the Nigerian offshore of the Dahomey Miogeocline and there may well be oil in combination structural-stratigraphic traps in this area, especially on the continental slope. Mobil and Tenneco have had areas onshore under lease at different times, and Union of California and Gulf have had Blocks A and B under lease also. None of the companies has been successful to date either onshore or offshore, but as has already been noted, few holes have been drilled.

2.5 CENOZOIC ERATHEM IN NIGERIA

2.5.1. General

Sediments belonging to the Palaeocene and Neogene Systems of the Cenozoic Erathem in Nigeria occur in six main areas (Figures 2, 9, 19 and 74):

1. Niger Delta Complex and Anambra Basin southwestern part
2. Eastern Dahomey Miogeocline, Nigeria
3. Iullemmeden Basin southern part known generally as the "Sokoto Basin" or "Sokoto Embayment"
4. Bornu and Chad Basins
5. Niger and Benue Valleys

The general distributions of rock units occurring in these areas is shown in Figure 9. The various stratigraphic successions are shown in Figures 77, 78, 87, and 93.

Except for the Cenozoic sediments of the subsurface section of the Niger Delta Complex and the Palaeocene Ewekoro Formation of southwestern Nigeria which have been described in detail, the stratigraphy of the Cenozoic rocks is poorly known. Outcrops are frequently poor, vegetation cover is extensive and we are dependent for our knowledge of lithology, thickness, facies etc. largely on subsurface data from water wells, oil and gas well, excavations etc.

Correlations between surface and subsurface formations for the Niger Delta Complex are shown in Figure 93. Figure 77 shows how stratigraphic nomenclature has developed for Cenozoic formations exposed at surface in western Nigeria in the Dahomey Miogeocline; in the Niger Delta Complex and in Eastern Nigeria.

The Cenozoic rocks of the Niger Valley above Onitsha outside the delta area and the Benue Valley are poorly known and consist at surface mainly of terrace deposits of Pleistocene to Recent age. They are very thin and are of little interest to the petroleum geologist except that they provide sedimentological data bearing on the Late Cenozoic evolution of the Niger Delta Complex (Burke, Durotoye and Whiteman 1971). The Cenozoic deposits of Nigeria are described below systematically and by basin. For convenience of presentation the Cenozoic deposits of the Nigerian parts of the Iullemmeden, Bornu and Chad Basins and of the Dahomey Miogeocline are described before the formations of the Niger Delta Complex.

Nearly all oil and gas produced to date in Nigeria (Chapter 5), has come from the Cenozoic Agbada reservoir sands of the Niger Delta Complex. Up to 1979 this has amounted to over 7.2×10^9 bbl and over 4.7×10^{12} cuf of associated gas. Furthermore, various observers think that it is highly probable that the greatest part of Nigeria's undiscovered oil and gas

Figure 77

Stratigraphic Nomenclature of the Surface and Subsurface Cenozoic Formations of Western Nigeria, Niger Delta Complex and Eastern Nigeria

	WESTERN NIGERIA			NIGER DELTA COMPLEX		EASTERN NIGERIA			
	JONES 1964 G.S.N. 1964	REYMENT 1965	ANTOLINI 1968	SHORT & STAUBLE 1967 FRANKL & CORDRY 1967		WILSON 1925 WILSON & BAIN 1928	SIMPSON 1954	G.S.N. 1957	REYMENT 1965
POST-EOCENE									
POST-EOCENE TO RECENT	Coastal Plains	Benin Formation	Coastal Plains Sands	Benin form-ation	Benin Sands	Benin Sands	Coastal Plains Sands	Coastal Plains Sands	Benin formation
					Lignite Group	Lignite Group	Lignite Series		Ogwashi-Asaba fm.
EOCENE	formation	Ameki formation	Ameki formation	Akata formation	Agbada Formation	Bende–Ameki Group	Bende–Ameki Series	Bende–Ameki Group	Ameki formation
		Illaro formation	Ilaro formation						
			Oshosun formation						
		Oshosun formation							
PALEOCENE	Ewekoro formation	Imo Shale / Ewekoro formation	Ewekoro formation	Imo Shale formation		Carbonaceous & pebbly Sandstone Group	Imo Clay-Shale	Imo Clay-Shale Group falsebedded sandstone and Upper Coal Measures	Imo Shale
									Nsukka formation

resources also lie in Agbada and related reservoir sands.

Prospects in Cenozoic rocks of the Iullemmeden, Chad and Bornu Basins are considered to be low or nil, but it should not be forgotten that the Cretaceous and older rocks of the Nigerian sections of the Chad and Bornu Basins which occur beneath the Cenozoic sediments may hold considerable potential. Oil has been disovered in the Cretaceous "Continental Intercalaire" sands of the Chad Basin in Chad (Figure 2), but none has been produced commercially to date; political unrest has largely prevented further developments.

2.5.2. Iullemmeden Basin, Sokoto Section, Northern Nigeria Cenozoic Formations

2.5.2.1. General

The youngest sediments, excluding the superficial deposits, encountered in the Sokoto Section of the Iullemmeden Basin are the Gwandu Formation (Figure 9). These rocks are extensively exposed on either bank of the Sokoto River southeast of the Niger-Nigeria Frontier and consist mainly of lacustrine sediments. In many areas the formation is capped by lateritic ironstone. Its age is uncertain but the formation has been assigned to the Pleistocene. However, palynological data provided by Shell-BP

Nigeria indicate that it is possibly Eocene. Beneath the Gwandu Formation are the Dange, Kalambaina and Gamba Formations (Figures 78 and 80). The oldest of these is the Dange Formation which consists of blue-grey, plastic shale with phosphatic nodules and thin limestones. This formation is thin and is only about 68 ft thick near Sokoto but thickens to 140 ft in the subsurface near Balle. The overlying Kalambaina Formation outcrops form a high level plain bordered on the east of the Dange scarp. The formation consists of white clayey limestone and marl, calcareous mudstone and slightly indurated shale. At Balle, the Kalambaina is about 160 ft thick. The beds are thought to be Eocene in age. The Dange, Kalambaina and Gamba Formations constitute the "Sokoto Group". The Gamba Formation is a thin shale and is really part of the Kalambaina Formation.

It is difficult to reconstruct the palaeogeography of Palaeocene and Eocene times for the southern section of the Iullemmeden Basin. To the north in Niger the Palaeocene is about 164 ft thick. It consists of limestones, atapulgites, phosphates, gypsum and ferruginous oolites. These rocks are thought to have been laid down in a warm shallow sea connected with the Atlantic via an ill-defined "north-west passage" generally located in the Gao area. Sediments were probably derived from deeply weathered Basement Complex and Mesozoic rocks. The sea is thought to have finally retreated from the Iullemmeden Basin in Ypresian (Late Palaeocene) time.

The "Continental Terminal" succeeds in Niger

Figure 78
Main Formations and Groups Nigerian Section of the
Iullemmeden Basin (Sokoto Basin of Literature)

	Age
GWANDU FORMATION	EOCENE-MIOCENE
Black to red mudstones with coarse grained white to brown quartz sand; lignite in bounds. Thickness 1000 ft	(? PLEISTOCENE)
GAMBA FORMATION	PALAEOCENE
Shales, grey with pliosphatic pellets. No thickness given Kogbe (1976)	
KALAMBAINA FORMA-TION	PALAEOCENE
Marine clayey limestone and shale. Thickness 60 ft in Nigeria	
DANGE FORMATION	
Bluish grey shale interbedded with yellowish limestone. Thickness 140 ft.	PALAEOCENE CENOZOIC CRETACEOUS MAESTRICHTIAN
WURNO FORMATION	
Sandstones, fine grained, friable; siltstones intercalated mudstones. Thickness 150 ft	
DUKUMAJE FORMATION	MAESTRICHTIAN – ? CAMPANIAN
Shales with limestones and mudstones. Thickness 100 ft	
TALOKA FORMATION	? MAESTRICHTIAN –
Mudstones and sandstones with carbonaceous shales and lignite. Thickness 400–600 ft	? CAMPANIAN– ? CONIACIAN– SANTONIAN?
ILLO FORMATION	PRE-MAESTRICHTIAN
Pebbly grits, sandstones and clay. Thickness 450–800 ft (Niger Valley)	
GUNDUMI FORMATION	PRE-MAESTRICHTIAN OR ? TURONIAN
Fine to medium grained sandstones, false bedded grits, clays, basal conglomerates and arkoses. Thickness 300–1000 ft	

(SOKOTO GROUP spans GWANDU through DANGE formations; RIMA GROUP spans DUKUMAJE through ILLO formations)

———————————————————— UNCONFORMITY

BASEMENT COMPLEX

Data from: Kogbe (1972 and 1976); Reyment (1965); Whiteman (1973).

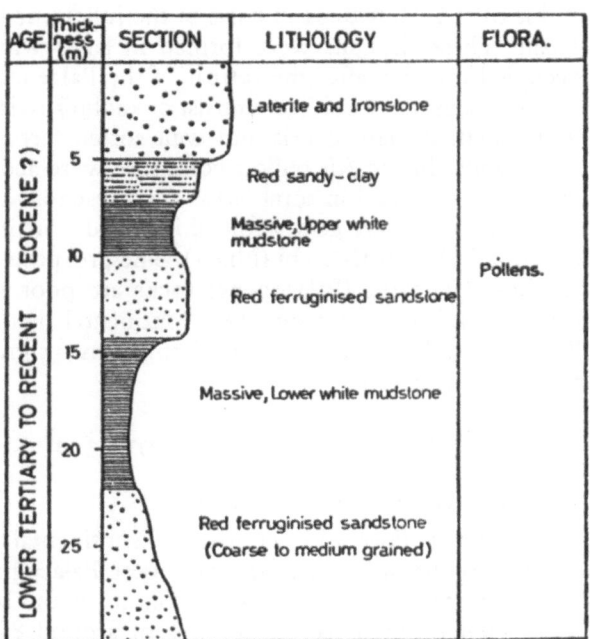

Figure 79. Type section Gwandu Formation, Birnin Kebbi, "Sokoto Embayment" Nigeria. Based on Kogbe (1976).

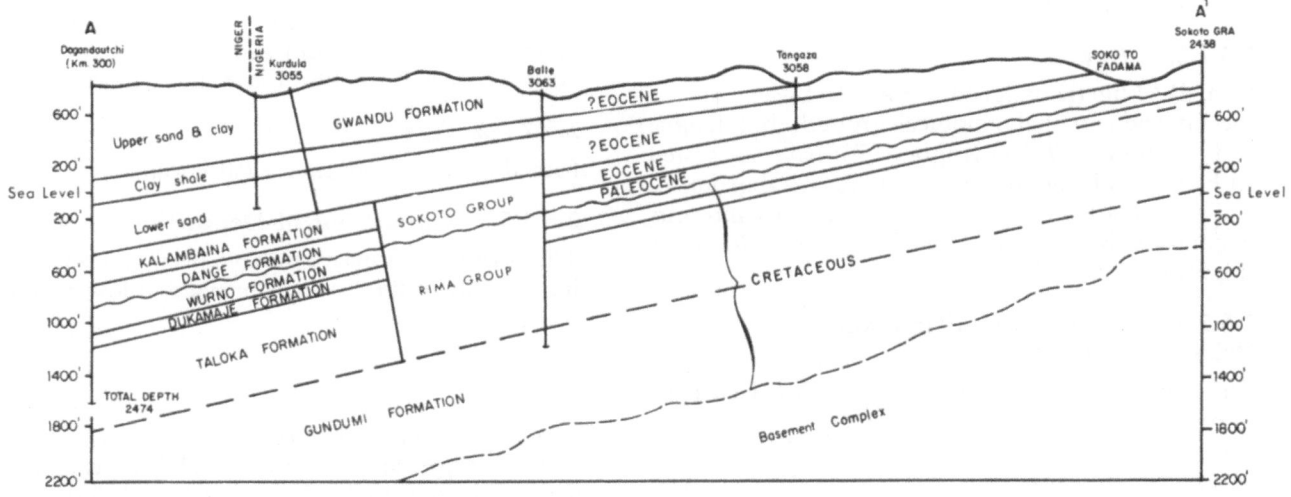

Figure 80. Section Cretaceous and Palaeogene Formations from Dogondoutchi (Niger) to Sokoto (Nigeria). Based on Kogbe (1972 and 1976). (Locations on Figure 73.)

and is over 1450 ft thick in parts of the Iullemmeden Basin. These deposits were formed in lakes which occupied the area after the retreat of the Palaeocene sea. The sediments in Niger consist of kaolinitic clays with calcified plant debris and well sorted ferruginous sands. In the Aïr region volcanoes were active in the Tertiary and in some areas the Mesozoic and Tertiary rocks were gently folded in broad synclines as around Dogon Doutchi (Niger). The prospects of all these Cenozoic (Palaeogene) rocks are poor because of lack of cover and their continental nature. The general distribution of these formations is shown in Figures 9, 81 etc.

2.5.2.2. Gwandu Formation

2.5.2.2.1. General Description

The Gwandu Formation consists of continental lacustrine sediments and is the youngest Palaeogene formation present in the Sokoto section of the Iullemmeden Basin. It covers an area of around 8500 square miles in northwestern Nigeria. Shell-BP Nigeria assigned a possible Eocene age to the formation, which consists of massive white clay interbedded with coarse grained red sandstone and mudstone with lignite or peat bands. Ironstone caps are common at outcrop. Individual mudstone sections may attain thicknesses of almost 200 ft, whereas the sandstones are usually thinner and around 100 ft. Kogbe (1972) proposed a type section for the formation at Birnin Kebbi in the Gwandu Emirate. Details of the type sections are shown in Figure 79.

2.5.2.2.2. Thickness Variations

Kogbe (1976) divided the formation into three members (unnamed): the basal member consists of fine to coarse poorly consolidated sand with thin beds of clay; overlain by semi-indurated clay-shale (up to 100–200 ft thick); and an upper member consisting of semi consolidated sand interbedded with whitish clay. The whole formation attains a thickness of 1000 ft or so near the Niger–Nigeria frontier (Kogbe 1976) although Dessauvagie (1974) gives a maximum thickness of 240 ft (75 m).

2.5.2.2.3. Age

Shell-BP Nigeria assigned a possible Eocene age to the Gwandu on palynological grounds but Kogbe and Sowunmi (1974 in Kogbe 1976) have suggested an Eocene to Miocene age on palynological grounds.

The formation oversteps Palaeocene beds and rests unconformably on Cretaceous beds at some localities (Figures 2 etc.).

2.5.2.2.4. General Conditions and Palaeogeography

The Gwandu sediments are continental and lacustrine in origin.

2.5.2.3. Gamba Formation

2.5.2.3.1. General Description

Kogbe (1973) proposed the name (new) Gamba Formation for the grey laminated shale which overlies the Kalambaina Formation (See below).

2.5.2.3.2. Thickness

The type section is given as Gamba Quarry where the thickness is around 12 ft and attains 30 ft in boreholes.

The shales are phosphatic and ironstone concretions occur in the upper part (Figure 82). Laterite overlies the Gamba Shales at the type locality.

2.5.2.3.3. Age

The age of the formation is given as Palaeocene.

2.5.2.4. Kalambaina Formation

2.5.2.4.1. General Description

The Kalambaina Formation underlies the Gamba Formation and consists of grey-yellow and white limestone, calcareous shale and laminated shale. The section, exposed in the Sokoto Cement Works Quarry is portrayed in Figure 82. The formation has yielded many fossils (Kogbe 1973).

Thickness Variations: The Kalambaina Formation is of variable thickness but is less than 70 ft.

2.5.2.4.2. Age

The Kalambaina Formation has yielded vertebrates as well as molluscs which indicate a Palaeocene age.

Conditions of Deposition: The Kalambaina Formation was laid down under marine conditions having yielded echinoids, corals, nautiloids, lamellibranchs and gastropods.

2.5.2.5. Dange Formation

2.5.2.5.1. General Description

The Dange Formation forms the base of the Sokoto Group and consists of blue grey, plastic shale with thin limestones (Figure 83). The Formation includes bands of fibrous gypsum and around Sokoto attains 70 ft. It thickens to 140 ft at Balle (Figure 53).

2.5.2.5.2. Age

Faunal evidence indicates a Palaeocene age.

2.5.2.5.3. Conditions of Deposition

The formation appears to have been laid down during periods of fluctuating normal and low salinity.

Figure 81. Geological sections northwestern Nigeria showing disposition of Cenozoic and Cretaceous formations "Sokoto Basin", Nigeria Geological Sheets 1959–1963 Scale 1 : 250 000. See Figure 72 for location of these sections.

Additional material from *Nigeria: Its Petroleum Geology, Resources and Potential*
ISBN 978-94-009-7363-3 (978-94-009-7363-3_OSFO4),
is available at http://extras.springer.com

AGE	FORM-ATION	(m)	SECTION	LITHOLOGIC DESCRIPTION	FAUNA
P A L E O C E N E	L A T E R I T E	5		Ferruginised sandstone and oolitic Ironstone or laterite	
	GAMBA FORMATION	10		Ironstone concretion in shales Grey laminated shale Phosphatic pellets and small coproliths Laminated shale with gypsium	Foraminifera Ostracods Lamellibranch.
	KALAMBIANA FORMATION	15 20		White limestone with crevices in-filled with non-calcareous clay Shally limestone white to whitish yellow White-yellowish limestone slightly shally Shally level within limestone bed Carbonaceous Limestone with shally inclusion 3 mm thick Shales, laminated within limestone, Carbonaceous	Foraminifera Ostracods Lamellibranch. Gastropods Crinoids Corals Nautilus Echinoderms

Figure 82. Section of Palaeogene Kalambaina and Gamba Formations, Cement Quarry near Kalambaina and Gamba 3 miles south of Sokoto, Northwestern Nigeria. Based on Kogbe (1976).

AGE	Thickness (m)	SECTION	LITHOLOGIC DESCRIPTION.	FAUNA
P A L E O C E N E	0 2 4 6 8 10 12 14 16 18 20		Laterite Clayey shale yellowish-white Greyish brown laminated clay shale sometimes with thin limestone intercalations limestone band shale (laminated) limestone Grey shale Limestone shale	Lamellibranch Nautilus Echinoids Ostracods Foraminiferas Vertebrates

Figure 83. Section Dange Formation at type section Sokoto-Gusau Road, Northwestern Nigeria. Based on Kogbe (1976).

103

2.5.2.6. Hydrocarbon prospects of the Gwandu Formation and Sokoto Group of the Nigerian Section of the Iullemmeden Basin

The petroleum prospects of these Cenozoic (Palaeogene) formations in Nigeria are considered to be nil because of lack of cover, immaturity of sediments, thinness, lack of structure and because of dissection (Figure 81).

2.5.3. Bornu Basin and adjacent Southern Chad Basin Cenozoic Formations

2.5.3.1. General

Little is known about Palaeogene rocks of the Bornu Basin and adjacent areas, mainly because the Neogene Chad Formation and superficial deposits cover the older sediments and lap onto the Cretaceous sediments and the Basement Complex. The only Palaeogene formation known to date in this area is the Palaeocene? Kerri-Kerri Formation. This is at least 650 ft thick.

The Chad Formation is of Plio-Pleistocene age and occupies some 40 000 square miles in northeastern Nigeria. Near Maiduguri the formation is more than 1800 ft thick. The Chad Formation as its name implies, extends into Chad and also into adjacent Niger. In Nigeria it dips gently eastward (Figures 2 and 9). The general relationships of the Kerri-Kerri and Chad Formations are shown in Figure 218. Both the Kerri-Kerri and Chad Formations are unfolded.

2.5.3.2. Kerri-Kerri Formation

2.5.3.2.1. General Description

The Kerri-Kerri Formation consists of loosely cemented coarse to fine grained sandstones, massive claystones and siltstone with bands of ironstone and conglomerate occurring locally. The sandstones are often cross bedded and lignites occur near the base of the formation. Plant remains are widespread and the age of the formation is thought to be Palaeocene. It is at least 650 ft thick and its distribution is shown in Figures 2 and 9. The petroleum prospects of the Kerri-Kerri are nil because of its thinness, lack of cover and the fact that parts of it are fresh water flushed.

2.5.3.3. Jos Sediments

This ill defined formation consists of kaolinitic clays said to be of Plio-Pleistocene age. They are of fluvio-volcanic origin and occur beneath the volcanics of the Jos area. Their petroleum prospects are nil.

2.5.3.4. Chad Formation

2.5.3.4.1. General Description

The Chad Formation consists of fine to coarse grained sand, yellow, blue grey with intercalations of sandy clay, clay and diatomite. The sandy sediments are often poorly sorted. Towards the centre of the basin lacustrine clays predominate interdigitating with sands, grits and gravels around the basin margin. The formation varies considerably in thickness and on the western shore of Lake Chad it may be over 2500 ft thick. Under parts of the lake basin outside the great beach ridge which is such a striking feature of Lake Chad geology (Figures 9 amd 218), the Chad Formation is less than 600 ft thick but within the Mega-Chad Beach Ridge the formation thickens rapidly to around 2500 ft and over 3000 ft perhaps in the centre of the basin (Burke 1976).

Again the petroleum prospects of the Chad Formation are nil because of flushing in three well developed aquifers and lack of cover. In addition the sediments are immature. The Chad Formation however may have acted as "roof" rock in helping to seal off hydrocarbons which may have accumulated in older sediments such as the Continental Intercalaire.

2.5.4. Dahomey Miogeocline, Western Nigeria Cenozoic Formations

2.5.4.1. General

The Dahomey Miogeocline which extends into western Nigeria as far east as the Okitipupa High or Ilesha Spur (Figure 84) and as far west as the Volta Delta Complex in Ghana, consists of an extensive wedge of Cretaceous, Palaeogene and Neogene sediments which thicken markedly from the onshore margin of the basin (where the predominantly clastic Cretaceous sediments rest on Basement Complex) into the offshore where thick finer grained Cenozoic sediments obscure the Cretaceous rocks developed in leptogeoclinal basins.

The oldest of the Cenozoic formations exposed in the Nigerian section of the Dahomey Miogeocline is the Imo Shale and the youngest is the Benin Formation. The Palaeocene Imo Shale was encountered between 2270–3145 ft in Ofowo-1 and the combined Mesozoic and Cenozoic thickness in Mobil's Ofowo-1, situated near the present day shore is around 7000 ft. The combined Mesozoic – Cenozoic thickness in Union Oil's Dahomey-1 is greater than 10 110 ft. Tops are not available for Union's wells but seismic data presented by Emery et al. (1975) indicate that the Cenozoic-Mesozoic sequence exceeds 12 000 ft not far from the shore and maintains this thickness for a considerable distance seaward (Figure 233).

Down to ocean faults, produced during continental margin adjustments processes affect Cretaceous

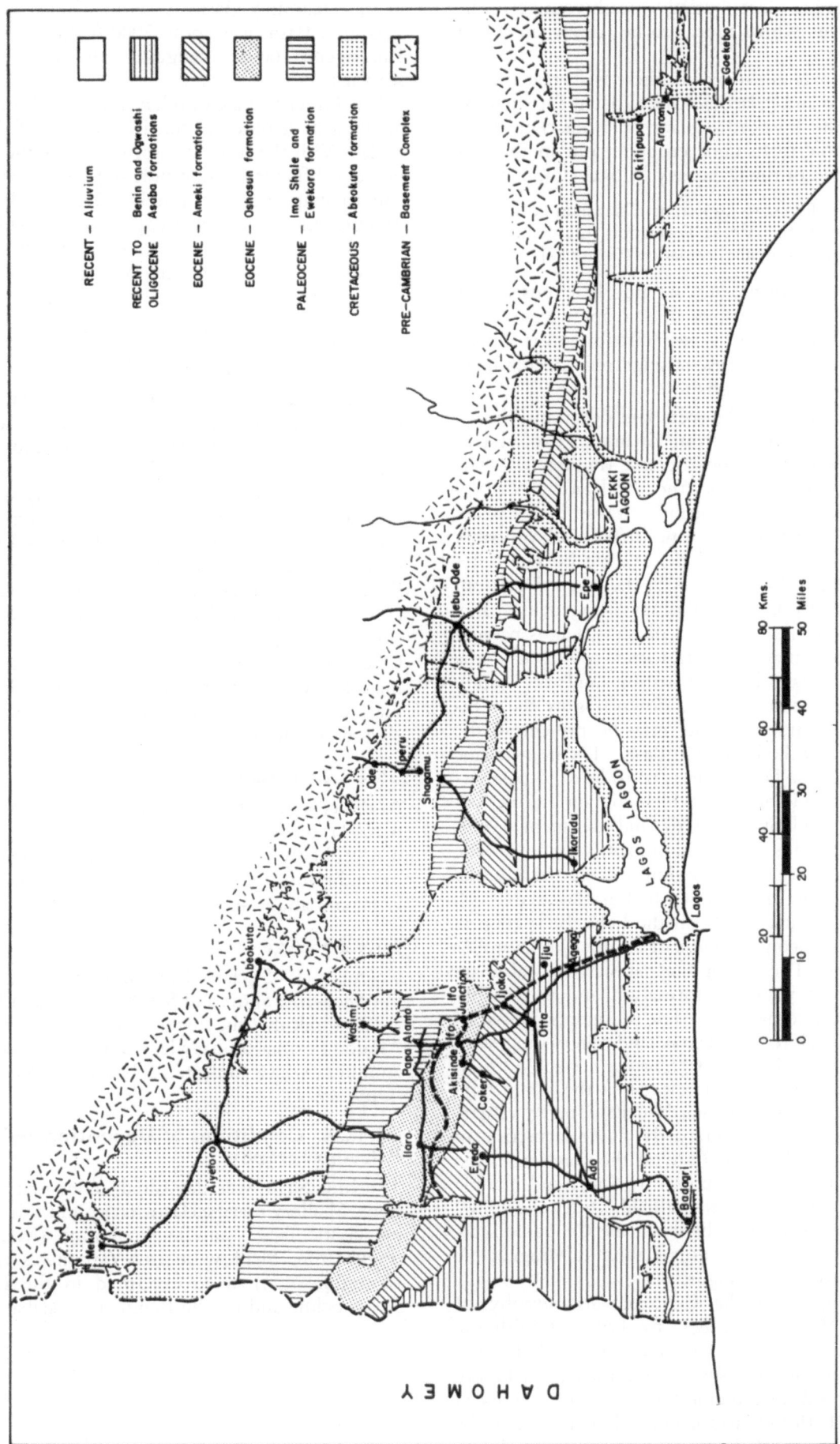

Figure 84.
Map showing distribution of Cretaceous and Cenozoic rocks, Western Nigeria. Based on Geological Survey of Nigeria Maps and Whiteman (1973).

and older Tertiary rocks and faulted? Palaeogene sediments are in places buried beneath younger Cenozoic deposits which appear to be largely unfaulted (Figure 268).

Onshore data is patchy but there is a great deal of stratigraphic and palaeontological detail for such localities as the Ewekoro Quarry and adjacent areas. The Cenozoic formations cropping out in the Nigerian section of the Dahomey Miogeocline include:

Imo Shale; Ewekoro Limestone; Oshoshun Formation; Ilaro Formation; Ameki Formation; Ogwashi–Asaba Formation; and Benin Formation (Figure 87).

The Cenozoic formations are poorly exposed and difficult to map because of thick tropical vegetation and superficial deposits. Their distribution is shown in Figures 9 and 84. Few stratigraphic details have been provided for the offshore.

The Palaeocene rocks were laid down in a broad embayment extending from Ghana on the west to the Okitipupa Ridge on the east (Figure 22). The Ewekoro Limestone and the "Calcaire à *Togocyamus*" were laid down during a transgressive phase and pass laterally both along strike and probably down dip into the Imo Shale. During Eocene times the sea regressed extensively in the Niger Delta but less extensively in the Dahomey Miogeocline. Bay or lagoonal facies developed north of Lagos passing into shallow marine clastics and southwards the phosphatic Oshoshun Formation and the sandy Ameki Formation were laid down (Figure 85). Oligocene and Miocene rocks may be present in the eastern part of the Dahomey Basin. The non-marine Ogwashi–Asaba Formation is considered to pass laterally into the Ijebu Beds which contain marine fossils (Short and Stauble 1967 and Reyment 1965). Both formations are predominantly sandy. The sandy beds alternate with lignite and clay beds in the Ogwashi–Asaba unit and with thin clay beds with occasional marine fossils on the Ijebu Beds (Figure 77). The latter contain traces of bitumen at a number of localities. The Benin Formation (Coastal Plain Sands) is the youngest formation and is possibly Miocene to Recent in age. The sediments consist of yellow and white continental sand, alternating with pebbly layers and clay beds (Short and Stauble 1967). Reyment (1965) holds the view that the Benin Formation is partly marine, partly deltaic, partly estuarine, partly lagoonal and partly fluvio-lacustrine in origin, which does not fit with the concept of the Benin Formation as proposed by Short and Stauble (1967) and now used in the oilfields. Adegoke (1969) attempted a correlation between the Cenozoic sediments exposed on land in the Dahomey Miogeocline and the Niger Delta (Figure 86).

In dealing with the Cenozoic stratigraphy of Southern Nigeria it is best to deal first with the outcrops and then the subsurface. A correlation between subsurface and surface is presented in Figure 93 and the subsurface units in general use are listed in Figure 91. Both brackish and truly marine deposits were laid down. Palaeogene strata thin markedly over the Okitipupa Ridge; the maximum thickness in the Dahomey Basin is about 1200 ft and in the Anambra Basin it is about 7000 ft. In both areas the Palaeogene (and Neogene) thickens markedly seaward.

The Niger Delta Complex and the Dahomey Miogeological sediments consist mainly of Neogene rocks. With the limited information publicly available it is difficult to subdivide these rocks at outcrop and to map them accurately. Detailed time-stratigraphic units have been recognized in the sub-surafce however and correlated throughout the Niger Delta Complex and the Dahomey Miogeocline.

2.5.4.2. Imo Shale

2.5.4.2.1. General Description

The Imo Shale is mainly a fine textured dark grey to bluish shale with thick sandstone bands and ironstones. The formation becomes more sandy towards the top where it consists of alternations of sandstone and shale. The Imo Shale grades laterally westwards into the Ewekoro Limestones of Western Nigeria. Type area for the formation is along the Imo River between Umuahia and Okigwi in Eastern Nigeria. Frankl and Cordry (1966) and Short and Stauble (1967) pointed out that the Imo Shale is the up-dip equivalent of part of the subsurface Akata Formation. Ogbe (1970) proposed a new name—Akinbo Formation—for the green laminated shale with bands of marl and glauconite exposed in western Nigeria and classed as Imo Shale.

2.5.4.2.2. Thickness

At the type locality in Eastern Nigeria the formation is 1600 ft thick but near the Ewekoro Cement Quarry it is probably less than 500 ft. The Imo Shale passed downdip into the Akata Shale which in its type area is around about 4000 ft thick.

2.5.4.2.3. Age

Reyment (1965) classed the Imo Shale as Palaeocene but Shell-BP, Short and Stauble (1967) and Berggren (1960) think that the upper part is Early Eocene. The Akinbo Formation has yielded many planktonic foraminifera which indicate a Thanetian – Ypresian age (Ogbe 1972).

2.5.4.2.4. Conditions of Deposition and Palaeogeography

The Imo Shale was deposited under marine conditions. Shallow marine clastic facies and a deeper marine clastic facies can be recognized (Murat 1972) (Figure 85). Shallow marine carbonates (Ewekoro Limestone etc.) were deposited in the Dahomey Miogeocline and on the Benin and Calabar Flanks.

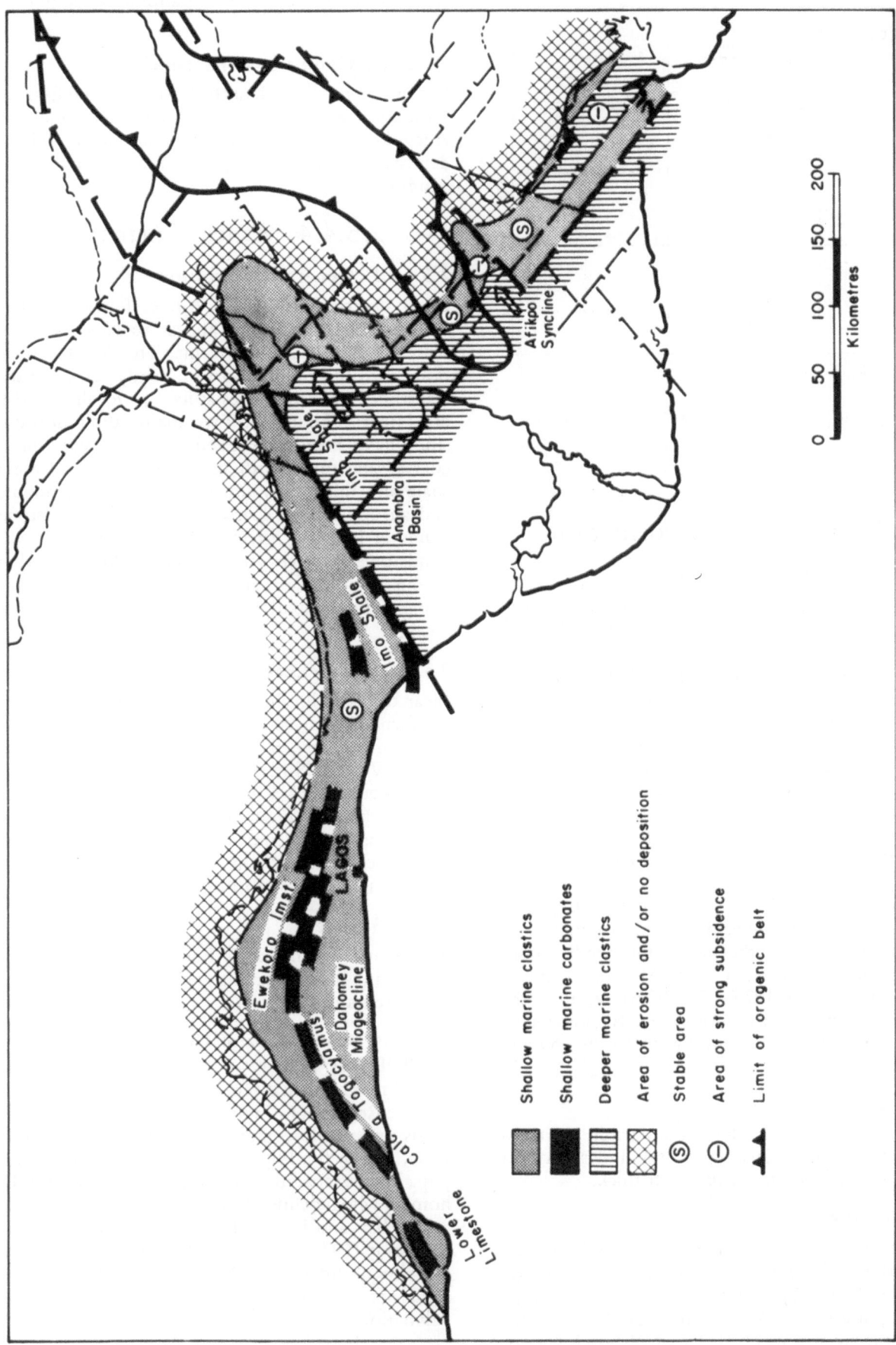

Figure 85. Palaeogeographic sketch map, Palaeocene Transgressive Phase, Dahomey Miogeocline, Anambra Basin, Afikpo Syncline and Calabar Flank. Based on Murat (1972).

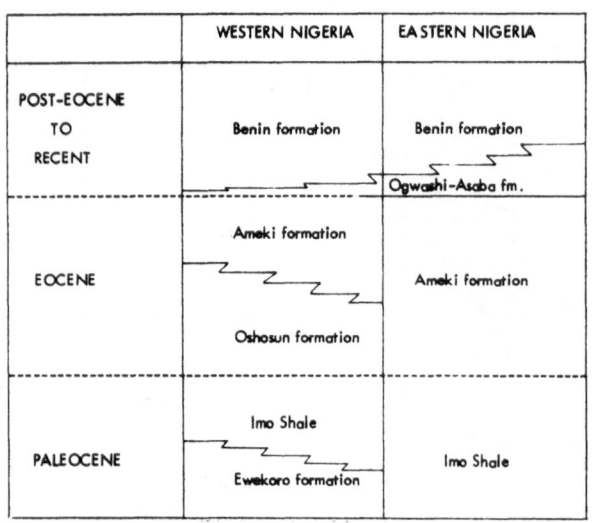

	WESTERN NIGERIA	EASTERN NIGERIA
POST-EOCENE TO RECENT	Benin formation	Benin formation
		Ogwashi-Asaba fm.
EOCENE	Ameki formation	Ameki formation
	Oshosun formation	
PALEOCENE	Imo Shale	Imo Shale
	Ewekoro formation	

Based on Adegoke (1969)

Figure 86
Stratigraphic Relationships of Surface Cenozoic Formations,
Southern Nigeria

2.5.4.3. Ewekoro Formation

2.5.4.3.1. General Description

The Ewekoro is a thin shelly limestone, in places glauconitic, which becomes sandy bear the base. It is a lateral equivalent of the Imo Shale. The shelly limestones are interbedded with bluish to white marls. The surface of the limestone is pot-holed, water worn and scoured. The cavities are infilled with shale and the limestone is reddened. The limestones tend to be sandy at the bottom; contain abundant shell fragments in the middle part and are poorly fossiliferous fine grained and algal in the upper part of the formation. The base of the formation grades down into the Abeokuta Formation (Adegoke *et al.* 1970).

Fayose and Asseez (1972) considered the Ewekoro Limestone to be a member of the Imo Formation in Western Nigeria which they divide into the Ewekoro Member and an unnamed Shale Member (Akinbo Formation of Ogbe 1970). The distribution of the formation is shown in Figures 9 etc. Regional stratigraphic relationships are shown in Figures 86 and 87.

2.5.4.3.2. Thickness

At the type locality in the West African Portland Cement Company's Quarry about 40 miles north of Lagos the formation is around 40 ft thick.

2.5.4.3.3. Age

The formation has yielded many fossil molluscs and echinoids which indicate a Danian – Montian age (Adegoke 1970). Microfossils collected from the type area suggest a Danian – Landenian age (Ogbe 1970).

Radiometric dating of glauconite from Ewekoro Limestone yielded an average age of 54 ± 2.7 million years (Adegoke *et al.* 1970) and most workers assign a Palaeocene age to the Ewekoro Limestone. Asseez (1970) from the study of foraminifera assigned an Ypresian age.

2.5.4.3.4. Conditions of Deposition

Like its lateral equivalent the Imo Shale, the Ewekoro Formation was laid down under marine conditions. Adegoke (1972) stated that the fauna clearly indicates deposition under tropical littoral to neritic conditions.

2.5.4.4. Ameki Formation

2.5.4.4.1. General Description

Local descriptions of the Ameki Formation from Western Nigeria do not appear to have been presented except in general terms. The Ameki Formation consists of grey-green sandy clays, sandy claystones and sandstones at its type locality in Eastern Nigeria (Figure 9). Two main lithological divisions have been recognized: a lower with fine to coarse sandstones with intercalations of calcareous shale and thin shelly limestone, limestone nodules; and an upper with coarse cross bedded sandstones, bands of fine grey-green sandstone and sandy clay. Sandstones in places attain 100 ft thickness. Carbonaceous plant remains and lignites occur at some horizons. The formation displays conspicuous facies changes.

2.5.4.4.2. Thickness

Near the type locality in Eastern Nigeria (Figure 2) the formation is 4800 ft thick. The Ameki Formation is thought to pinch out over the Okitipupa Ridge and in Western Nigeria is much thinner than in Eastern Nigeria where it passes laterally into the Oshoshun Formation (Figure 87). Down dip in the Niger Delta it is in part equivalent to the Agbada Formation. Locally at outcrop the Ameki is only around 300 ft thick in Western Nigeria.

2.5.4.4.3. Age

The Ameki Formation is Eocene (Lutetian to Early Bartonian) in age, having yielded *Pappocetus lugardi* whale remains.

2.5.4.4.4. Conditions of Deposition and Palaeogeography

The Ameki Formation was deposited during the Eocene Regression (Figure 45) and consists of deltaic sands and shallow marine clastics in the Anambra Basin and in the Niger Delta Complex. The Okitipupa Ridge separated the Niger Delta from the Dahomey Miogeocline during the Eocene regression where bay and lagoonal facies developed according to Murat (1972) and the Oshoshun Formation was laid down. Adegoke (1969) however challenged the idea of a lagoonal origin for the Oshoshun Formation.

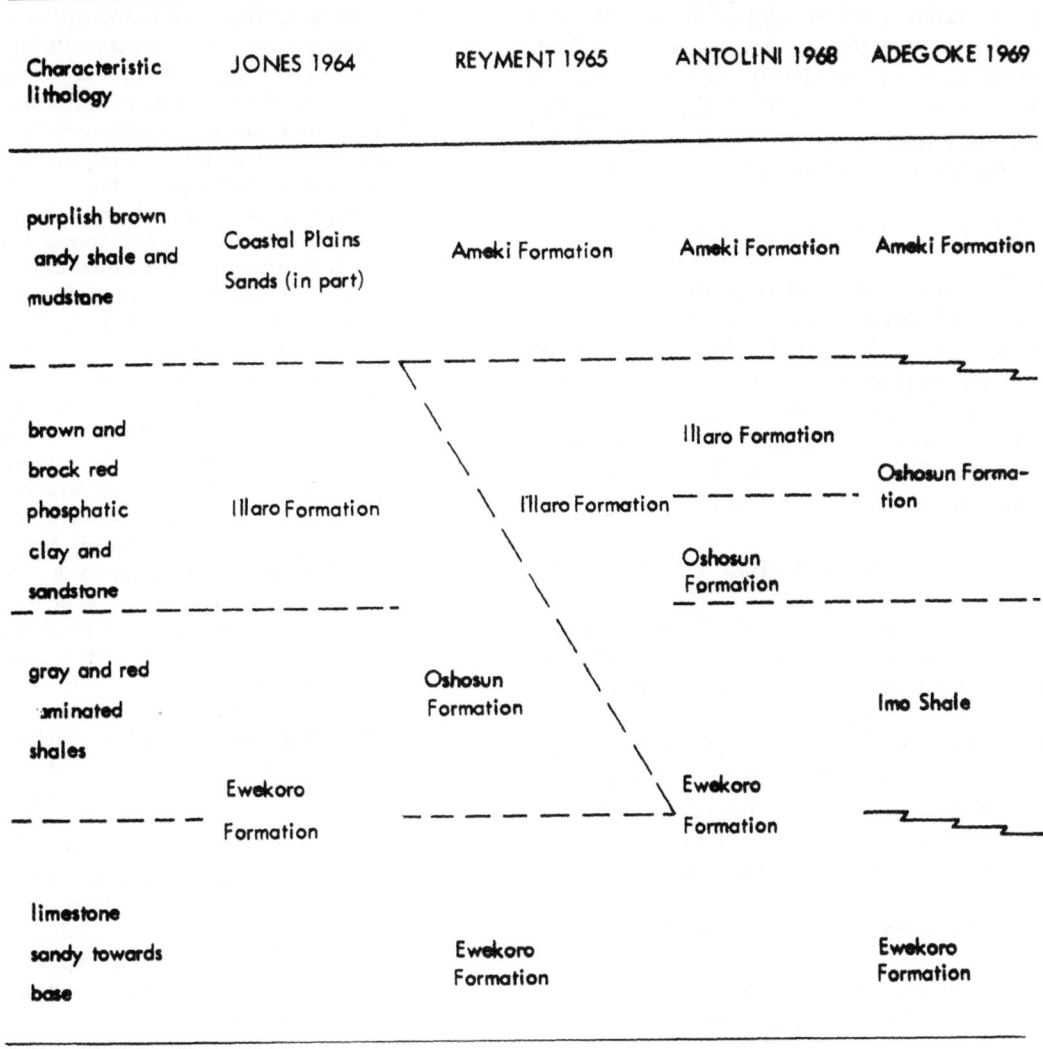

Characteristic lithology	JONES 1964	REYMENT 1965	ANTOLINI 1968	ADEGOKE 1969
purplish brown andy shale and mudstone	Coastal Plains Sands (in part)	Ameki Formation	Ameki Formation	Ameki Formation
brown and brock red phosphatic clay and sandstone	Illaro Formation	Illaro Formation	Illaro Formation / Oshosun Formation	Oshosun Formation
gray and red aminated shales	Ewekoro Formation	Oshosun Formation	Ewekoro Formation	Imo Shale
limestone sandy towards base		Ewekoro Formation		Ewekoro Formation

Figure 87
Characteristic Lithologies and Stratigraphic Nomenclature
Palaeogene Formations, Western Nigeria

2.5.4.5. Oshoshun Formation

2.5.4.5.1. General Description

The Oshoshun Formation is the lateral equivalent of part of the Ameki Formation in Southern Nigeria (Figure 86). It consists in the lower part of dull brown to brick red, mottled sandy mudstone. Thin pebble beds and coarse pebbly sandstones occur and the unit contains glauconite and phosphate. The middle part of the formation contains much more phosphatic material in three major phosphate horizons. Pebbly mudstone and mudstones occur between the phosphate horizons which are traceable laterally into workable deposits in Togoland. The upper part of the Oshoshun is much more sandy. The Oshoshun Formation only occurs west of the Okitipupa Ridge (Figure 86). The Illaro Formation (Jones 1954) was erected to include in part the phosphatic beds of the Oshoshun Formation.

2.5.4.5.2. Thickness Variations

These are shown in Figure 86. The formation is of variable thickness but is probably less than 200 ft.

2.5.4.5.3. Age

The Oshoshun Formation is assigned a Medial Eocene (Lutetian) age (Reyment 1965 and Adegoke 1969).

2.5.4.5.4. Conditions of Deposition and Palaeogeography

The Oshoshun Formation was deposited in a marine environment probably in fairly deep water (Adegoke 1969).

2.5.4.6. Illaro Formation

The Illaro Formation (Jones 1964) is in part a lateral equivalent of the Oshoshun Formation (Figure 87).

109

It consists of massive, yellowish poorly consolidated sandstone. Sometimes the sandstones are cross bedded according to Reyment (1965). The formation, as defined by Jones, is about 200 ft thick. Adegoke (1969) considers that the Illaro Formation is ill defined as it embraces both Oshoshun and Ameki beds.

2.5.4.7. Ijebu Formation

The Ijebu Formation consists largely of sands, locally impregnated with bitumen and clays. Their thickness is greater than 30 ft. The type locality is in the Ijebu area, east of Lagos (Figures 9 and 84) and adjacent to Lekki Lagoon. The beds contain marine fossils and are in part the marine equivalent of the non-marine Ogwashi-Asaba Formation. The Ijebu Formation grades upwards into the Benin Formation.

2.5.4.8. Ogwashi-Asaba Formation

This formation consists of alternations of lignite seams and clays. It is found mainly in Benin, Onitsha and Owerri Provinces (Figure 9) but extends into eastern and western Nigeria (Figure 77). Some of the lignite seams are more than 20 ft thick and more than 62 million tons are said to have been proven overall. The age of the formation is ? Oligocene to Miocene. Dessauvagie (1974) did not map the Ogwashi – Asaba Formation in Western Nigeria (Figure 9).

2.5.4.9. Benin Formation (Surface)

Reyment (1965) reinstated the name Benin Formation (Parkinson 1907) for the outcropping yellow and white sands and clays which occur in coastal Nigeria especially in Western and Mid-Western Nigeria. Pebble beds occur in the sands and clays and sandy clays form lenses. Apart from a plant remain few fossil have been found. Foraminifera, ostracods, mollusc fragments, corals, fish remains have been found from boreholes, according to Reyment (1965) and he interpreted these to indicate that the Benin Formation was deposited in partly marine, deltaic, estuarine, lagoonal and fluvio-lacustrine environments. The Benin Formation is described in detail below in Section 2.5.5.2. dealing with the subsurface Cenozoic of the Niger Delta Complex. Its distribution in Western Nigeria is shown in Figure 9.

2.5.5. Niger Delta Complex Cenozoic Formations

2.5.5.1. General

As we have already noted three major depositional complexes can be distinguished within the Cenozoic and Cretaceous formations of Southern Nigeria in the Abakaliki Trough, Benue Depression, Anambra Basin, the Benin and Calabar Flanks and the Niger Delta area. These are the:

3. **Cenozoic Niger Delta Complex** which developed as a regressive offlap sequence during Cenozoic time. The delta complex built out across the Anambra Basin and Cross River margins and eventually extended onto the Late Cretaceous continental margin. These sediments form part of the West African Miogeocline and eventually spread onto cooling and subsiding oceanic crust which had been generated as Africa and South America spread apart. A constructive high energy depositional environment appears to have prevailed only since Miocene times. From Eocene to Early Miocene times three separate depocentres have been identified and lobate-elongate deltas (Figure 244) may have been formed. The Miocene – Present Day delta complex is a fine example of a high energy constructive lobate-arcuate delta in which the ratio of deposition (Rd) to the rate of subsidence (Rs) is considerably greater than 1 ($Rd/Rs>1$) although there were periods when $Rd/Rs\sim1$; (Figure 271). Since Miocene times the Niger-Benue Delta Complex has formed a united system. The delta complex is deformed by well developed growth faults and large scale mud diapirs and its growth is closely related to the development of the diapirs.

2. **Late Cretaceous – Palaeocene Post Santonian Anambra Delta Complex** which developed largely in the Anambra Basin situated on the northwest flank of the Abakaliki Fold Belt. This regressive offlap sequence, began to develop in Campanian times, prograded extensively in Maestrichtian times and and the delta development ended in Palaeocene time when the Imo Shale Palaeocene Transgression extended into the Lower Anambra Basin (Figure 85). The Anambra Delta Complex may be thought of as a Late Cretaceous – Palaeocene proto Niger – Benue Delta Complex which developed wholly upon continental crust. It was somewhat smaller than the Present Day Niger Delta Complex and does not appear to have developed large growth faults or mud diapirs as the Cenozoic Delta Complex did.

1. **Early to Late Cretaceous Pre-Santonian Transgressive – Regressive Complexes** which began with sediments being laid down during the Medial Albian Transgression in the Abakaliki Trough and adjacent flanking areas and in the Benue Depression. Three cycles have been recognized ending with the Coniacian – ? Early Santonian Regression. The Albian – Cenomanian Bima Delta Complex developed at the head of the Benue Depression some 600 miles from the present day Niger shore, as did various smaller deltas deriving sediments from the sides of the Benue Depression. Large thicknesses of sediment accumulated in the Benue – Abakaliki Troughs in Pre-Santonian times and in places exceed 15 000 ft.

Cretaceous formations belonging to the first two major depositional units have been described in Section 2.4 and the structural framework in which these rocks were laid down is described in Chapter 3.

The Cenozoic formations, which constitute the Niger Delta Complex in which most of Nigeria's oil and gas have been located to date, form the subject of this section. The stratigraphic framework in which the complex developed is described systematically below in Sections 2.5.5.2. – 2.5.5.4.

Published information about the lithological units

which constitute the Niger Delta Complex is derived from both surface and subsurface exposures and as often is the case when this happens and because of confidentiality rules, methods and motives for publication, the different methods by which the basic data has been obtained etc., considerable problems arise in trying to reconcile the two kinds of data. The two different approaches to publication of data held by the Geological Survey of Nigeria, (the official government body for publication of maps, memoirs etc.) and the Federal Ministry of Petroleum (and now NNPC) and the various oil companies are in many ways irreconcible especially in the current political, geopolitical and economic environment prevailing in Nigeria.

Knowledge about surface formations derives mainly from field surveys undertaken by the Geological Survey of Nigeria and by oil companies, mainly Shell-BP Nigeria, who at one time held all of colonial Nigeria under concession (Figure 298). Other companies such as Gulf, Esso and Mobil have completed considerable amounts of geological and geophysical exploratory work in Nigeria but relatively little of their findings have been published.

Shell-BP Nigeria and the Geological Survey of Nigeria in 1957 published 1:250 000 Scale Geological Sheets covering the Niger Delta, Dahomey Basin, Anambra Basin and the Lower Benue Depression (Figure 89) and much of the information incorporated in these geological sheets has been assimilated on Dessauvagie's 1972 Edn and 1974 Edn Coloured Geological Map of Nigeria, presented here in part as an uncoloured map (Figure 9).

Unfortunately these Shell-BP and Geological Survey of Nigeria maps were not accompanied by memoirs and the numerous cartographic, stratigraphic, nomenclatural and structural problems which have arisen because of the choice of lithological units, the ways in which the units were mapped, correlations of surface units and subsurface units etc. have remained unanswered for twenty years (Figure 90). Additional stratigraphic nomenclatural problems have arisen because of the ways in which Reyment (1965) named but did not map certain units (e.g. the Ogwashi – Asaba Formation); and because of the different names adopted by Dessauvagie (1972 and 1974) and Allen (1965) for similar lithological units of mainly Recent and Pleistocene age.

Problems have arisen also in determining the surface limits of the Present Day and Pleistocene portions of the subsurface units designated Benin, Agbada and Akata formations. Something of the complexity of the stratigraphic-cartographic problems existing can be gained from studying Figure 90 in which the terminology proposed by Dessauvagie (1972 and 1974) for the surface Cenozoic deposits of the Niger Delta Complex is compared with that proposed by Shell-BP and the Geological Survey of Nigeria (1957) and Short and Stauble (1967) and from studying the distributions shown on Figures 9 and 92. Dessauvagie's 1:1 000 000 Geological Map and Correlation Chart (1974) appears to be the best

compromise available to date.

Very little specific subsurface data, i.e. detailed well data, logs, cross-sectional data has been published for the Niger Delta Complex and only a few regional structural sections of the delta complex showing how surface formations actually pass down dip into subsurface formations and how the surface and subsurface units are disposed structurally have been published (Evamy et al. 1978). In fact, frequently oil company working sections only show sectional data to a mean sea level datum, so effectively avoiding the problem of surface-subsurface correlation.

As is often the case with a developing oil province much more information is held "confidential" by oil companies and by government bodies such as the Federal Ministry of Petroleum and NNPC than has been published or is available to the Geological Survey of Nigeria, or to individual investigators and so the pictures which have emerged showing the stratigraphic and structural evolution of the Niger Delta Complex are of necessity sometimes generalized and sketchy ones. To some extent the scarcity of published data is understandable because of confidentiality rules, strategic and economic value of data and of course the competitive proprietary nature of the basic information. Students of deltas can only hope that someday a little more of the great wealth of geological and geophysical data available for the Niger Delta Province will be published. The Niger Delta certainly stands among the world's best studied delta complexes but except for its superficial deposits it is among the poorest known publicly in terms of specific detail. In this respect it is important to note that since the Shell-BP 1:250 000 geological maps were published in 1957, that the Geological Survey of Nigeria has not published any geological maps of the delta province; and in fact the most comprehensive map available to date, Dessauvagie's 1:2 000 000 Geological Map (1974) was compiled privately at Ibadan University and published. Shell-BP Petroleum Development Company's account (Evamy et al. 1978) dealing with the hydrocarbon habitat of the Tertiary Niger Delta however has done much to unravel the mysteries of the delta dynamics and petroleum generation, migration and accumulation and this paper must become a classic among the world's delta complex studies.

Three main formations have been recognized in the subsurface of the Niger Delta Complex (Short and Stauble 1967; Frankl and Cordry 1967 etc.). These are the Benin, Agbada and Akata Formations. The three formations were laid down under continental, transitional and marine environments respectively (Short and Stauble 1967). The Benin Formation was deposited in a continental-fluviatile environment and mainly consists of sands, gravels and back swamp deposits which vary in thickness from 0 to 7000 ft. The Agbada Formation was laid down in paralic, brackish to marine fluviatile, coastal and fluvio-marine environments and consists of interbedded sands and shales. Many sub-environments have been recognized within these major units. The

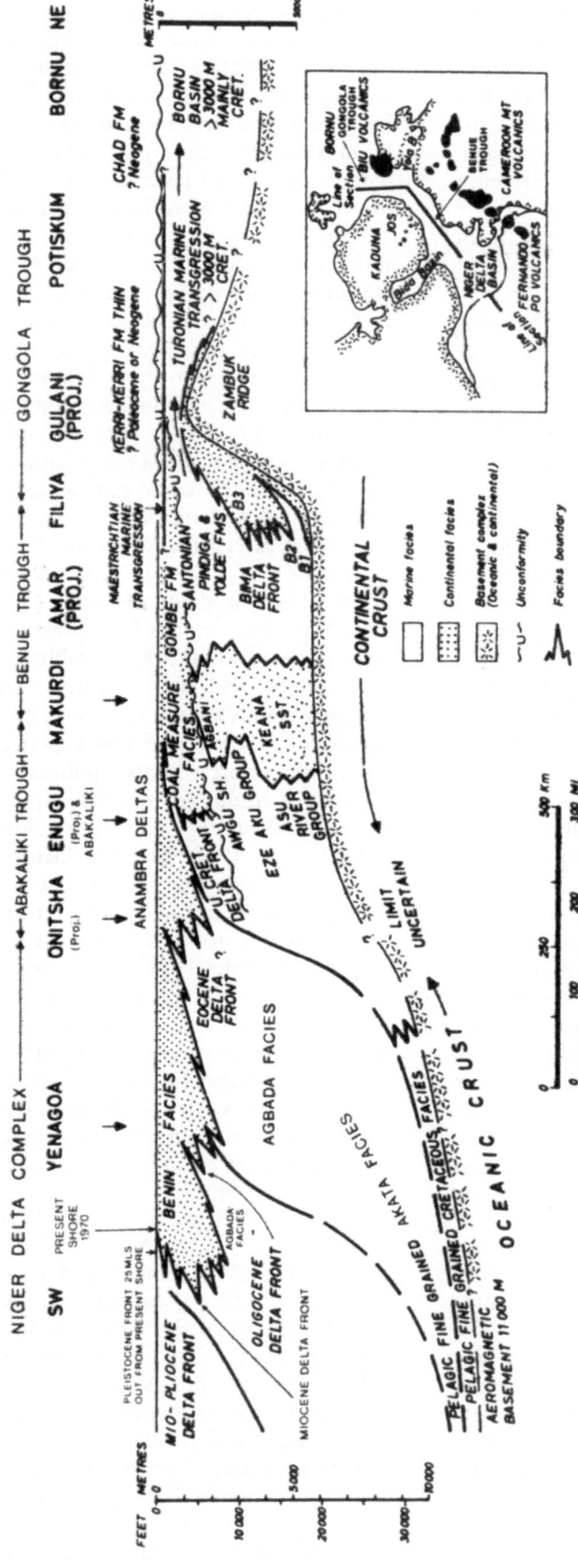

Figure 88. Reconstructed section, highly schematic, Niger Delta Complex, Abakaliki Benue and Gongola Troughs showing evolution of delta complexes. Based on Whiteman 1973.

Figure 88. Reconstructed section, highly schematic, Niger Delta Complex, Abakaliki Benue and Gongola Troughs showing evolution of delta complexes. Based on Whiteman 1973.

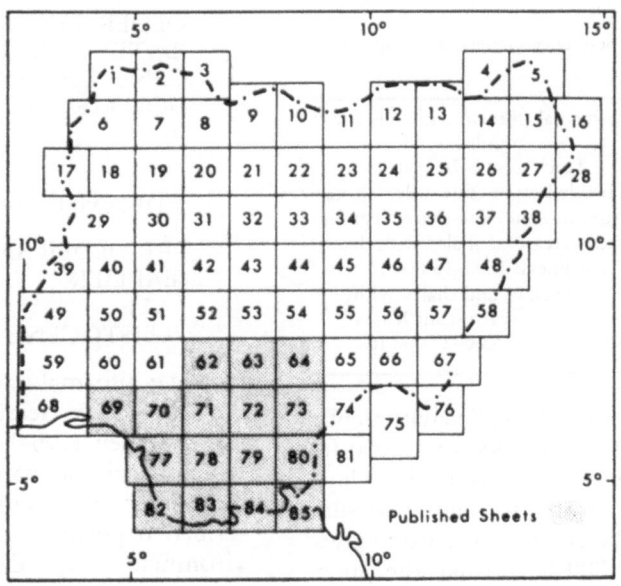

Figure 89. Index map showing coverage 1 : 250 000 Shell–BP Nigeria and Geological Survey of Nigeria Geological Sheets 1957 Edition, Niger Delta Complex Anambra Basin, Abakaliki Trough, Afikpo Syncline, Calabar Flank and Lower Benue Trough.

	SHELL–B P & GEOLOGICAL SURVEY OF NIGERIA 1957			DESSAUVAGIE 1972 & 74		SHORT & STAUBLE 1967	
HOLOCENE	13 Alluvium in General	UPPER DELTAIC PLAIN	—	Alluvium			Includes 13a & 13b in O/R Benin Formation
	13a Meander Belts		Qmb	Meander Belt Deposits			
	13b Wooded Back Swamps Fresh Water Swamps		Qfs	Fresh Water and Back Swamp Deposits			
	13c Mangrove Swamps	LOWER DELTAIC PLAIN	Qm	Mangrove Swamp Deposits			R Agbada Formation (Recent Outcrop)
	13d Lagoonal Marshes			Not Differentiated Including — Alluvium			Includes 13c, 13d & 13e in Agbada
	13e Abandoned Beach Ridges			Abandoned Beach Ridges			
EARLY HOLO TO LATE PLEISTOCENE	12,13 Sombreiro – Warri Deltaic Plain Deposits Invaded by Mangrove Swamps		Qd	Deltaic Plain Deposits			Includes 12/13 as O/R Benin Formation
PLEISTO- OLIGOCENE	12 Coastal Plain Sands		Tqb Too	Benin Sands Ogwashi – Asaba Formation	RECENT- OGLIO		O/R Benin Formation Ogwashi – Asaba Formation & Equivalents
U–M EOCENE	11,11a Bende Ameki Group Incl. Nanca Sands		Ta	Ameki Formation	EOCENE		Ameki Formation & Equivalents
L. EOCENE PALAEOCENE	10,10a Imo Shale Group Incl. Igbaku Sandstone Ebenebe Sandstone Umuna Sandstone		Ti	Imo Shale	L. EOCENE PALAEOCENE		Imo Shale & Equivalents

Compiled by A.J. WHITEMAN 1976

Figure 90
Comparison of Stratigraphic and Environmental Nomenclature of the Niger Delta Complex

Figure 91
Niger Delta Complex Offlap Sequence General Description

Benin Formation	Continental, fluviatile gravels and sands. Thickness: 0–7000 ft.
Agbada Formation	Interbedded sands (fluviatile, coastal and fluviomarine) and shales. Paralic, marked increase of shales with depth. Thickness: 1000–15 000 ft?
Akata Formation	Marine clays and shales with turbidite sands and silt lenses. Thickness: 2000–20 000 ft?

Agbada formation becomes much shalier with depth and varies in thickness from 0 to 1000 to 15 000 ft. The Akata Formation consists of marine silts, clays and shales with occasional turbidite sands and silts forming sinuous lenses. The Akata Formation varies in thickness from 0 to 20 000 ft and like the other two formations age varies from Palaeocene to Recent. Deposits belonging to these three formations are being laid down today and so thickness estimates should be viewed as arbitrary and are dependent on location.

The Benin Formation, and its equivalent, form extensive outcrops inland from the Agbada Formation and south of the outcrops of the Ameki Formation and Imo Shale (Figure 92). The Afam Clay Member has been differentiated within the Benin Formation. In addition to being a subsurface unit the Agbada Formation outcrops a few miles inland from the present shore, south of the outcrop of the Ogwashi–Asaba Formation–Benin Formation (O/R Figure 92).

The Akata Formation outcrops subsea on the delta slope and open continental shelf, and is not exposed onshore, unless we view the deeper water facies of the Imo Shale as Akata Formation laid down in the front of the Palaeocene Anambra Delta Complex.

Surface units mapped as part of the Niger Delta Complex and given "formal" formation names include: the Imo Shale Formation (Palaeocene); Ameki Formation (Eocene); the Ogwashi–Asaba Formation with its so called "Lignite Group" and the Agbada and Benin Formations (Figure 93).

Using less formal and more descriptive terms Dessauvagie (1974) has designated (Figure 9): Meander Belt Deposits (Qmb); Freshwater and Backswamp Deposits (Qfs); Mangrove Swamp Deposits (Qm); Abandoned Beach Ridges; and Deltaic Plain Deposits (Qd). The distributions of the units shown in Figure 9 are based mainly on Allen (1965). Correlations between subsurface and surface units according to Short and Stauble (1967) are presented in Figure 93.

Allen (1964, 1965 etc.) recognized the following deposits of Pleistocene and Holocene age in the surface Niger Delta:

YOUNGER SUITE (HOLOCENE)	Sands, silts and clays of modern delta and associated barrier island – lagoon and estuary complexes.
OLDER SANDS (LATE PLEISTOCENE TO EARLIER HOLOCENE)	Quartzose sands with shell debris and glauconite deposited during eustatic rise of sea level following the last glaciation

DEPOSITIONAL BREAK (REGRESSION)

PRE-OLDER SANDS (EARLIER THAN LATE PLEISTOCENE)	Plant bearing clays, silts and sands. Some soils, especially at top. Poorly consolidated.

He gave informal names to these units however and the deposits are clearly part of the Agbada and Benin Facies. Allen (1964) also named the "Lagos Sand" which despite its name occurs within the delta area. A number of present day subenvironments and described deposits being laid down within these environments were described by Allan (1964). The afore mentioned formations etc. are described below systematically under the headings: General Description; Thickness Variations; Age; and Conditions of Deposition.

2.5.5.2. Benin Formation (Subsurface)

2.5.5.2.1. General Description

The Benin Formation was first used by Parkinson (1907, p. 311) and fell into disuse. It was reinstated by Reyment (1965, p. 104) replacing "Coastal Plain Sands" which outcrop in Benin, Onitsha and Owerri Provinces and elsewhere in the delta province. Over the years the "Benin Sands Series" etc. acquired a variety of meanings some of which are listed below:

Parkinson (1907, p. 311)	*Benin Sands Series*
Tattam (1944, p. 33)	*Coastal Plain Sands*
Simpson (1955, p. 29)	*Coastal Plain Sands*
Reyment and Barber (1956, p. 42)	*Coastal Plain Sands*
Reyment (1965, p. 104)	*Basin Sands*

Needless to say a formation such as the Benin, deposited in a continental fluviatile environment, has a highly variable lithology and Short and Stauble (1967) of Shell-BP attempting to regularize stratigraphic terminology for the Niger Delta Complex within the terms of the Stratigraphic Code of the American Association of Petroleum Geologists, designated a type locality and type description. Parkinson (1907) did not designate a type locality.

Benin Formation

Type Section:	Elele–1, new field wild cat drilled 24 miles NNW of Port Harcourt.
Interval:	0–4600 ft below derrick floor
Location:	Lat. 5°4'12"N; Long. 6°50'4"E.
Derrick Floor Elevation:	98 ft above MSL
Total Depth:	10 635 ft.
Samples:	Ditch cuttings every 30 ft. Stored with Shell-BP Petroleum Development Company of Nigeria Ltd. and with the Geological Survey of Nigeria, Kaduna. Well drilled by Shell-BP Ltd.

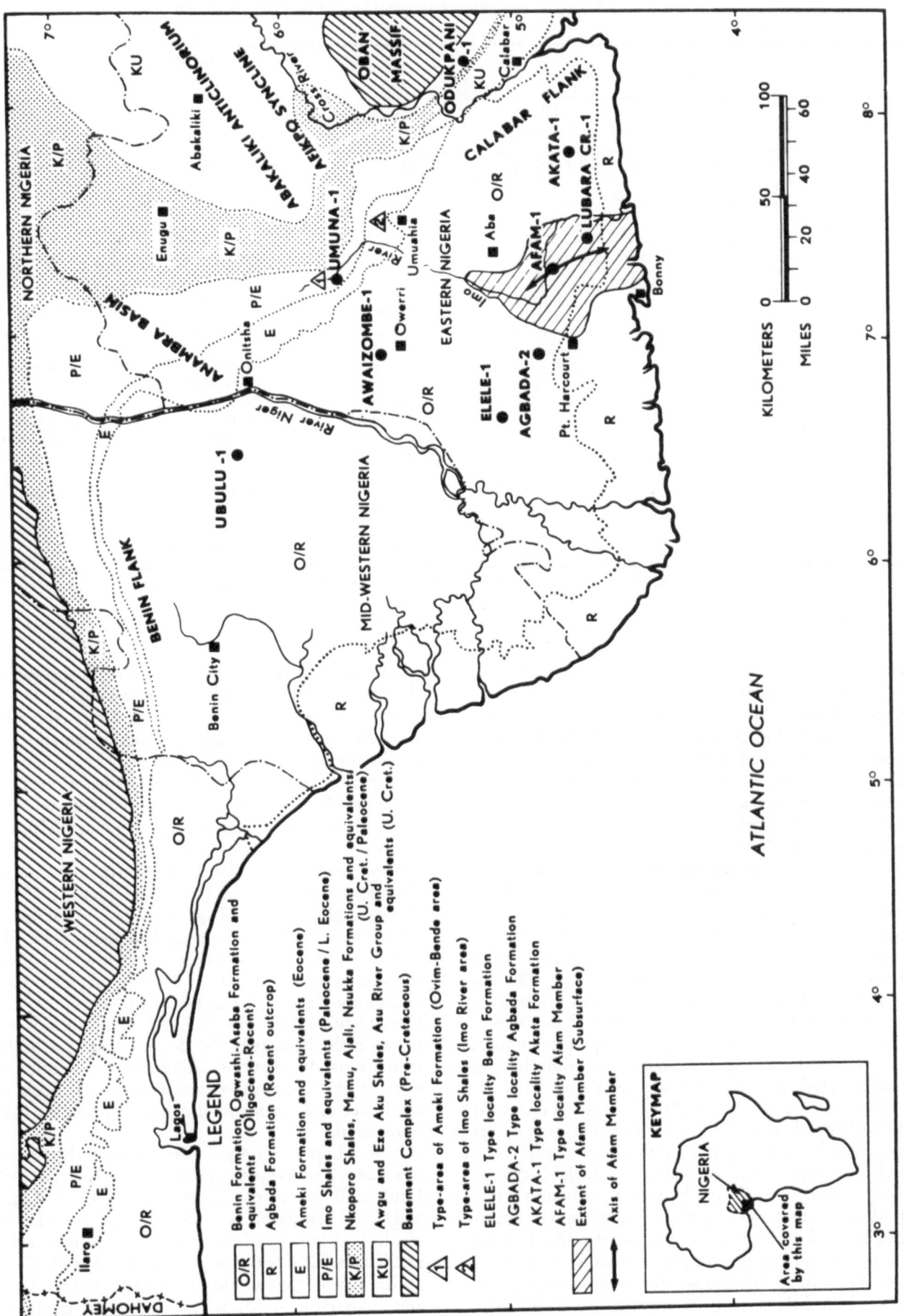

Figure 92. Geological Sketch Map Niger Delta Complex, Dahomey Miogeocline and Calabar Flank. Based on Short and Stauble (1967). Reprinted with permission of the American Association of Petroleum Geologists.

Figure 93
Correlations of subsurface and surface formations of the Niger Delta Complex

Subsurface			Surface Outcrops		
Youngest known age		Oldest known age	Youngest known age		Oldest known age
Recent	Benin Formation	Oligocene	Plio/Pleistocene	Benin Formation	Miocene ?
	Afam Clay Member				
Recent	Agbada Formation	Eocene	Miocene	Ogwashi-Asaba Formation	Oligocene
			Eocene	Ameki Formation	Eocene
Recent	Akata	Eocene	L. Eocene	Imo Shale Formation	Palaeocene
			Palaeocene	Nsukka fm	Maestrichtian
Equivalents not known			Maestrichtian	Ajali formation	Maestrichtian
			Campanian	Mamu Formation	Campanian
			Camp/Maest.	Nkporo Shale	Santonian
			Coniacian/ Santonian	Awgu Shale	Turonian
			Turonian	Eze-Aku Shale	Turonian
			Albian	Asu River Group	Albian

Based on Short and Stauble (1967) Reprinted with permission of the American Association of Petroleum Geologists.

The sequence encountered in Elele–1 is more than 90% sand and contains a few shaly intercalations. Shale content increases towards the base as is shown by the electric log (Figure 94). The sands and sandstones are coarse to fine grained and commonly granular in texture. Grains are subangular to well rounded. Pebbles are common and the sands and sandstones are poorly sorted. Parts of the section are unconsolidated. The sands and sandstones are white or yellowish brown because of limonitic coats. Lignite occurs in thin streaks or as finely dispersed fragments. Haematite and feldspar grains are common. Benin shales are greyish brown, sandy to silty and contain plant remains and dispersed lignite. Shales form only a very small part of the sequence. The top of the type section is at the surface and the base of the type section is at a depth of 4600 ft in Elele–1. The base is drawn arbitrarily at the top of the highest marine shale with foraminifera, the point having been selected from a gradational sequence.

The type section of the log of Elele–1 (Figure 94) shows a break at a depth of 3700 ft but only Short and Stauble (1967) assign significance to this. The highest marine shale at 4600 ft in Elele–1 can be recognized delta wide as a boundary, although within these boundaries there are fundamental stratigraphic problems in differentiation of units. In brief the formation can be recognized because of its high sand percentage (70–100%), few minor shale intercalations and the absence of brackish water and marine faunas. To date very little oil has been found in the Benin Formation (mainly minor oil shows) and the formation is generally water bearing.

In the eastern part of the Niger Delta, the Afam Clay Member of the Benin Formation separates the Benin Formation into two parts.

The lateral extent of the Benin Formation is shown in Figures 9, 92 etc. It occurs almost delta wide and has a seaward limit just within the Plio-Pleistocene Shore.

Short and Stauble (1967) did not describe general variations and Parkinson (1907), Reyment (1965) and Dessauvagie (1974) all gave general descriptions which presumably can be applied delta wide. The sediments are said to consist of yellow and white, sometimes crossbedded sands with pebbles. Clays and sandy clays occur in lenses (Reyment 1965 and Dessauvagie 1974). Neither of these two writers mention the lignites in the Benin Formation in their formational descriptions, although both mention lignites in the Ogwashi–Asaba Formation. Both disseminated and bedded lignites occur within the Benin Formation.

If the "Lignite Group", as mapped by the Geological Survey of Nigeria (in an Unpublished Report, Geological Survey of Nigeria 1971 by C. N. Okezie and S. A. Onuogo), is plotted on the Shell-BP and Geological Survey of Nigeria 1957 Edition Geological Sheets 1:250 000, the rocks of the Lignite Group apparently occur within the Bende–Ameki Group and "Coastal Plain Sands" as mapped i.e. the lignites are part of the Benin Formation. If this is accepted

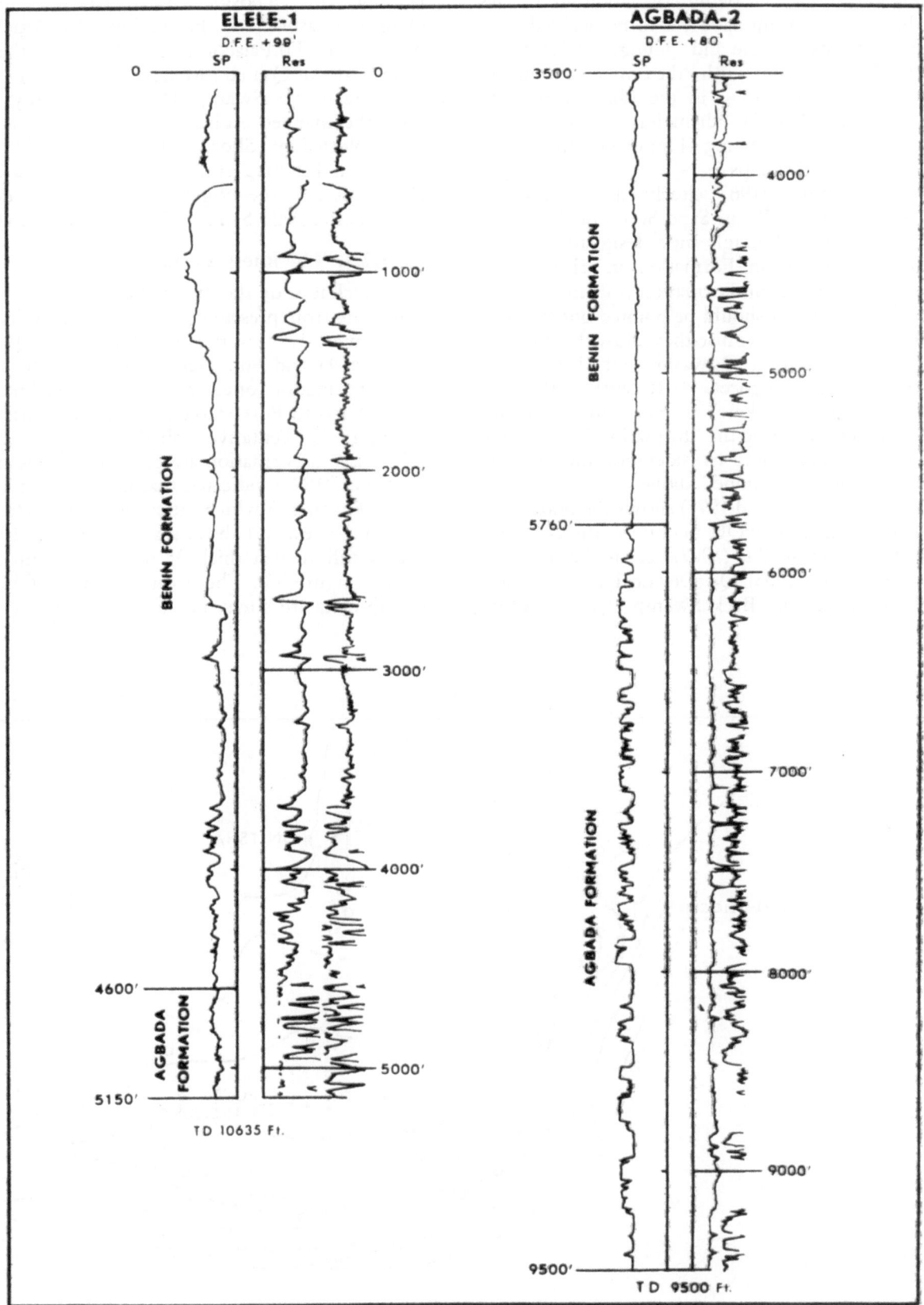

Figure 94. Summary SP-Logs, Resistivity logs showing petrophysical characteristics of the Benin Formation, type sections in Elele-1 and Agbada-2. Based on Short and Stauble 1967. Reprinted with permission of the American Association of Petroleum Geologists.

the relationships of the Benin Formation sand facies and the Lignite Group have yet to be decided even in general terms. Okezie and Onuogo (1971) correlated the Lignite Group with "the only oil producing strata" at present known in the Niger Delta Oil Fields i.e. the Agbada Formation. Cross-sections were not presented showing "Lignite Group"–Benin and Agbada relationships.

Short and Stauble (1967) clearly equate the Benin Formation with the Benin Sand Series of Parkinson (1907). However although they designated a type section for the Benin Formation in Elele–1 and claimed that "the Benin Formation is delimited here for the first time", it should be pointed out that this is not strictly correct because they show the limits of O/R (Benin Formation, Ogwashi–Asaba Formations and equivalents – Oligocene to Recent) on their locality and geological sketch map (Figure 92 this report) without separating the "formations" within the formation or taking into consideration the "Lignite Group" problem mentioned above.

Dessauvagie (1972 and 1974) shows the outcrop of the Benin Formation (TQb) in quite a different manner to Short and Stauble (1967) because he separates (following Allen 1965) Qd Deltaic Plain Deposits; Qfs Fresh Water and Back Swamp Deposits; Qmb Meander Belt Deposits; with the landward limits of (Tog) Ogwashi–Asaba Formation, (Ta) Ameki Formation and (Ti) Palaeocene Imo Shale and the down delta limits of Qm etc. Mangrove Swamp Deposits. The Mangrove Swamp Deposits roughly equate with the mapped Agbada Formation (Recent outcrop) plotted by Short and Stauble (1967). Murat (1969) on his map simply generalized the surface distribution of the Niger Delta Complex deposits into Coastal Plain Sands (12) and Alluvium (13).

2.5.5.2.2. Thickness Variations

The thickness of the Benin Formation is variable extending from present day depositional limits where it is obviously very thin to around 6000–10 000 ft (Figures 95 and 96). Generalized isopachs for the Benin Formation for an area seawards of a line drawn from Warri to Port Harcourt to Calabar are shown in Figure 95. Tentative isobaths drawn on the base of the Benin Formation are presented in Figure 96.

Dailley (1976) indicated that in the Benin area the Benin Formation which outcrops at the surface situated only a few feet above sea level exceeds 10 000 ft and is still 1000 ft thick 35 miles out from Bonny Beach (Figure 96). The great thickness of Benin in the up-delta areas may be due to two causes:

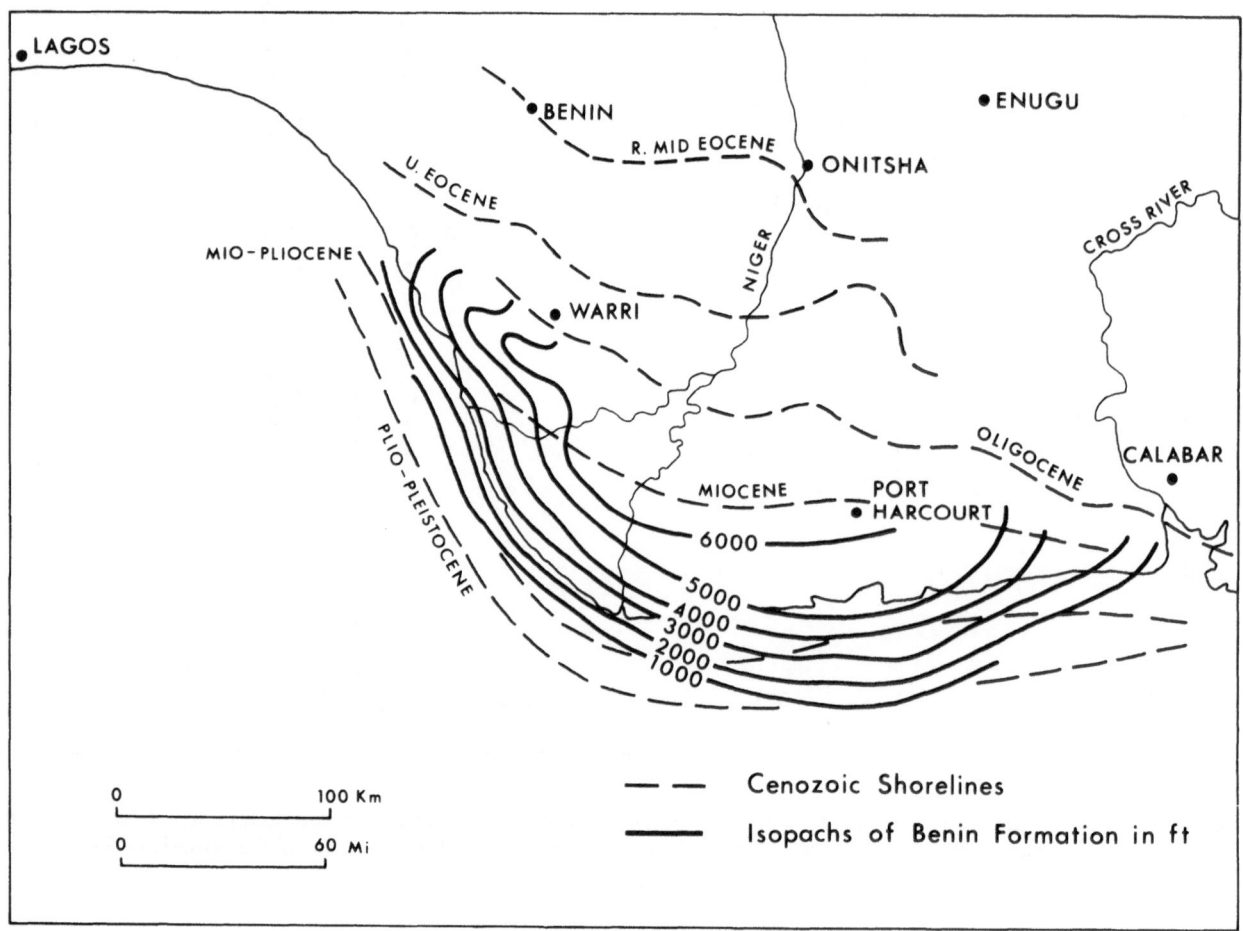

Figure 95. Map showing isopachs of the Benin Formation. Based on Knapp (1971).

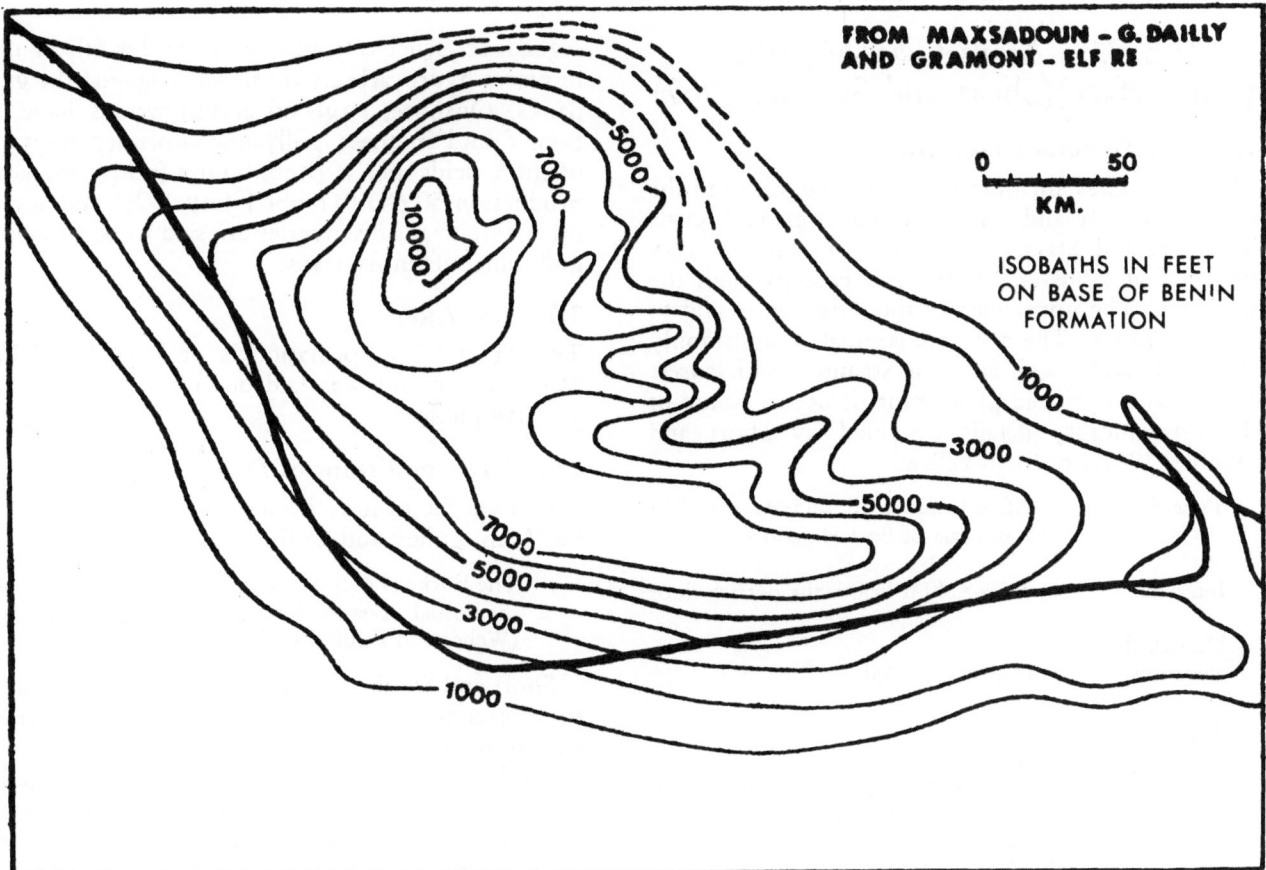

Figure 96. Map showing approximate isobaths on the base of the Benin Formation. Based on Dailly (1976).

1. greater subsidence of transitional oceanic to continental basement compared with less subsidence of continental crust and/or
2. mass seaward movement of the Akata Shale to form the great diapir zone so creating a buoyant frontal zone to the delta in which sedimentation is and was less than in the subsiding areas up-delta.

Knaap's map (Figure 95) shows that the Benin Formation is around 1000 ft thick some 10 miles north of the southern boundary of OPL 90 (Occidental's relinquished permit, 1976) more than 30 miles out from the Present Day Shore, implying that the Plio–Pleistocene shore line position plotted by Short and Stauble (1967) some 25 miles out from the mouth of the Bonny River, must be situated seaward of this 30 mile from shore 1000 ft Benin Sand isopach. Delta growth in the Miocene must have been much faster along the southern face of the Niger Delta Complex than off the southwestern face of the Delta westwards of Nun River where the position of Plio-Pleistocene shore plotted by Short and Stauble (1967) lies some 10 miles outside the 1000 ft Benin Formation isopach plotted by Knaap (1971).

2.5.5.2.3. Age

In the type section of Elele-1, the highest dated shale in the sandy Benin sequence has yielded an Early Miocene assemblage and Short and Stauble (1967, p. 767) gave the total span of the formation as Mio-

cene to Recent. However in Short and Stauble (1967, Figure 1) (Figure 93) the age of the Benin Formation is given as Oligocene to Recent. In a constructive high energy delta environment Benin facies could have been generated as soon as delta formation started in the Eocene (Chapter 3).

2.5.5.2.4. Conditions of Deposition and Palaeogeography

As we have described above clearly there is considerable confusion as to what constitutes the Benin Formation overall. Different maps show different limits, and to some extent this has arisen because of the confusion which exists about the conditions under which the Benin Formation were laid down. Reyment (1965, p. 104) clearly thought that the Benin Formation is partly marine;

> The Benin Formation is thus partly *marine*, partly deltaic, partly estuarine and partly lagoonal and fluviolacustrine in origin

Whereas Short and Stauble (1967, p. 167) hold that the Benin Formation was laid down in a continental, probably upper deltaic environment. They picked the base at the appearance of the first marine incursion. Genetically the Benin Sands and sandstones are mainly deposits of the continental upper deltaic plain environment, according to current company terminology.

119

2.5.5.3. Afam Clay Member (Subsurface)(Short and Stauble 1967)

2.5.5.3.1. General Description

The Afam Clay member has been recognized in the subsurface only and was described in general terms by Short and Stauble (1967) as consisting of clay with a few sandstone bodies scattered through the sequence. The upper part of the sequence is said to be more sandy. The clays are laminated, commonly silty and sandy with siltstone stringers and lenses. Lignite and pyritized plant remains occur. Some of the type locality details provided by Short and Stauble (1967) are listed below:

Type Section:	Afam I, new field wildcat, drilled 22 miles east of Port Harcourt
Interval:	3420–5540 ft below the derrick floor.
Derrick floor elevation:	86 ft above MSL
Total Depth:	10 207 ft
Samples:	Ditch cuttings every 20 ft. Stored with Shell-BP Petroleum Development Company of Nigeria Ltd. and Geological Survey of Nigeria, Kaduna. Well drilled by Shell-BP Ltd.

The distribution of the Afam Clay Member in the subsurface is shown diagrammatically in plan and section in Figures 92 and 97. It is restricted to the subsurface of the south east part of the delta in the lower Imo River area.

The Afam Formation appears to have been deposited in a gully or valley. The base is said to be marked by an unconformity which overlies the lower part of the Benin Formation. A basal sand bed occurs just above the unconformity. The top of the Afam Clay Member is drawn at the top of the uppermost clay intercalation in a sequence which grades up into the Upper Benin Formation. On electric logs, the Afam unit is recognized as the first continuous shale with occasional sand bodies within the Benin Formation over much of the area. The Afam is the most extensive of a group of "clay fill" structures known from the Niger Delta Complex. It occupies an area of around 1350 square miles in the southeastern part of the delta complex and the gully in which it was deposited appears to have been more than 20 miles wide and over 60 miles long. The base of the formation is unconformable and sharp where it rests on the Benin Formation but is more difficult to determine where it contains basal sand intercalations and rests on the underlying paralic Agbada Formation.

2.5.5.3.2. Thickness

The formation thickens from north to south where it is over 2500 ft thick in Lubara Creek. It thins rapidly east and west away from the channel axis. It is interesting to note in passing that where the paralic Agbada sequence is truncated by the clay filled gullies in places hydrocarbons are trapped against the gully flanks. Onshore, large oil accumulations have not been found in these gully-unconformity traps but offshore fields such as Mobil's Ubit Field (cumulative production 50 683 011 bbl 1 July 1976) and several other important discoveries are said to be located in gully unconformity traps.

2.5.5.3.3. Age

The Afam Clay Formation is of Medial and Late Miocene age and was laid down during the transgressive phase.

2.5.5.3.4. Conditions of Deposition

Two hypotheses have been proposed to account for the Afam gulley and its fill:

1. The Burke (1972) submarine canyon hypothesis; and
2. the initial river valley – transgression hypothesis (Weber and Daukoru 1975)

Short and Stauble (1967) suggested that deposition took place in a very restricted marine to transitional environment which does not have a modern counterpart in Nigeria. Burke (1972) thinks that gullies were formed at the head of submarine canyons and that the fills are essentially upper canyon deposits. He holds the view that the gullies are located where they are because of the divergent longshore drift pattern caused by the southwesterly wind generating drift away from the delta nose (Figure 142).

Weber and Daukoru (1975) do not appear to agree with Burke's explanation and have postulated that the Afam Gully was formed during a major transgression and that initially existing river valleys were responsible for the locations of the gullies. They stated that there are possibly five or more clay filled gullies cut into the paralic and continental facies of the delta complex in the southeastern part of the delta. Names have only been given publicly to the Afam Clay Member and the Kwa Ibo Clay Member, although the latter is undescribed. Canyon fills have been described from Mobil's MD-1 and from Shell-BP's Opobo South-1 as the Kwo Ibo and Opobo Members of the Agbada Formation (Frankl and Cordry 1967). The gullies are said to be of Miocene age and to be entrenched first into continental sediments and then to the south into paralic sediments. Like the Afam Gully the fill is said to be clay (Weber and Daukoru 1975).

2.5.5.4. Agbada Formation (Subsurface)

2.5.5.4.1. General Description

The Agbada Formation (Short and Stauble 1967; Frankl and Cordry 1967) underlies the Benin Formation and forms the second of the three strongly diachronous Niger Delta Complex formations (Figure 246B).

As with the Benin Formation, Short and Stauble

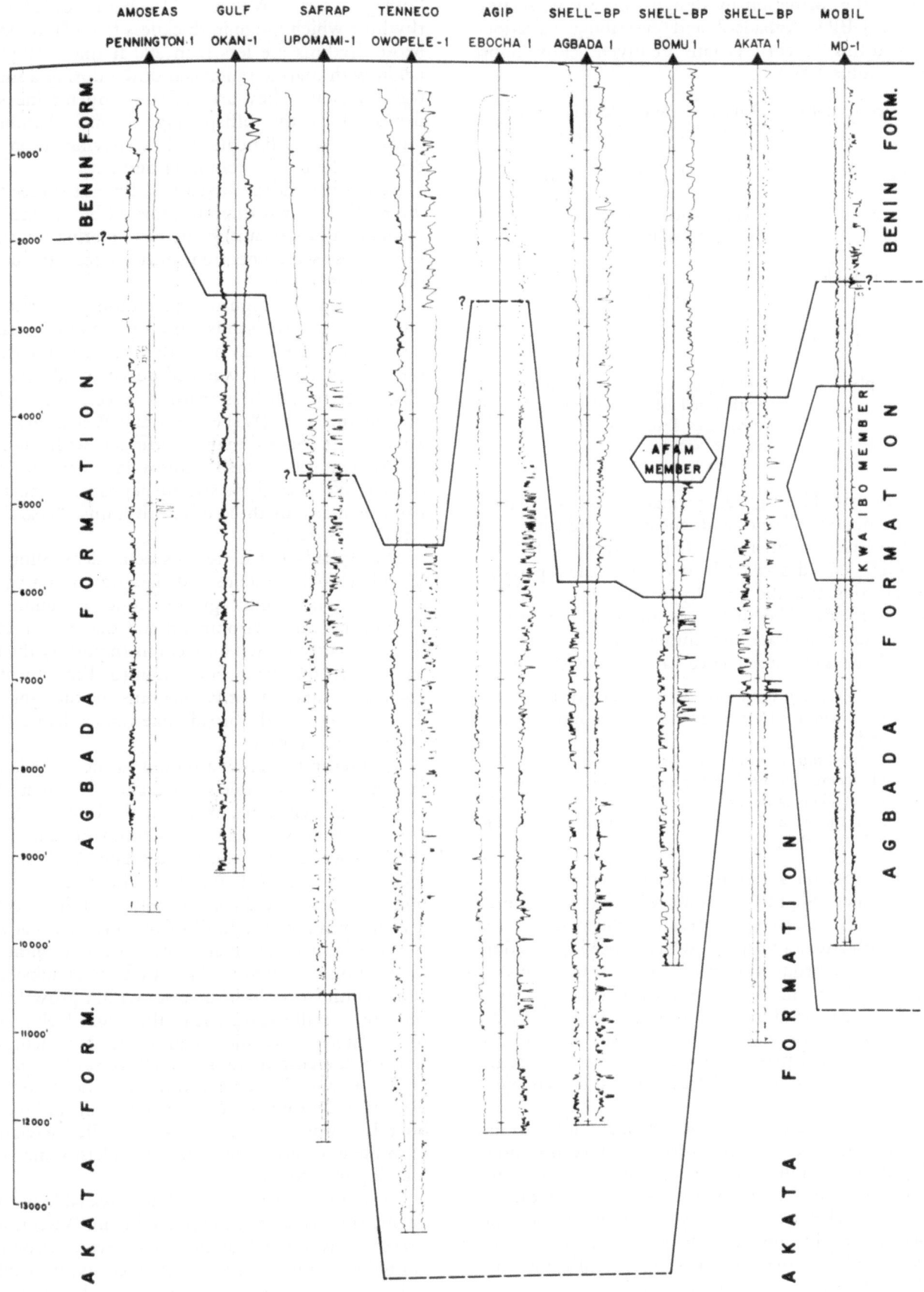

Figure 97. Generalized electric logs showing the relationship of the Afam and Kwa Ibo Members, Niger Delta Complex. Based on Frankl and Cordry (1967).

(1967) decided to deal with lithological variability within the formation by designating a type section in Shell-BP's Agbada-2 and describing variations around this "norm". Details are given below (Short and Stauble 1967);

Type section:	Agbada-2, new field wildcat, drilled 7 miles NNW of Port Harcourt
Interval:	5760–9500 ft below derrick floor
Location:	Geographical coordinates 4°55'39.94" N; 7°1'50.92"E
Derrick Floor Elevation:	80 ft above MSL
Total Depth:	9500 ft
Samples:	Ditch cuttings every 15 ft. Stored with Shell-BP Petroleum Development Co. Ltd. and Geological Survey of Nigeria, Kaduna.

Lithology: The Agbada Formation consists mainly of alternations of sands and sandstones and siltstones. It consists of numerous offlap rhythms (ABCDE, ABCDE etc. described below), the sandy parts of which constitute the main hydrocarbon reservoirs in delta oilfields. The shales constitute seals to reservoirs and as such are very important

In Agbada-2 the sequence has been divided into:

2. *an upper unit* consisting of sandstone-shale alternations with the former pre-dominating over the latter; and

1. *a lower unit* in which the shales predominate and in places are thicker than the intercalated sandstones or sands. The sandstone percentage ranges from 75% near the upper limit of the formation to 50% and below in the lower part of the unit.

The sandstones or sands are very coarse to very fine grained. Slightly consolidated sands have a predominantly calcareous matrix but the majority are unconsolidated, a property which caused great completion and production problems in early days of field development. The sands are often poorly sorted except where sand grades into shale. Lignite streaks and limonite are common but limonite coated sand grains and feldspars, which characterize the Benin Formation are rare. Shell fragments and glauconite occur.

The shales are denser at the base than higher in the column, because of normal processes of compaction. The shales overall have been described as dense and at 12 000 ft bulk shale densities of 2.6 g/cc can be expected. The shales are grey in colour and become silty and sandy upwards. Shaliness increases downwards and the formation passes gradually into the Akata Shale Formation (described below).

The structural elevation of the base of the Agbada Formation fluctuates widely throughout the delta because of synsedimentary diapirism largely within the Akata Shale and growth fault development.

The paralic Agbada sequence consists of a series of "offlap cycles" (Weber 1971), better called offlap rhythms which range in thickness from 50 to 330 ft. Most rhythms are less than 200 ft thick. Rhythms begin with marine sands laid down during a marine transgression. They are followed by marine shale deposited as the offlap stage begins. Laminated fluviomarine sediments follow having been laid down in the barrier foot environment.

Barrier bar and/or fluviatile sediments succeed and the rhythm unit is often terminated by marine sediments laid down during the next marine transgression. These sediments frequently truncate the barrier sand deposits.

As is shown in Figure 98 in addition to barrier bar sands, point bar sands and distributory channel sands, sandy beds can be distinguished in the rhythmic sequence which was laid down as tidal channel fills, river mouth bars, natural levées, and shallow marine sand bars (Weber 1971 and Weber and Daukory 1975). Sands may interfinger with lagoonal clayey, deposits, marsh, oxbow lake deposits, tidal flat deposits, mangrove swamp deposits. In the upper parts of many of these offlap rhythms, "coaly" intercalations occur.

The Agbada Formation, as defined by Short and Stauble (1967), contains beds laid down in a variety of sub-environments groups together under the heading paralic environment; and the zig-zag facies line frequently shown on sections to portray the limits between the Benin Formation (or Facies) and the Agbada Formation (or Facies) must be thought of as demarcating a fairly broad zone rather than a sharp depositional contract.

At outcrop the landward limit of the Agbada Formation on Dessauvagie's (1972 and 1974) map (Figure 9) is drawn more or less at the limit of Allen's (1965) mangrove swamp environment (marine to brackish water) and Short and Stauble's (1967) (R-) Agbada Formation (Figure 92). The position of this boundary has shifted considerably with time because it roughly marks the limit of the intertidal flat zone (i.e. the limit of the marine brackish environment). Its position is related to the net effects of subsidence and accumulation in local areas within growth fault cells; to overall subsidence of the Niger Delta Complex related to oceanic – continental basement subsidence, and eustatic rises and falls of sea level during Holocene and Pleistocene times and delta complex diapirism. During the Wisconsin low of sea level the boundary was situated some 40–50 miles seaward of its present position near the Plio-Pleistocene shore line (Figure 246A).

It is not clear from data publicly available whether during the overall rise in sea level from Wisconsin to Present Day times that the mangrove environment existed as a continuous migrating entity; probably it did not, so we should not expect to find a continuous sequence of Agbada deposits extending from the Pleistocene shore to the Present Day Shore. Where the boundary of the mangrove zone lay during Pleistocene highs of sea level is not clear and there may well be Agbada deposits defined in terms of the mangrove limit up delta and beneath the margin of the Benin Continental fluviatile sand facies.

On the seaward side, the present day limit of the Agbada Formation and the Agbada facies lies in front of the delta at the landward edge of Holomarine environment of the open continental shelf, where it meets the pro-delta slope deposits (Weber 1971). Fine silts and clays and admixtures predominate in this environment. The transition from Agbada (sand-shale facies) to Akata (shale) facies takes place on the outer delta slope of the Present Day delta front. Again the limit is best thought of as fairly broad zone rather than as a line. As we have pointed out its Present Day position is anomalous because of the Late Pleistocene – Recent rises and falls in sea level and overall Late Pleistocene – Recent marine transgression. In the absence of borehole data from the outer shelf where the Plio-Pleistocene contact is located we can but surmise that the conditions of deposition were similar then to those prevailing now and that the facies and lithofacies units and their distributions were similar also. The Pre-Pleistocene limits of the Agbada Formation drawn say on the top Pliocene depositional surface might have occupied a belt some 25–40 miles wide if we are to judge from the width of the belt in which the Present Day Agbada Facies is being laid down. Sectional data available on the onshore delta indicate that 10–30 miles deep indentations of the lower boundary of the Benin Formation have taken place in the Pre-Pleistocene section of the Niger Delta Complex, so indicating that periods existed when net subsidence predominated and allowed the sea to flood the delta complex.

The Agbada Formation occurs almost delta wide beneath the Benin Formation. On the landward side of the Niger Delta Complex regional subsidence and warping have resulted in exposure and erosion of the older parts of the delta complex. The Ogwashi–Asaba Formation of Oligocene to Miocene age and the Ameki Formation of Eocene age differentiated at outcrop probably pass into the Agbada Formation in the subsurface (Figure 93). The two fold division of the Agbada Formation into an upper sandy, sand-shale alternation predominantly unit and a lower shaley unit (mentioned above) probably corresponds roughly with Ogwashi–Asaba and Ameki divisions at outcrop (Short and Stauble 1967). However, these two divisions are "up dip delta" units and cannot be differentiated delta-wide because the Agbada Formation becomes more marine down-delta; it is for this reason (among others) that only one subsurface formational name Agbada is employed.

2.5.5.4.2. Thickness Variations

The top of the Agbada is drawn on the highest occurrence of shale yielding a brackish or marine fauna (Short and Stauble 1967); consequently different tops may be picked in different places and so the position of the formational boundary may vary considerably being dependent on the recognition of brackish water

or marine microfossils in ditch cuttings. Preservation, collection, treatment, presence or absence of microfossils etc. will all affect the "pickability" of tops for the formation. However, in practice the formation has been defined lithologically using wire line log data etc. at the incoming of the first major sand of the Benin Formation. The formational boundary then is in reality a broad belt rather than the crisp arbitrary line shown on sections.

In the type section in Agbada-2 the base of the Agbada Formation is not exposed. However by convention the base of the formation is taken at the first significant shale body i.e. the top of the Akata Shale. The log "signature" of the lower Agbada Formation in Akata-1 is shown in Figures 94 and 97.

The thickest known section of the Agbada at the time Short and Stauble (1967) presented their description was around 10 000 ft. Merki (1970) gave 15 000 ft as a maximum thickness and Weber and Daukoru (1975) gave a range of thickness of 9600 ft to 14 000 ft. Obviously thickness will vary from place to place dependent on structural and depositional position and criteria adopted for definition. Isopach data have not been presented for the Agbada Formation largely because of the considerable synsedimentary deformation it has been subjected to as a consequence of Akata Shale diapirism and growth faulting and the time transgressive nature of the formation. The formation is thinnest on the present day shelf because of well developed Akata diapirism. It thickens inland so that the Agbada is thickest beneath the zone where the Benin Formation is thickest. The zone of maximum subsidence (Benin, Agbada and Akata) overlies the transition zone between oceanic and continental crust (Figure 246). The thickest section is probably of Miocene-Pliocene age and thins up delta into the Oligocene and Eocene sediments where the complex is developed over Cretaceous sediments resting granitic crust.

2.5.5.4.3. Age

The Agbada Formation gets younger down delta from NE to SW in general terms. It is of Eocene age in Awaizombe-1 and Ubulu-1 and Agbada facies is being laid down at the Present Day on the inner conshelf and within the landward limits of the mangrove swamp interval brackish environment. Positions of the Late Eocene and Medial Eocene shore lines are shown in Figure 244B.

2.5.5.4.4. Conditions of Deposition and Palaeogeography

The Agbada Formation as defined by Short and Stauble (1967) was laid down under a variety of environmental conditions and the composition of an Agbada rhythm encountered in a well will depend mainly on the location of the well with respect to positions of the coastlines prevailing at the times of deposition considered, growth faulting, regional subsidence etc. As we have pointed out deposition locally is controlled among other things by the rate of subsidence within a growth fault cell. In wide cells, wells drilled

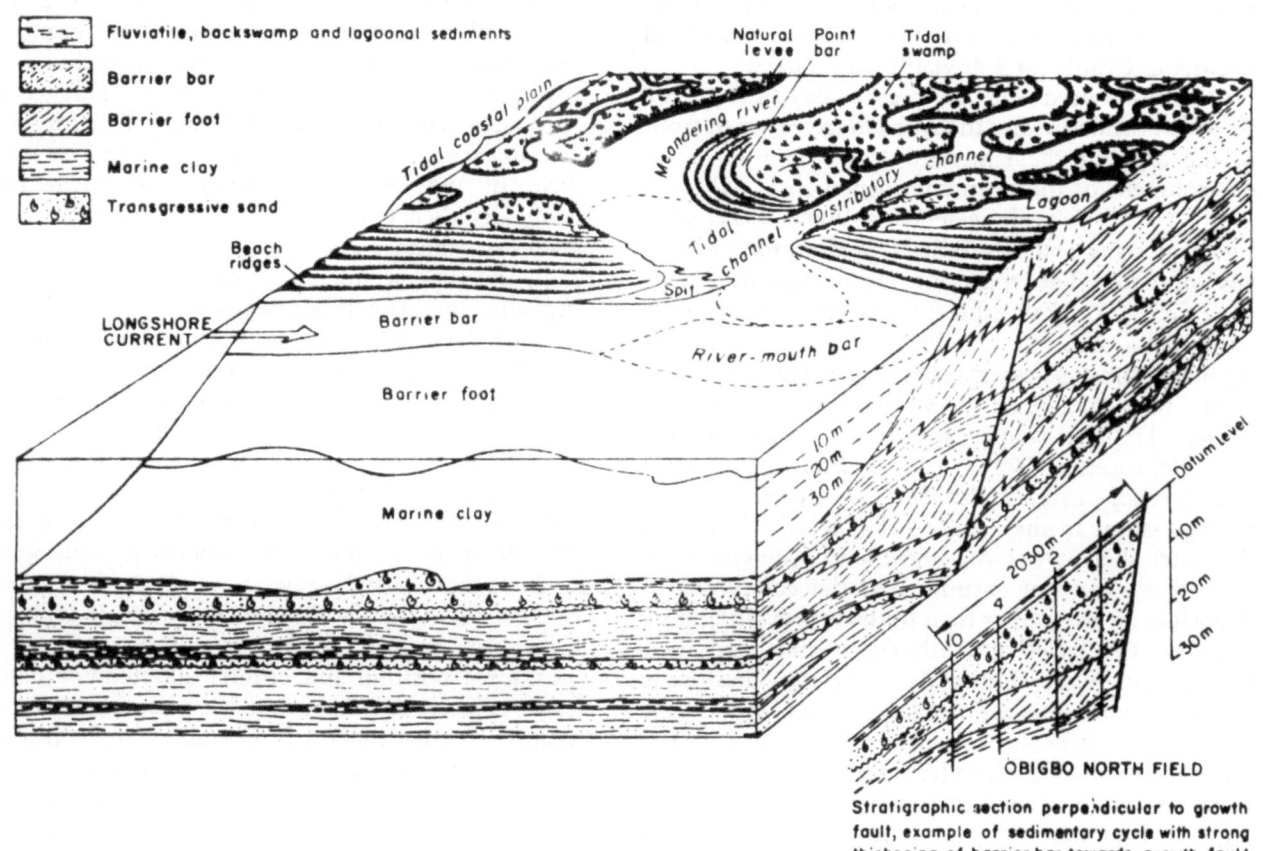

Figure 98. Block diagram showing the distribution of rhythmic units (transgressive and offlap sequences) environments of deposition within a developing growth fault cell. Note the marked thickening of the sedimentary rhythm into the Obigbo North Field growth fault. Based on Weber (1971).

across the cell will tend to show lateral facies changes better than wells drilled across a narrow growth fault cell.

Gross formational details are of general interest but if we are to understand variation within Agbada reservoirs we must study the details of the rhythmic sequences which constitute the Agbada Formation because the separate sections of these rhythms were laid down under a variety of conditions.

The facies distribution within the F-1 sands of the Bomu Field are shown in Figure 99 (N–S section, length of 5 km). Four rhythms are recognizable within this section. A transgressive sand is developed only in the topmost cycle. The two lower cycles are only partly developed with the field area. A basal conglomerate of clay ironstone pebbles is developed only at the top of the third cycle.

A typical offlap rhythm is illustrated in Figure 100 and consists of:

TRANSGRESSIVE MARINE SAND (ONLAP)

E. Fluviatile backswamp and lagoonal sediments (offlap)
D. Barrier bar and bar foot deposits (offlap)   Typical
C. Laminated fluvio marine sediments (offlap)   Agbada   ABCDE
B. Marine shale (offlap)   Rhythm
A. Transgressive marine (on-lap) sand

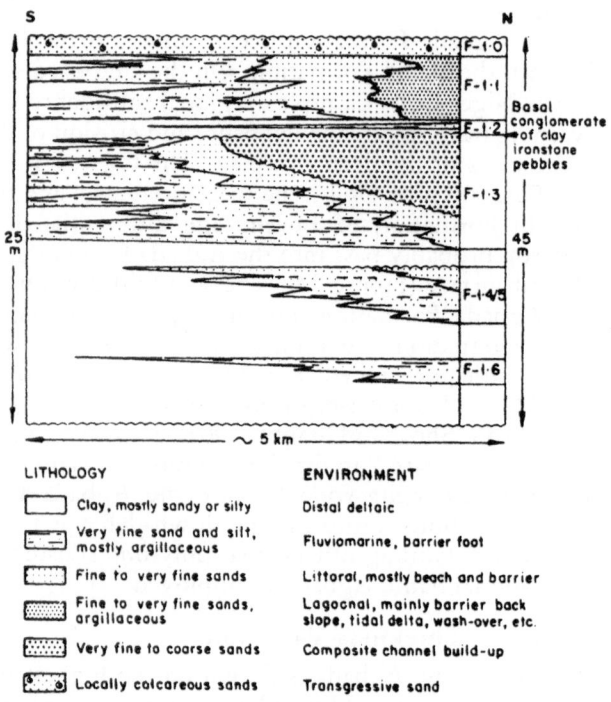

Figure 99. Section of F–1 sands of the Bornu Field, Nigeria showing lithologies and environments of deposition of rhythmic units F–1.0 to F–1.6. Based on Weber (1971) after B.V. Rossum.

124

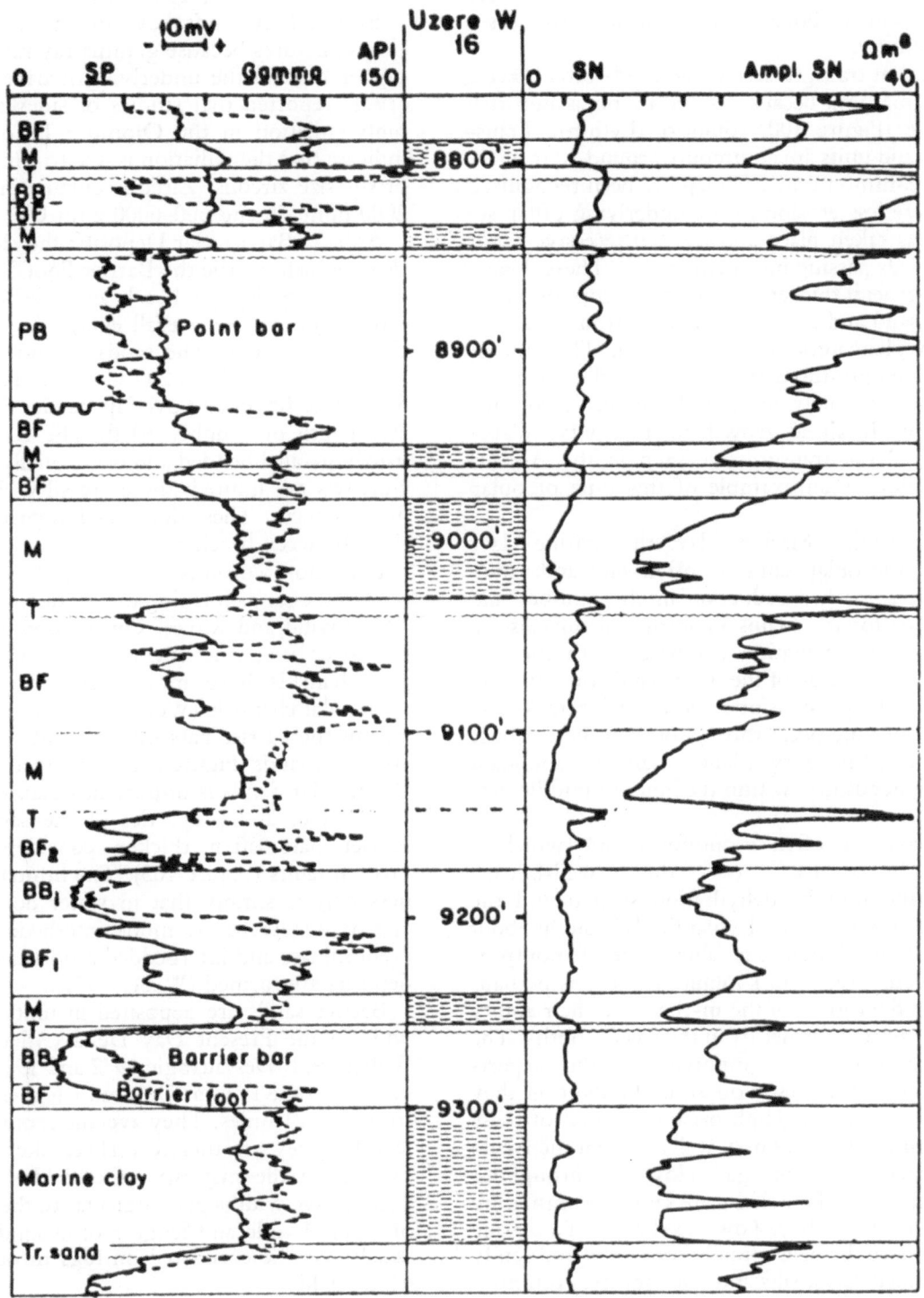

Figure 100. SP, Gamma Ray and SN responses of offlap sequences encountered in Shell-BP's Uzere west field. Based on Weber (1971).

Agbada Onlap or Transgressive Marine Sands: Most Agbada rhythms began with the erosion of underlying sand or silt units and as the sea transgressed a thin fossiliferous transgressive marine sand was formed. Transgressive sands were formed mainly by erosion and winnowing of the substratum and often contain shale fragments, foraminifera, bryozoa and fish otoliths as such deposits do today (Weber 1971). The onlap sands often contain burrows which extend down into the preceding offlap sequence. Glauconite is common, pointing to slow deposition and indicating that fluvio-marine sources of sediment were fairly distant and that overall deposition was slow. Transgressive sands were deposited in the Holomarine Zone in waters over 100 ft deep. As Weber (1971) pointed out these transgressive onlap

125

sands can be identified on logs because their pores are filled with carbonate cement which produces a high resistivity.

In addition transgressive onlap sands often gave a high gamma peak because caused by potassium rich glauconite (Figure 100, topmost rhythm). Transgressive sand units are commonly around 10 ft thick.

A less common type of onlap has been recognized also, where less erosion of the underlying offlap sequence has taken place and the transgressive sands fine upwards passing into marine clay. These deposits are in general thicker than the more common type of onlap sand and average around 30 ft in thickness. Streaks of glauconitic clay are common. They appear to have been produced in areas where the onlap has taken place gradually and may have been associated with strong localized growth fault activity. Weber (1971) cited the transgressive sand in the Agbada Field rhythm as an example of this type of onlap sand.

Agbada Offlap Marine Clay: the marine clays overlying the onlap sand are often silty and sandy and show a marked reduction in the number and diversity of fossil remains indicating an increase in the rate of sedimentation compared with the rate during the formation of the onlap sand. Burrows are common and layering is often destroyed. Streaks and lenses of fine sand occur throughout but there is little interconnected porosity. Plant remains are abundant indicating deposition within the inner to middle neritic zones.

The clays are often montmorillonitic which is important because of the well-known montmorillonite-illite dehydration system and the extrusion of pore and molecular fluids brought about with depth and burial. The abundance of montmorillonite compared with kaolinite in clays is perhaps due to the finer grain of the m-clays and their ability to be transported further from the river mouths. The offlap marine clays are important in other aspects because they act as reservoir seals. In addition they may be source rocks. High organic content of these montmorillonitic shales may well augment the Akata Shale sources of oil and gas. However throughout much of the delta these shales are said to be immature (See Evamy et al. 1978). Towards the top of a marine shale sequence the beds often become silty and sandy and laminated clays, silts and fine sand are common.

Agbada Fluvio-marine Barrier Foot deposits: These consist of sands in which grain size and laminae increase upwards. Plant remains are very common and occasionally occur as thin lignitic streaks. There is a gradation from the marine clays below through laminated clay to silt and sand. Burrows are not conspicuous perhaps because of rapid sedimentation. These deposits were laid down in the barrier foot environment, i.e. in the proximal fluviomarine frontal part of the coastal barriers. By analogy with the modern delta the sediments were probably laid down in water depths ranging from 30 to 100 ft. The sands and silts were transported along the coast, as they are today, by longshore currents redistributing river mouth loads (Figure 138).

Barrier foot sands have distinctive Gamma Ray Log signatures because gamma ray radiation is often higher than in the underlying marine clays. Weber (1971) reported that studies of sidewall cores from high radiation in the Olomoro Field (Figure 101) indicate that the radiation is due to a high percentage of silt size zircons. Zircon contains in general 200–2000 ppm Th and 600–6000 ppm U.

Agbada Barrier Bar Deposits: the cleaner, coarser sands which overlie the Barrier Foot Sands were laid down in the barrier beach and washover sand environments where overall energy is greater than in the deeper water. The sands are fine grained with average size of 250 to 260 microns. The Barrier Bar Sands interfinger and overlap Barrier Foot sediments at a depth of roughly 30 ft. The Barrier Bars are usualy parallel bedded and occasionally crossbedded. Burrows are limited; there are silty claybreaks and lignites occur. These developed in peat filled depressions between beach ridges.

Longshore currents are responsible for the transport of the sands forming the barriers which are built up by wind and waves. On the down current ends of barrier bars, sand spits develop extending into the tidal channels. River mouth bars form at the front of these tidal channels (Weber 1971). On the landward side of the barrier bars tidal currents in some places erode the sands (Figure 102). The overall size of these Barrier Bar Sands is important because of the reservoir shapes and properties of the sands. Sands in barrier bars often thicken considerably towards growth faults (Figure 103). The limitation on thickness here is simply that in water depths of below approximately 35 ft, muds interbedded with sand predominate and interbedded sand-shale barrier foot deposits are formed (Weber 1971).

Barrier sands are deposited in units over 3 miles wide in the Present Day Delta (Allen 1964, 1965; Weber 1971; Dessauvagie 1972 and 1974) and extend parallel to the Present Day coast for lengths ranging from 3 to 21 miles. They average around 11 miles in length. They are therefore sheet like when viewed overall. In the majority of oilfields, Agbada sand bodies have widths perpendicular to the coastal trend of about 1 mile and lengths of around 5 miles and usually can be correlated on logs throughout much of the field.

In the Oweh and Olomoro field area barrier bar sands at the base of the cycle deposited into the growth fault can be correlated over a distnce of 13 miles from the eastern end of the Olomoro Field up to the Afiesere Field. The barrier sand is only 20 ft thick, giving a length/thickness ratio of several thousand (Weber 1971).

In fields that extend a good distance from the coastline clean barrier sands can be seen to thin seawards. The Barrier Bar Sand of the Agbada Field is a good example of this (Figure 103) (Weber 1971). The upward coarsening of the barrier foot to barrier bar sediments is clearly shown in this example.

The SP log and gamma ray responses show

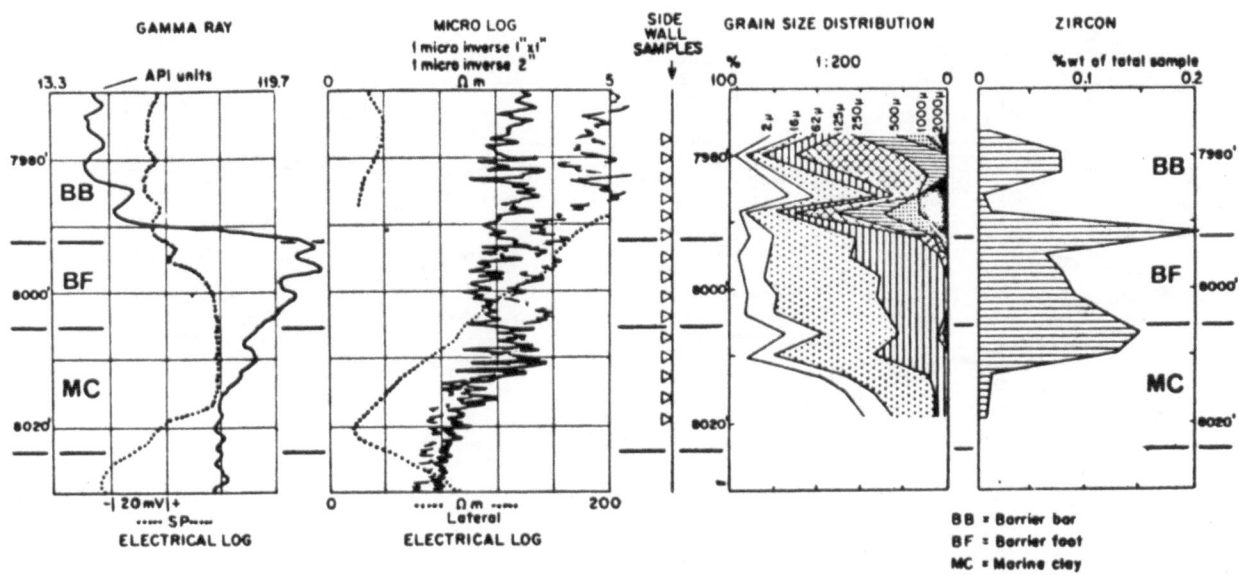

Figure 101. Sections showing the gamma ray and microlog responses, grain size distributions and zircon concentrations of side wall cores from barrier foot sediments from Shell-BP's Olomoro-5. Based on Weber (1971) after M.B.K. Lutz.

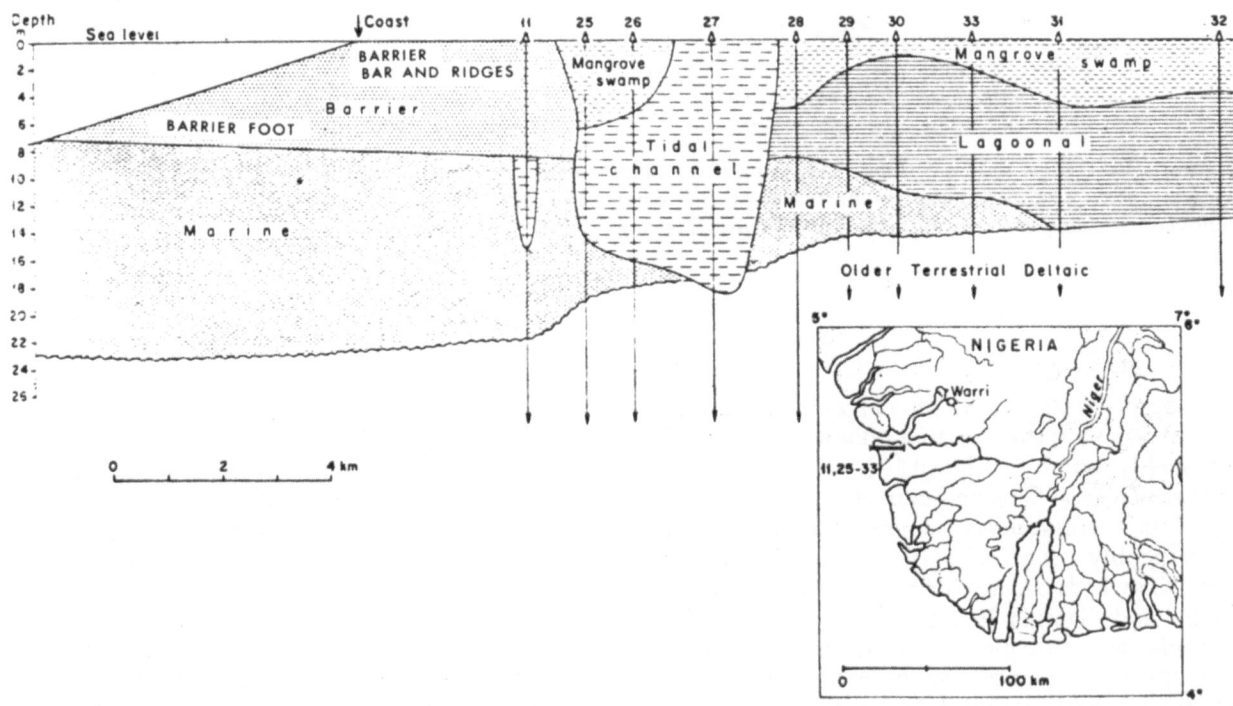

Figure 102. Section showing facies distributions (Agbada paralic facies) encountered in coreholes 11, 25–33 (See accompanying map for locations), Present Day portion of the Niger Delta Complex. Based on Weber (1971) after J.P. v.d. Sluis (KSEPL)

changes related to the upward coarsening of barrier sands. Coarsening is nearly always accompanied by an increase in permeability so in oil bearing sands of this type resistivity gradually increases upwards. The vertical permeability of Barrier Bar Sand reservoirs is often broken up by clay streaks, and this is important as far as the overall development of a reservoir is concerned because barrier sands are often deposited on top of one another separated by marine clay and/or dirtier, thin barrier foot sands.

Agbada Tidal Channel Deposits: The Present Day Barrier Bar Zone is breached by many tidal channels and estuaries (Figures 9, 124 etc). Some of these tidal channels exceed 1.5 miles in width and are more than

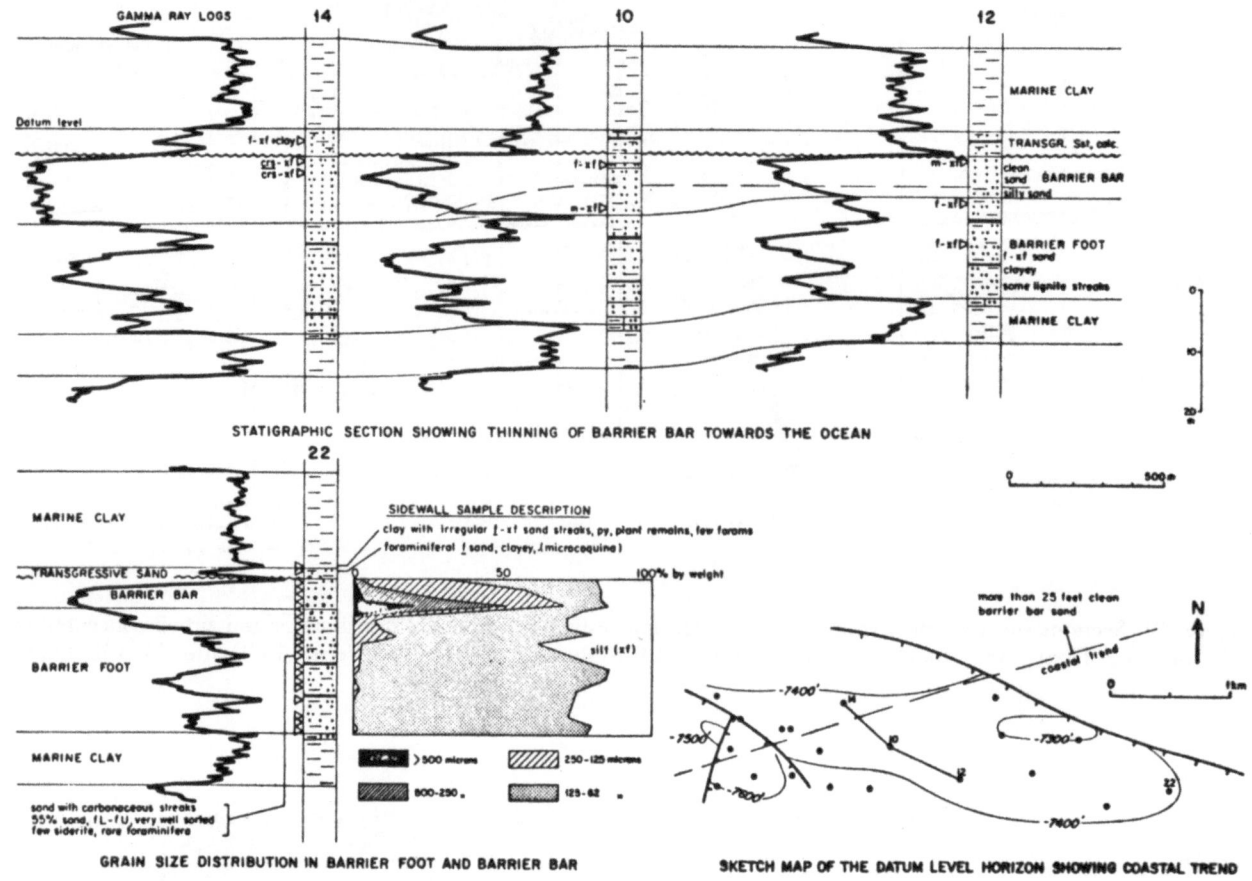

Figure 103. Sections showing the thinning of barrier sand towards the ocean (sections 14, 10 and 12) and a section (22) showing grain size distribution in Barrier Foot and Barrier Bar deposits in the Agbada Field. Map shows coastal and rollover associated with the Agbada growth fault. Based on Weber (1971).

70 ft deep. Width/depth ratio ranges in the subsurface are similar. A section cutting obliquely across a tidal channel is shown on Figure 102.

Agbada Tidal Channel sediments consist frequently of thin crossbedded sequences fining upwards. Clay pebbly or gravelly lag deposits occur at the base. Clay beds occur which give a serrated character to SP and Gamma Ray logs. The clay breaks are irregular and difficult to correlate individually. Fining upward sequences can be recognized on some micrologs. River mouth bars are difficult to spot on log evidence alone but have been identified.

Agbada Fluviatile Deposits: Agbada Fluviatile Deposits were laid down in the river flood plains and coastal distributary channels of the Lower Deltaic Plain as a "brickwork" of point bars. They were laid down in brackish and marine environments. Sometimes thick sheets of medium to coarse sand occur with clayey beds developed between individual point bars.

In the paralic Agbada sequence the maximum thickness of a point bar complex is usually less than 300 ft and is built up out of point bar sand units 30 – 45 ft thick. The fluviatile sand bodies have sharp bases on logs but occasionally basal conglomerates in sand limits blur this log effect.

Back swamp deposits consisting of silty clay, thin sand beds (small creek infills), and peats occur between point bar sands and these are especially developed on the flanks of the river channels. Crevasse and levée sands have been recognized in wells. Width to thickness ratios vary from 400 to 800 for point bar sands. Distributary channel fills often cross crests of rollover anticlines indicating perhaps a relationship between growth faulting and channel location.

An Agbada tidal channel sequence in the Egwa Field is shown in Figure 104. The tidal channel developed at an angle to the coastal trend and the barrier bar, with back swamp and lagoonal deposits flanking the channel (Weber 1971). Tidal channel deposits clearly show the effects of differential compaction. They are often about 100 ft thick and 1 mile wide.

From the foregoing detailed descriptions it is evident that the Agbada Formation constitutes a complex series of deposits laid down under at least 5 sub-environments of deposition:

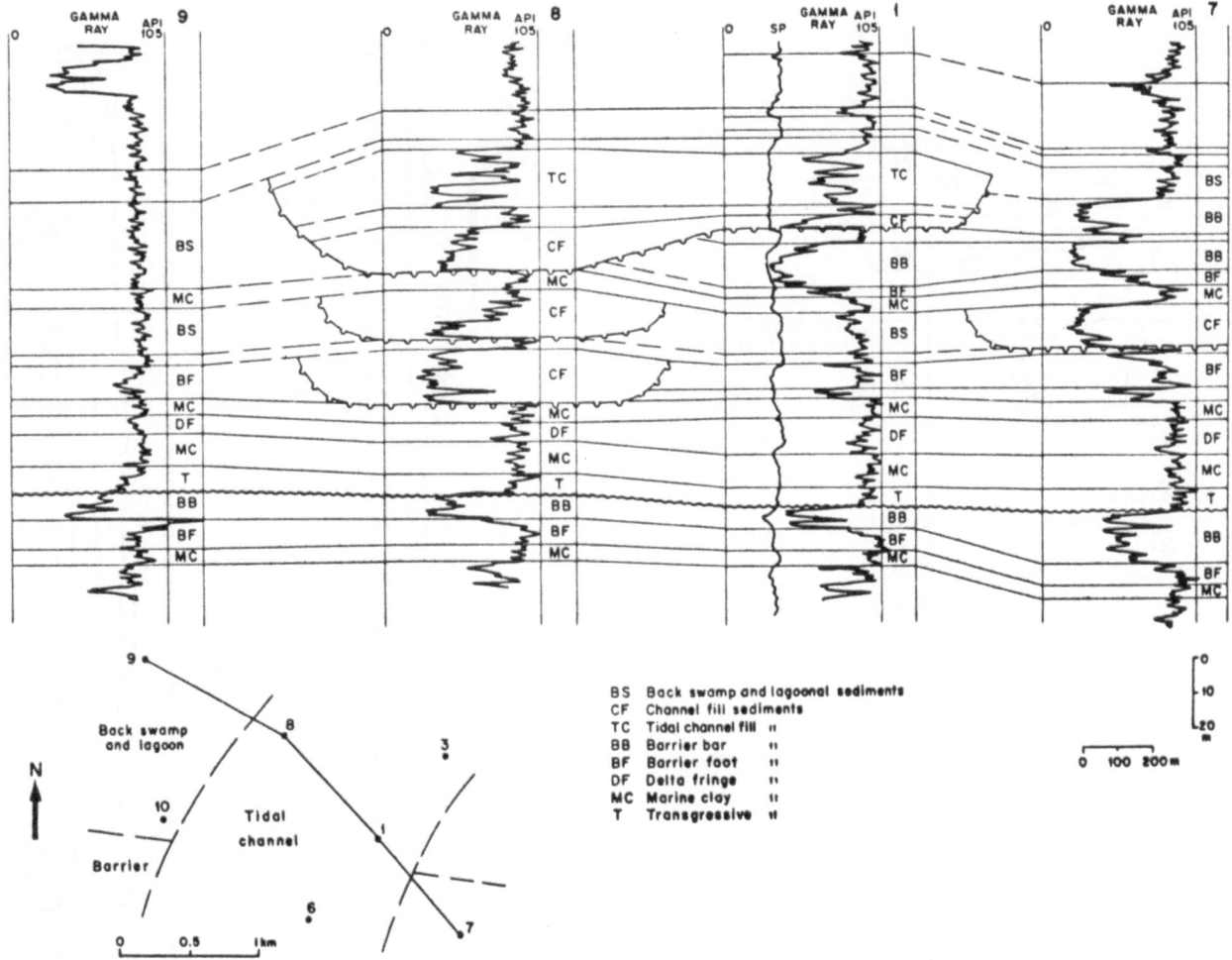

Figure 104. Stratigraphic sections in part of the Egwa Field paralic sequence showing a tidal-channel fill. Based on Weber (1971).

Environment	Deposit or Lithofacies
5. Lower Deltaic Flood plains	Agbada Fluviatile Deposits, Point Bar Sands and Backswamp Deposits etc.
4. Tidal Coastal Plain	Agbada Tidal Channel Deposits
3. Barrier Bars	Agbada Barrier Bar Sands
2. Barrier Bar Foot	Agbada Fluvio-Marine Barrier Sands Foot Sands
1. Holomarine	(iii) Agbada Offlap Marine Clays
	(ii) Marine Transgressive
	(i) Onlap Sands

Although the limits of individual lithofacies are distinct and recognizable on wire line logs the overall limits of the Agbada Formation are more difficult to define and as we have said because of the questions of picking faunal and lithological environmental indicators the limits as plotted in individual wells may be somewhat arbitrary. A formation which has as its defined top the last appearance of brackish or marine shale and its defined base taken at the deepest significant sand body can only be classed as loosely defined but is nevertheless a practicable and acceptable litho-unit.

2.5.5.5. Akata Formation (Subsurface)

2.5.5.5.1. General Description:

The basal major time-transgressive lithological unit of the Niger Delta Complex is the Akata Formation. It is composed mainly of marine shales but contains sandy and silty beds which are thought to have been laid down as turbidites and continental slope channel fills. The type locality of this formation has been selected in Akata-1, a Shell-BP well situated east of Port Harcourt (Figure 106).

Details of the type section are given below (Short and Stauble 1967):

Type Section: Akata-1, new field wildcat, drilled 50 miles east of Port Harcourt.

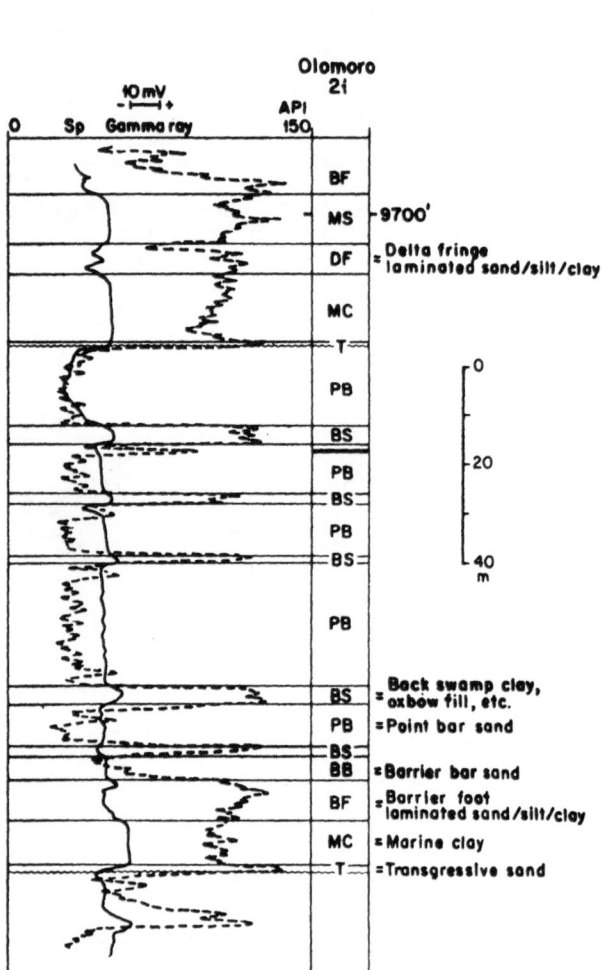

Figure 105. Section showing SP and gamma ray log response in a thick predominantly fluviatile rhythm in Olomore 21, Niger Delta Complex. Based on Weber (1971).

Interval:	7180 – 11 121 ft below derrick floor
Location:	Lat. 4° 41′ 50.5″N and Long. 7° 46′ 58.6″E
Derrick Floor Elevation:	106 ft above MSL
Total Depth:	11 121 ft
Samples:	Ditch cutting samples every 20 ft. Stored with Shell-BP Development Co. Ltd. of Nigeria and with the Geological Survey of Nigeria, Kaduna.

The Akata Formation consists of dark grey uniform shales, especially in the upper part. In some sections it is sandy or silty in the upper part of the formation where it grades into the Agbada Formation. An SP-Resistivity-log of the part of the Akata

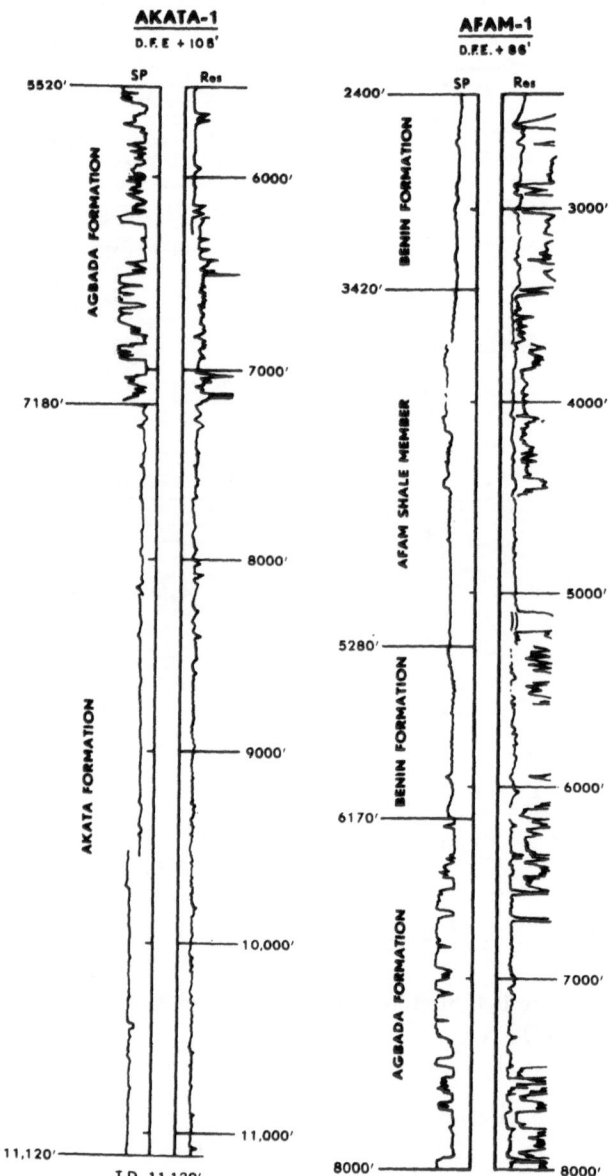

Figure 106. Summary SP and Resistivity logs showing petrophysical characteristics of the Akata Formation in type well Akata-1 and type Afam Clay Member in Afam-1, Niger Delta Complex. Based on Short and Stauble (1967). Reprinted with permission of the American Association of Petroleum Geologists.

Formation drilled in Akata-1 is presented in Figure 106. The Agbada Fauna is generally very rich in planktonic foraminifera which make up more than 50% of the microfauna.

The Akata Formation is thought to be the main source rock for Niger Delta Complex oil and gas. The formation probably underlies the whole of the Niger Delta Complex south of the Imo Shale outcrop which itself was probably deposited under similar conditions of deposition and may be considered an up-dip equivalent of the Akata Facies (Figure 92).

2.5.5.5.2. Thickness Variations

The top of the Akata Formation is taken arbitrarily

at the deepest development of deltaic sandstones at 7810 ft in the type section (Figure 106). The base of the formation was not reached at a total depth of 11 121 ft in Akata-1 but the base has been penetrated in wells situated on the delta flanks. Short and Stauble (1967) provisionally drew the base of the Akata at the top of the first major sandstone body or unconformity found below the shale.

In the type sections of the Akata Shale are over 4000 ft thick but Merki (1970) gave the thickness as 2000–20 000 ft. Weber and Daukoru (1976) gave the range in thickness as 1900–14 400 ft, although in their diagram they show that it is over 19 000 ft in the zone of maximum subsidence overlying the transition zone between continental and oceanic crust beneath the delta. Thickness drilled will depend on position within the delta complex and the amount of lutokinetic deformation (diapiric movement) and flowage to which the formation has been subjected.

2.5.5.5.3. Age

The age of the formation ranges from Eocene to Present Day (Short and Stauble 1967) but conceptually deep water Palaeocene Imo Shale and even Late Cretaceous Nkporo Shale (Late Cretaceous) could be classed as Akata Facies.

2.5.5.5.4. Conditions of Deposition and Palaeogeography

The Akata fauna is rich in planktonic foraminifera which indicate deposition on a shallow marine shelf. According to Short and Stauble (1967) faunal assemblages indicating deep water or bathyal deposition have not been found in wells drilled into the Akata Formation. This has a simple explanation in that the majority of wells drilled onshore and in the shallow offshore have only penetrated shallow water deposits. Wells drilled near the outer limit of the conshelf such as in Occidental's OPL 90 and OPL 85, 87, 88 and Deminex's OPL 76 must have penetrated deeply into the Akata Formation. Quite clearly the Akata Facies extends into deep water and must contain deep water assemblages. Information about the "Stratpalaeo" of such sediments has yet to be published. Also the Akata Formation must grade laterally and vertically into the deep water turbidites of the Avon-Mahin Fan, the Niger Fan, and the Calabar Fan (Burke 1972) (Figure 142).

Fans and turbidites may well have developed from time to time as the Niger Delta Complex grew and we can think of such deposits as extending "backwards" either continuously or intermittently through the delta complex; perhaps more or less occupying the same areas because of delta dynamics and the patterns of longshore drift which must have prevailed for the greater part of the Cenozoic. The Afam Clay Member, the Kwa Ibo Member and other channel infills as yet unnamed and undefined are thought by Burke (1972) to be submarine canyon infills; however other views on origin have been presented.

The arbitrary nature of the boundary between the Agbada and Akata facies has been commented upon

(Section 2.3.5.5.4). At the Present Day, Akata facies are being deposited on the continental shelf and slope and perhaps on the lower part of the pro-delta slope. The Present Day outcrop of the Akata then is completely submarine. The upper boundary is markedly time-transgressive and has been deformed structurally (synsedimentary) on a large scale. The Imo Shale (Palaeocene) and Nkporo Shale (Upper Cretaceous) may represent up-dip subaerial outcrops of Akata facies. We do not have much data on the Akata Formation deep beneath the delta. Diapirs and high pressure zones are developed on a grand scale but details are limited.

2.5.5.6. Imo Shale Formations (Surface)

2.5.5.6.1. General Description

The Imo Shale Formation is the oldest of the Niger Delta Complex surface formations and outcrops on the Benin Flank and in the Anambra Basin where its outcrop widens and then narrows down along the north eastern flank of the delta (Figure 74). The outcrop extends into the Imo River area and beyond where it is overlapped on the Calabar Flank by the Benin Formation and Alluvium. Frankl and Cordry (1967) and Short and Stauble (1967) pointed out that the Imo Shale is the up dip equivalent of part of the subsurface Akata Formation.

Only general descriptions are available for the Imo Shale outcrops which rim the delta area (Reyment 1965; Short and Stauble 1967; and Dessauvagie 1974). Originally the name was used by Tattam (1944 p. 30). He also used Imo-Anambra Shales (Tattam 1944, p. 34). Grove (1951 p. 5), Simpson (1955 p. 24) and Reyment and Barber (1956 p. 48) used the name Clay-Shales. Reyment (1965 p. 90) referred to Imo Shale.

The type locality of the Imo Shale is on the Imo River in outcrops near the Umuahia–Okigwi Road. Reyment (1965) gave only a limited description of the formation describing it as a thick clayey shale with a thickness of around 3200 ft. According to Reyment (1965) the name Imo River Shales of Tattam (1944) was first used in an unpublished report made by the Shell-BP Petroleum Development Company.

On the northwest flank of the Niger Delta Complex the formation rests apparently conformably on the Cretaceous Abeokuta Formation (Kab, Figures 9 and 74) and equivalents and on the Late Cretaceous Nsukka Formation around the nose of the Abakaliki Fold Belt. Dessauvagie (1974) described the formation as a fine textured, dark grey and bluish grey shale with occasional admixture of clay ironstone and thin sandstone bands, which occur especially towards the top of the unit.

2.5.5.6.2. Thickness Variations

Again only general data are available for the surface outcrop of the Imo Shale. Reyment (1965) gave a thickness of 3200 ft with the formation thinning down to 640 and 570 ft in wells in the Araromi and Gbekebo areas of western Nigeria (Figures 9 and 74). Dessauvagie (1974) gave a thickness of 1600 ft for the type area of the Imo River.

2.5.5.6.3. Age

Berggren (1960) and Shell-BP Nigeria think that the Imo Shale may be of Early Eocene age but Reyment (1965) classed it as Palaeocene. Dessauvagie (1974) classed it as Palaeocene to Eocene on his map (Figure 9). Stolk (1963) stated that the Imo Shale ranges up into the Lower Eocene and Short and Stauble (1967) gave the age range as Early Eocene to Palaeocene.

2.5.5.6.4. Conditions of Deposition and Palaeogeography

Very little specific information is available about the Imo Formation exposed on the Niger Delta Complex rim. Murat (1970) described the general palaeogeography of Palaeocene times differentiating shallow and marine clastics in an Imo Shale Embayment situated east of the Okitipupa or Ilesha High, and extending into the Anambra Basin and beyond into the horst and graben area of the Calabar Flank.

The Imo Shale was deposited during the Palaeocene Transgressive Phase and, as we have noted, Short and Stauble (1967) regard the Imo Shale as an up dip equivalent of the subsurface Akata Formation. Regressive deltaic offlap sequences of Imo age, if they existed, must have been eroded away. We can but speculate how these early Cenozoic deltas evolved.

Figure 107
Corelation of Eocene Formations, Eastern and Western Nigeria

	Western Nigeria	Eastern Nigeria	
Middle Eocene	Ameki Formation (normal marine facies)	Ameki Formation (fossils indicate normal marine origin)	
	Ilaro Formation		Nanka Sand
Lower Eocene	Oshosun Formation	Lower Ameki Formation	

Based on Reyment (1965)

2.5.5.7. Ameki Formation (Surface)

2.5.5.7.1. General Description

The Ameki Formation succeeds the Imo Shale at surface on Benin Flank and in the Anambra Basin but southeast of Umuahia it is involved in the Ogwashi–Asaba facies problem and is absent at surface along the Calabar Flank. The Ameki Formation is thought to pinch out on the Okitipupa Ridge on the west of the Niger Delta Complex (Figure 108). Over the years the Ameki Formation has been referred to as the "Bende–Ameki Group" (Wilson and Bain 1925, p. 71 and 1928 p. 32; Tattam 1944, p. 33; Du Preez 1947, p. 25; Jones 1948, p. 50; Reyment and Barber 1956, p. 38) or Bende-Ameki Series (White 1926, p. 7; Grove 1951, p. 5; and Simpson 1955, p. 25). Reyment (1965) selected Ameki as the type locality because the best exposures occur in the railway section between Mile 73 and Mile 87 along the Eastern Railway near Ameki Station.

The Ameki Formation consists of green, sandy clays with calcareous concretions and white, clayey sandstones. In places two lithological units are discernible:

2. an upper unit consisting of coarse, cross bedded sandstones, fine grey green sandstone and sandy clay. Interbedded sandstones may be as much as 320 ft thick; and
1. a lower unit with fine to coarse sandstones and intercalations of calcareous shale and thin shelly limestones.

The name Bende was used by Reyment (1965) for the "Bende Sandstone member". This is a white calcareous sandstone, in an informal sense. The Nanka Sand is thought to be a lateral equivalent of the Ameki Formation in Onitsha Province and is found in gully between Nanka and Oko (Reyment 1965). In general terms west of the Niger River the Ameki Formation is predominantly shaley but north of Lagos in Western Nigeria it passes into the sandy Ilaro Formation and the lagoonal clay of the Oshoshun Formation (Figure 108). East of the Niger the Ameki Formation consists of alternations of sandstone and shale, sandy or calcareous shale, marl and limestone.

The surface Ameki Formation is in part equivalent to the subsurface Agbada Formation. Together with the Ogwashi–Asaba Formation, the Ameki Formation probably is roughly equivalent to the Upper Sandstone-Shale alternating unit with thin shales and the lower predominantly shaley unit of the Agbada-2 type section (Short and Stauble 1967).

2.5.5.7.2. Thickness Variations

Reyment (1965) gave a thickness of nearly 4600 ft and Dessauvagie (1974) of nearly 5000 ft. Both pointed out that the Ameki thins over the Okitipupa High into western Nigeria (Figures 77 and 108).

2.5.5.7.3. Age

Reyment (1965) gave a Lutetian (Early to Medial Eocene) age to the Ameki Formation (Figure 107). Dessauvagie (1974) assigned an Eocene age, as did Short and Stauble (1967). The correlation adopted for Western Eastern Nigeria by Adegoke (1969) is presented in Figure 86.

2.5.5.7.4. Conditions of Deposition and Palaeogeography

As we have noted the Ameki Formation around the

132

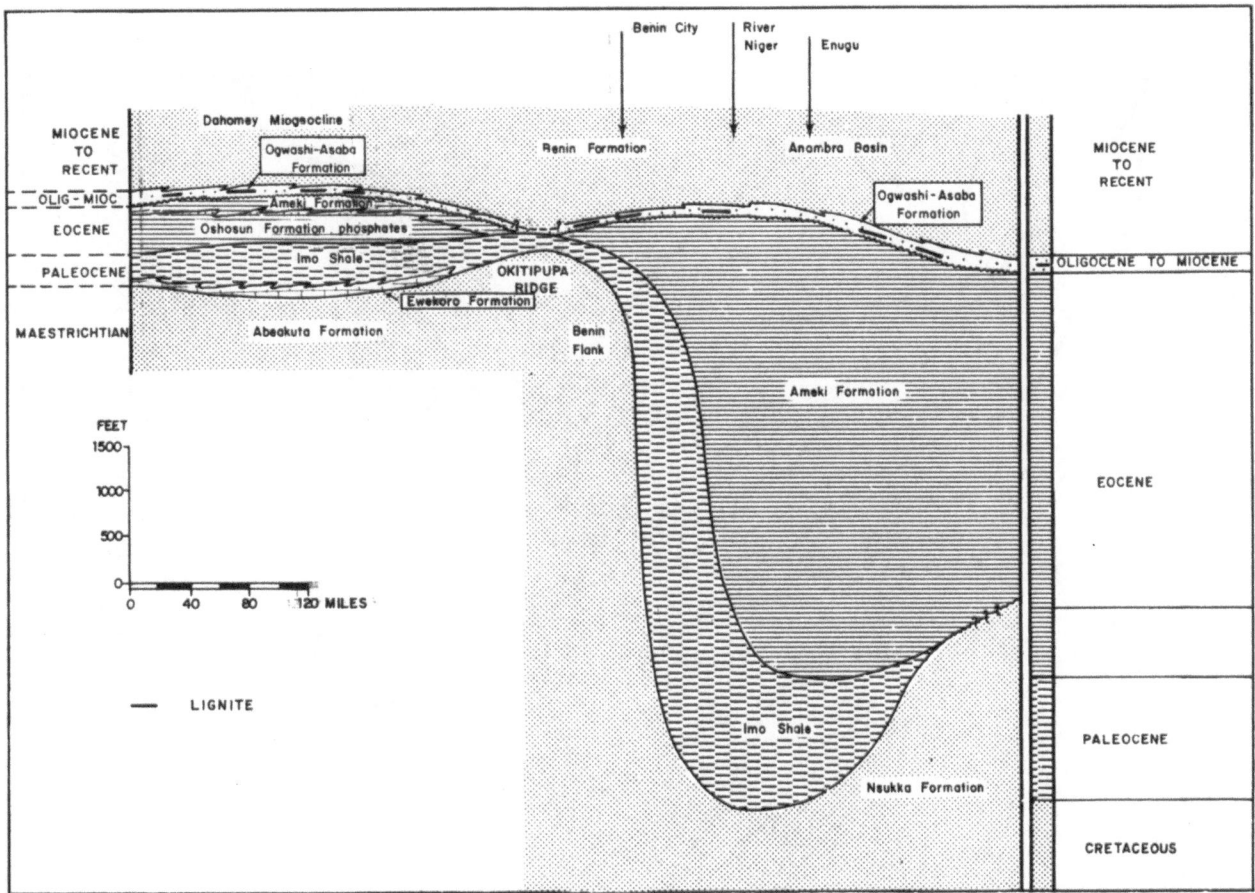

Figure 108. Section, highly diagrammatic, showing the correlation of Palaeocene to Miocene formations and Maestrichtian Formations exposed in the Dahomey Miogeocline and on the Benin Flank of the Niger Delta Complex. Based on Adegoke (1969).

delta rim becomes increasingly sandy because the Ameki was developed during an extensive regressive phase. A shallow marine environment was established giving a considerable variation in the pattern of sedimentation.

The formation of the Niger Delta Complex began in Eocene time and the shore line in Early-Medial Eocene time lay along a line just north of Benin City and south of Onitsha, swinging down towards Orlu situated at the head of the Ihuo Embayment (Figure 109). By Late Eocene time the shore line extended from just south of Ethiope-1, through the Aboh-1 area towards Orlu. The Ameki Formation consists of deltaic deposits laid down as the delta complex advanced over the shallow Anambra Shelf onto the continental margin. By Oligocene time the shore line was situated just below Onitsha. Whether the Ameki Formation is developed in the Cross River area of the delta complex is not known but it is thought that the Cross River Delta Complex was less developed than the Benue-Niger Complex at that time. The Ameki Formation has not been traced at surface southeast of Umuahia (Figure 9).

2.5.5.8. Ogwashi–Asaba Formation (Surface)

2.5.5.8.1. General Description

The Ogwashi–Asaba Formation overlies the Ameki Formation and extends from just west of the Siluko River on the eastern flank of the Okitipupa High in a steadily widening outcrop towards Onitsha, where the outcrop pattern is complicated by a fault (Figure 9). The outcrop then extends southeastwards via the Umuahia, Ikot Ekpene and Uyo areas to near Calabar where it is "overlapped" by the Benin Formation. The Ogwashi–Asaba Formation has not been mapped in Cameroun and the formation was not differentiated on the 1:2 000 000 Geological Map of Nigeria published 1964 by the Geological Survey of Nigeria. Reyment (1965) described the Ogwashi–Asaba Formation as a new formation. He proposed the name to include the:

Lignite Series	Parkinson (1907, p. 31)
Lignite Group	Wilson (1925, p. 79)
Lignite Group	Wilson and Bain (1928, p. 37)
Lignite Series	Simpson (1949, p. 8 and 1955, p. 26)
Lignite Series	Reyment and Barber (1956, p. 49)

133

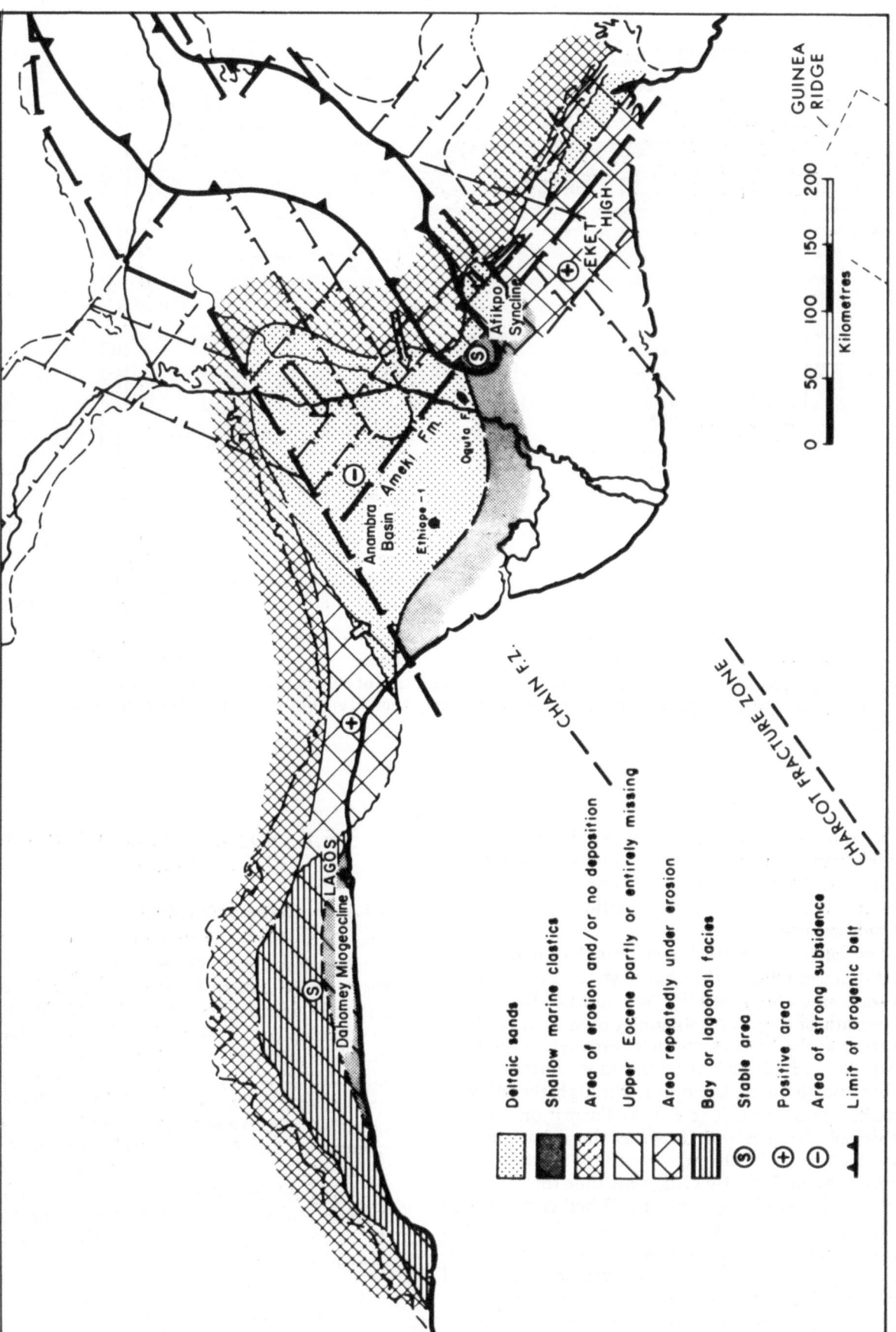

Figure 109. Palaeogeographic sketch map for Eocene times Anambra Basin, Niger Delta Complex and Dahomey Mio–geocline. Based on Murat (1972).

The Ogwashi–Asaba Formation consists of clays, sands, grits and seams of lignite alternating with gritty clays. The lignites were discovered by the Mineral Survey of Nigeria 1908–1909. Within the Ogwashi–Asaba Formation the lignites are confined to a narrow belt about 10 miles wide and 150 miles long trending northwest–southeast from the Niger in the west to the Nigeria–Cameroun frontier, east of Calabar. Within this belt the lignites have not been fully explored and are only known with certainty between Oba, 10 miles southeast of Onitsha on the northwest and Ikot Ekpene in the southeast. The lignites occur in the Lignite Group now called the Ogwashi–Asaba Formation (Reyment 1965) who assigned them to the Miocene? Oligocene (Figure 110).

Figure 110
Tentative Correlation of West African Miocene Formations

Western Nigeria	Gabon	Angola		Camerouns
Ogwashi-Asaba Formation Ijebu Formation	Mandorové and Animba Formations	beds with benthonic foraminifers	Missellele R. area with miogypsinids	Bomono and Kwa-Kwa areas with benthonic fora-minifers and miogypsinids.

Based on Reyment (1965)

The Lignite Group of Okezie and Onuogu (1971) presumably equivalent to the Ogwashi–Asaba Formation of Reyment (1965) occupies part of the mapped "Bende–Ameki Group" and "Coastal Plain Sands" on the Shell-BP Nigeria Ltd and Geological Survey of Nigeria 1:250 000 Geological Sheets published 1957. The Ogwashi–Asaba Formation of Reyment (1965) apparently was not mapped in the field and its definition is based on discontinuous section geology. Because of this Okezie and Onuogu (1971) seemed reluctant to use Reyment's term and used the general name Lignite Group in their report and on their map.

The lignites of the Nnewi–Oba area, near Onitsha occur in the "Bende–Ameki Group" whereas in the Umuahia–Ikot Espere–Uyo areas the lignites occur in the "Coastal Plain Sands" of the GSN–Shell-BP 1:250 000 Sheet Classification. The Lignite Group and Ogwashi–Asaba Formation are considered to be in part surface equivalents of the subsurface Agbada Formation (Figure 93; however cf. Figure 86).

According to Okezie and Onuogu (1971) the Lignite Group consists of a succession of cross bedded sandstones and grits; carbonaceous mudstones and shales; lignites and white blue and pink clays. Exposures are scarce except in deep valleys and in railway cuttings near Umuahia. Dips are low 2–3° and the beds are undisturbed. The thickest seams of lignite occur in the Oba and Orlu areas. A 12½ ft seam is known from the Oba area and a 14 ft seam has been recorded from Nnewi. Thicknesses vary considerably with seams splitting and jointing. The lignites vary from woody to earthy varieties and contain a great deal of palm debris. They break with a conchoidal fracture and are well jointed. The fact that the lignites contain large quantities of palm debris accounts for the high resin and wax content. The lignites were probably formed in a tropical forest swamp environment.

Because only part of the lignite bearing sediments has been drilled, Okezie and Onuogu (1971) could not estimate reserves or undiscovered resources. Gray-King Analyses of Oba and Nnewi Lignites together with Asaba lignites (west of the Niger) and Ultimate and Rational Properties are presented in Figures 111 and 112.

2.5.5.8.2. Thickness Variations

As we have mentioned above because Reyment's Ogwashi–Asaba Formation is only section defined and the boundaries have not been mapped because Okezie and Onuogu (1971) used "Lignite Group" as the mappable unit and because it is not clear where Dessauvagie (1974) obtained data for his upper limit of the Ogwashi–Asaba Formation it is difficult to give a figure for the overall thickness of the Ogwashi–Asaba Formation. No data have been given about the extension of the Ogwashi–Asaba Formation into the subsurface but wells like Ubula–1, west of Onitsha and east of the Niger, Ihuo–1 (TD 11 228 ft), Umuowo–1 (TD 6803 ft), Ngwu–1 (TD 8828 ft), Ikpe (TD 8935 ft) Olumbe (TD 11 005 ft), etc. must have penetrated parts of the Ogwashi–Asaba Formation or its equivalents passing down to the deep into the Agbada Formation. Adegoke (1969) shows the Ogwashi–Asaba Formation as a thin unit resting unconformably on the Ameki and interdigitating with the Benin Formation (Figure 108).

2.5.5.8.3. Age

As with thickness data we have very little positive age data. Reyment (1965) assigns Oligocene–Miocene age and correlated the Ogwashi–Asaba with the Ijebu Formation of the Lagos area of the Dahomey Miogeocline (Figure 108). Short and Stauble (1967) and Dessauvagie (1974) all assigned an Oligocene–Miocene age to the formation.

2.5.5.8.4. Conditions of Deposition and Palaeogeography

Okezie and Onuogu (1971) cited evidence from palynological studies of the lignites made by Powell Duffryn Technical Services Ltd who said that pollen grains from the lignites were derived from tropical and semi-tropical plants consisting mainly of palms. This would account for the high wax and resin content of the lignites. Powell Duffryn concluded:

> that the Tertiary coal forming forests of Nigeria were probably dominated by palms and closely resembled the tropical forest swamps of today. (Okezie and Onuogu 1971).

The southern and top boundary of the Ogwashi–Asaba Formation along the River Niger lies 80 miles within the Miocene Shore plotted by Short and Stauble (1967) (Figures 9 and 92) and some 60 miles within the Oligocene shore. Using the Present Day

Figure 111
Gray-King Assays at 600°C of Oba and Nnewi Lignites Compared With Those of Iyioshi River Lignite, Asaba Division

Origin of Sample	Yields/100 gm. d.a.f.					Gas analysis % (air free basis)								
	Gas Litres	Residue	Tar and Light Oil	Liquor	Gas gm.	CO_2	C_nH_m	CO	H_2	C_2H_6	CH_4	N_2	Sp. gr.	C.V. B.Th.U/ cuf
Onitsha. Oba. 7'0" seam														
5'10"–7'0" top	13.1	43.6	30.4	10.2	15.8	33.2	7.6	9.4	12.5	4.4	25.9	7.0	0.936	555
4'8"–5'10"	13.9	45.3	27.8	10.2	16.7	30.0	5.7	15.3	13.8	8.6	16.5	10.1	0.926	481
1'2"–2'4"	14.5	56.5	13.6	8.9	21.0	48.9	3.9	16.2	8.0	3.1	5.2	14.7	1.040	254
1–1'2" bottom	17.5	56.7	8.4	13.7	21.2	—	—	—	—	—	—	—	—	—
Onitsha. Nnewi. 12'5"														
10'5"–12'5" top	16.8	51.2	18.0	11.2	19.6	—	—	—	—	—	—	—	—	—
8'5"–10'5"	16.1	50.4	17.8	12.9	18.9	29.8	6.0	14.8	13.5	6.9	22.8	6.2	0.909	532
0–2'2" bottom	13.0	45.3	16.1	18.5	20.1	—	—	—	—	—	—	—	—	—
Iyioshi River. 9'6"														
Bulk sample	13.8	46.9	26.1	10.3	16.7	35.8	3.4	18.3	8.9	4.3	24.8	4.5	0.970	459

Analyst: Powell Duffryn Technical Services Ltd.

Origin of Sample	Gas Litres	Residue	Tar and Light Oil	Liquor	Gas Wt.	Wt. Balance	Gas cu.ft. at STP	Coke Cwt.	Tar and lt. oil galls.	Liquor gall.
Onitsha. Oba. 7'0" seam										
5'10"–7'0" top	10.5	54.5	24.4	8.2	12.7	99.8	3760	11.0	54.7	18.4
4'8'–5'10"	11.6	53.7	23.2	8.6	14.0	99.5	4160	10.7	52.0	19.2
1'2"–2'4"	14.0	57.1	13.2	8.6	20.3	100.2	5030	11.4	29.6	19.2
0–1'2" bottom	16.5	59.0	7.9	12.9	20.0	99.8	5930	11.8	17.7	28.9
Onitsha. Nnewi. 12'5" seam										
10'5"–12'5" top	14.8	56.2	15.9	9.9	17.3	99.3	5140	11.2	35.6	22.2
8'5"–10'5"	14.6	54.9	16.2	11.7	17.2	99.9	5250	11.0	36.2	26.2
0–2'2" bottom	10.6	55.5	13.1	15.0	16.4	100.0	3810	11.1	29.3	33.6
Iyioshi River. 9'6" seam										
Bulk sample	13.0	49.0	24.6	9.7	15.7	99.0	4670	9.8	55.0	21.7

Analyst: Powell Duffryn Technical Services Ltd.

Figure 112
Ultimate and Rational Analyses of Oba and Nnewi Lignites

Origin of Sample	Ultimate analyses			Rational analyses				
	C %	H %	N+S+O %	Resins and waxes	Plant remains structured	Opaque matter	Fusain	Ulmins
Onitsha. Oba. 7'0" seam								
5'10"–7'0" top	71.3	6.7	22.0	11.2	7.4	Nil	Nil	81.4
4'8"–5'10"	72.2	6.2	21.6	10.9	—	—	—	—
1'2"–2'4"	71.4	4.4	24.2	5.4	6.8	Nil	Nil	87.8
0–1'2" bottom	68.4	4.5	27.1	2.9	—	—	—	—
Onitsha. Nnewi. 12'5"								
10'5"–12'5" top	72.2	6.0	21.8	9.6	—	—	—	—
8'5"–10'5"	70.6	4.9	24.5	11.1	6.2	Nil	Nil	82.7
0–2'2" bottom	71.4	4.8	23.8	7.8	—	—	—	—
Iyioshi River, second face								
Bulk sample	72.9	6.4	20.7	18.6	7.2	Nil	Nil	74.2

Analyst: Powell Duffryn Technical Services Ltd.

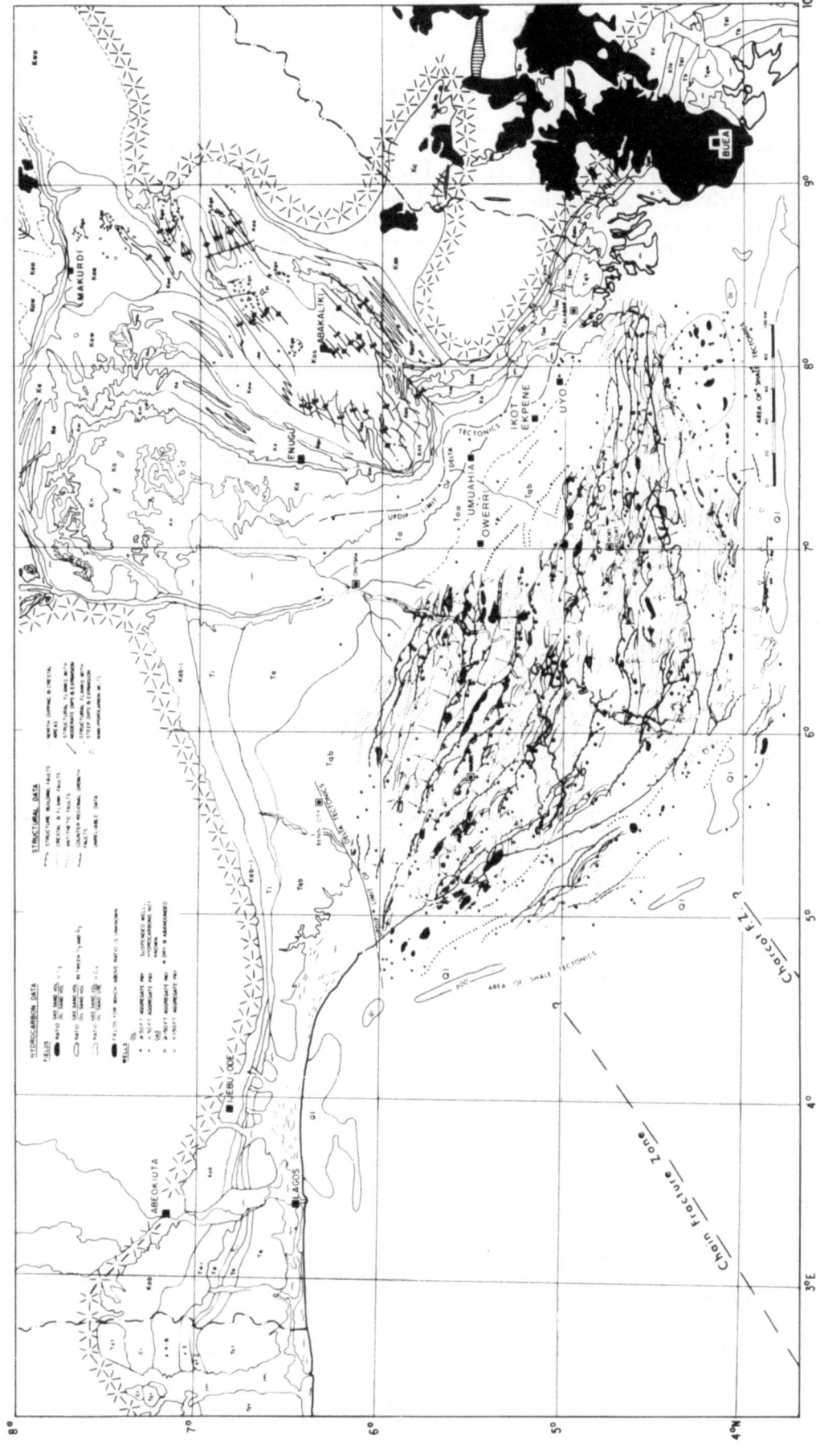

Figure 113. Map of the Niger Delta Complex showing present outcrops, shore lines, growth faults and major structural features. Part of key in Figure 8. Based on Short and Stauble (1967); Weber (1971); Merki (1972); Whiteman (1973); Dessauvagie (1974) Weber and Daukoru (1975), Evamy *et al.* 1978. Reprinted with permission of the American Association of Petroleum Geologists.

Sedimentary Environments of the Niger Delta for scale (Allen 1965) this could imply the Ogwashi–Asaba Formation was formed within the Upper Flood Plain Environment.

2.5.5.9. Benin Formation (Surface)

2.5.5.9.1. General Description

In addition to being a subsurface formation the Benin Formation crops out widely at surface across the delta province. Its limits shown on Figure 92 based on Short and Stauble (1967) are much more extensive than those shown by Dessauvagie (1974). Short and Stauble (1967) mapped the surface Benin Formation along with the Ogwashi–Asaba Formation and equivalents and used the top of the Ameki Formation as the base of the Ogwashi–Asaba Formation. Both Short and Stauble (1967) and Dessauvagie (1974) in part use the same boundary for the top of the Benin Formation. Short and Stauble (1967) used the base of the Agbada (Recent) outcrop as the top of the Benin Outcrop and in both cases this is the same as Allen's (1965) Mangrove Swamp limits. However, within the upper and lower limits of the Benin Formation, Dessauvagie (1974) using limits mapped by Allen (1965), differentiated Qd Deltaic Plain Deposits, Qmb Meander Belt Deposits and Qfs Freshwater and Back Swamp Deposits and assigned a Holocene age to these deposits differentiating the TQb Benin Formation as Mio–Pliocene.

Only general descriptions are available from the surface Benin Formation (Geological Survey of Nigeria 1956 in *International Stratigraphic Lexicon* 1956; Reyment 1965; Short and Stauble 1967; Frankl and Cordry 1967; Dessauvagie 1974). As Reyment (1965) pointed out the Benin Formation replaces the Benin Sands of Parkinson (1907), and the Coastal Plain Sands of Tattam (1944) and Geological Survey of Nigeria (1956). The surface Benin Formation has been variously described as yellow and white, sometimes crossbedded sands, largely unconsolidated with pebble beds, clays and sandy clays occurring in lenses.

2.5.5.9.3. Thickness Variations

In the subsurface of the Benin Formation the rocks are thought to attain 6000 ft (Knaap 1971) but according to Dailly (1976) the Benin may be as much as 10 000 ft beneath the Benin area (Figures 95 and 96) (Section 2.3.5.5.2.). Thicknesses cannot be calculated from surface outcrops in the up-dip delta area because of cover and poor exposure.

2.5.5.9.4. Age

The Geological Survey of Nigeria (1956) gave the age of the "Coastal Plain Sands" as "Miocene and the Upper part Pliocene and younger". Reyment (1965) gave the age as Pliocene to Pleistocene and Dessauvagie (1974) gave the age as Miocene to Pliocene and younger, although on his map (Figure 9) he assigned a Mio–Pliocene age to the formation. The Benin Formation or Facies could have started to

develop within the Niger Delta Complex as soon as delta formation started in Eocene times. The formation is being formed at the present day within the subaerial part of the delta complex.

2.5.5.9.5. Conditions of Deposition and Palaeogeography

This topic has been discussed in Section 2.3.5.5.2. The Benin facies of Reyment (1965) differs from that of Short and Stauble (1967) because Reyment holds that the Benin Formation is partly marine whereas Short and Stauble (1967) hold that the Benin Formation was laid down in a continental probably upper deltaic environment (upper floodplain Figure 115). They pick its base at the top of the first marine incursion. The down delta boundary of the Benin Formation/Facies is drawn where the brackish and marine Mangrove Swamp environment develops (Figures 114 and 115). During the Pleistocene the position of this boundary fluctuated widely through a zone around 50 miles wide.

2.5.5.10. Agbada (Surface) and Akata (Subsea) Formations

2.5.5.10.1. General

These formations outcrop at the surface and on the sea floor. They were laid down in a series of different environments under brackish and marine conditions (Figures 114 and 115).

Figure 114
Classification of Sedimentary Environments in the Present Day Niger Delta Complex.

First order: super-environments	Second order: environments	Third order: sub-environments
I. SUBAERIAL DELTA	1. Upper floodplain	
	2. Lower floodplain	a. Channel
		b. Point-bar
		c. Levee
		d. Backswamp
		e. Cut-off channel
	3. Mangrove swamp	a. Main channel
		b. Gullies
		c. Inter-creek flat
		d. Point-bar
		e. Intra-swamp delta
	4. Beach	a. Active
		b. Beach ridge
		c. Transverse channel
II. MARGINAL BARRIER ISLAND-LAGOON	1. Beach	
	2. Channel and creek	
	3. Lagoon	
	4. Lagoon delta	
	5. Marginal swamp	
III. MARGINAL ESTUARY	1. Marginal swamp	
	2. Open water	
IV. CONTINENTAL SHELF UNDER DELTA INFLUENCE	1. River mouth bar	Sub-environments not distinguished
	2. Delta-front platform	
	3. Pro-delta slope	
	4. Open shelf	
V. CONTINENTAL SLOPE	Environments not distinguished	
VI. NON-DEPOSITIONAL		

Based on Allen (1965). Reprinted with permission of the American Association of Petroleum Geologists.

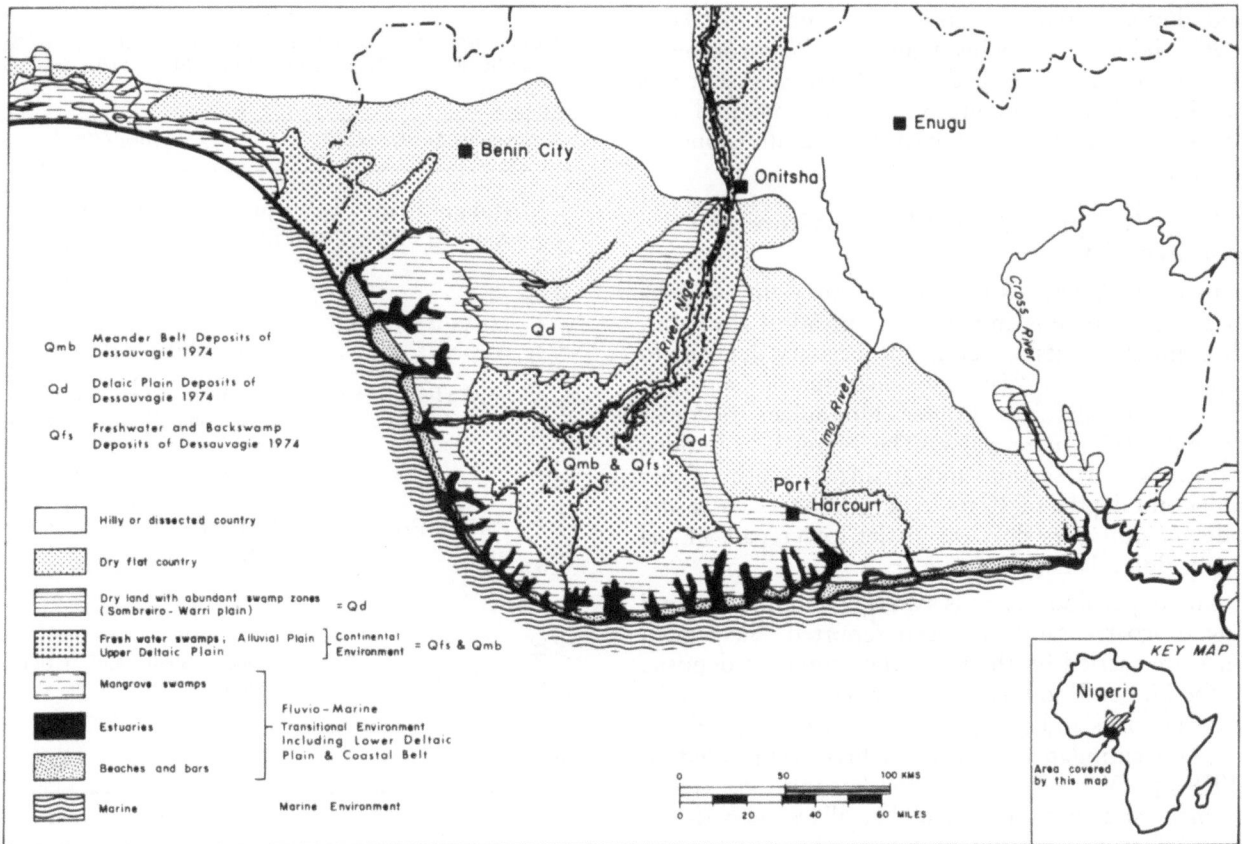

Figure 115. General map showing main sedimentary environments and morphological features Present Day Niger Delta Complex and adjacent areas. Qmb symbols etc. refer to deposits laid down in the environments shown and are taken from Dessauvagie's 1 : 1 000 000 Scale Map of Nigeria. Environments and morphological features from Allen (1965). Reprinted with permission of the American Association of Petroleum Geologists.

2.5.5.11. Formations and Environments of the Late Quaternary Niger Delta

Our knowledge of sedimentary processes, environments and deposits forming today and which formed the Late Pleistocene in the Niger Delta and adjacent areas is largely based on work done by Pugh (1954); Hill and Webb (1958); NEDECO (1954, 1959 and 1961); Allen and Wells (1962); Allen (1964 and 1965); Sluis in Weber (1971); Weber (1971); Oomkens (1974) and Weber and Daukoru (1975). Information about processes, environments etc. are highly relevant to log interpretation, reservoir and production studies etc. and are therefore discussed in detail in this section.

The most comprehensive papers are those of NEDECO (1959), Allen (1965 etc.) and Oomkens (1974) and most of the descriptions which follow are based on their writings. Allen (1965) recognized six super environments:

I   Subaerial delta
II   Marginal barrier island–lagoon complex
III Marginal estuary complex
IV   Continental shelf under delta influence
V   Continental slope
VI   Non-depositional

These super-environments were divided further into Second Order Environments (Figure 114):

I   Subaerial Delta
     1. Upper Floodplain
     2. Lower Floodplain
     3. Mangrove Swamp
     4. Beach
II   Marginal Barrier Island–Lagoon
     1. Lagoon
     2. Lagoon delta
     3. Channel and creek
     4. Marginal swamp
     5. Beach
III Marginal Estuary
     1. Marginal Swamp
     2. Open Water
IV   Conshelf under Delta Influence
     1. River Mouth Bar
     2. Delta Front Platform
     3. Pro Delta Slope
     4. Open Shelf
V   Continental Slope (Environments not distinguished)
VI   Non-depositional (Environments not distinguished)

Third Order Subenvironments were distinguished and are shown in Figure 114. Short and Stauble (1967) regrouped Allen's categories and the deposits laid down therein as shown in Figure 116. Allen (1965) did not classify his deposits and environments in terms of Akata, Agbada and Benin facies but this has been done in Figure 116. Environmental and depositional details for each of these subdivisions are given below.

2.5.5.11.1. Upper Flood Plain Environment and Deposits

This environment extends from above the Onitsha Gap where there is an extensive spread of alluvium, through the Onitsha Gap in the Tertiary deposits to where the Nun and Forcados distributaries separate. Below Onitsha the Upper Floodplain occupies around 3243 square miles. The Niger is a braided stream in this section and attains a width of 1½ miles and depths of 40 ft. Large meanders enclosing ridged point bars develop in the lower part of the upper floodplain.

The upper flood plain is swampy and forested and fine to coarse sand has been reported. Suspended fines are carried by the river. The topmost deposits of the Upper Floodplain are coarse sands, fine sands, silts and clays and swamp deposits. Both drifted and *in situ* accumulation of plant debris occur. Nedeco (1959, p. 309) stated that the medium to coarse grained sands with intercalated argillaceous beds exceed 150 ft in borings in the Upper Floodplain near Onitsha.

2.5.5.11.2 Lower floodplain environment and deposits

This environment extends from the Forcados–Nun bifurcation of the Niger to the levee indented margin of the brackish–marine mangrove swamp environment. More and more distributaries develop in this zone and stream depths rarely exceed 32 ft. River channel deposits, point bars, levees back swamp and cut off channel deposits develop. NEDECO (1961, p. 42) reported low stream flows and sediment structures reflect low stream activity. Large scale and small scale cross bedded strata occur in sand bodies. Levees (6 ft) border the stream channels and fall away into fresh water backswamp deposits. Swamp facies consist of layered sediments, mainly clays or silts with mica and abundant plant debris. Cut off deposits occur and resemble swamp deposits.

Dessauvagie (1974) divided the Upper Floodplain and Lower Floodplain deposits of Allen (1965) into Qd Deltaic Plain Deposits; Qfs Fresh Water and Swamp Deposits and Qmb Meander Belt Deposits (Figures 9 and 115).

The Upper Floodplain and Lower Floodplain deposits of Allen (1965) occupy around 3243 square miles and the Qd Deltaic Plain Deposits, Qfs Freshwater and Swamp Deposits and Qmb Meander Belt Deposits of Dessauvagie (1974) laid down in the Upper Deltaic Plain and Alluvial Valley Environment all constitute part of the Benin Facies (Figure 116).

The Present Day area occupied by Benin Facies is much less than the area occupied by the mapped

Figure 116
Comparison of Classifications of Environments used by Short and Stauble (1967) and Allen (1965), Niger Delta Complex

	Short and Stauble (1967)	Allen (1965)
BENIN FACIES CONTINENTAL	UPPER DELTAIC PLAIN AND ALLUVIAL VALLEY	UPPER FLOOD PLAIN LOWER FLOOD PLAIN
	LOWER DELTAIC PLAIN	Marginal Barrier Island
AGBADA FACIES FLUVIO-MARINE	COASTAL BELT	Marginal Barrier Island – Lagoonal Complex and Marginal Estuaries
	Fluvio-Marine	River Mouth Bar Delta Front Platform, Pro Delta Slope
AKATA FACIES MARINE	Holomarine	Open Shelf and Continental Slope

Based on Whiteman (1973)

Benin Formation as a whole which extends from Lagos Lagoon in the west via the Benin City area almost to the foot of Cameroun Mountain in the southeast. Probably the older parts of the formation are polygenetic and in the past river systems such as the Cross River System, the Imo River and the smaller rivers of Western and Southeastern Nigeria must have been more powerful and supplied more sandy sediment than they do now. The Present Day spread of Benin deposits is related to the Niger-Benue River System and may be a later delta feature probably related to Late Neogene development of the complex.

2.5.5.11.3 Mangrove swamp environment and deposits

The intertidal mangrove swamp covers around 1900 square miles fronting 440 miles of delta coastline (Oomkens 1974) or around 3474 square miles (Allen 1965). Barrier Beach islands broken through by tidal channels separate the swamps from the open sea. In the central area the mangrove swamps are 5–10 miles across and interfinger with lower flood plain alluvium. On the flanks of the delta the mangrove swamps are 20–25 miles wide, especially in those areas where the swamps are developed on Tertiary and Cretaceous rocks (Figures 117, 118 and 120).

Mangrove swamps occur within the Lagos Lagoon Complex in the Dahomey Basin and fringe sections of the coast extending into the delta area proper. In the Cross River, Calabar area and Rio del Rey Basin mangrove swamps are developed on the Nkporo,

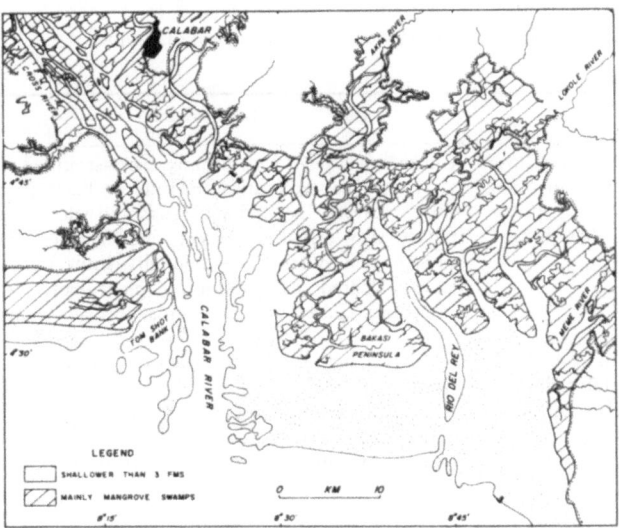

Figure 118. Map showing distribution of Present Day Agbada Facies Mangrove Swamp sub facies in the Brass and St. Nicholas River areas, Nigeria. Based on Allen (1965) and aerial photographs. Reprinted with permission of the American Association of Petroleum Geologists.

Figure 117. Sketch map showing the main features of the high energy tidal destructive Cross River Delta Complex at Present Day. Based on Allen (1965); Geological Survey of Nigeria Sheet 85 and British Admiralty Charts 1860 and 3423. Reprinted with permission of the American Association of Petroleum Geologists.

Ogwashi-Asaba and Benin outcrops and extend almost to the base of the Cameroun Volcanic pile. Dessauvagie (1974) mapped these deposits as Alluvium and curiously Allen (1965 Figure 7) labelled them as Prodelta Slope. Short and Stauble (1967) mapped these deposits as Mangrove Swamps overlapping onto dry flat country. Within the delta area the inland limit of the Mangrove Swamp facies is by implication the inland limit of the Agbada facies but in the southeastern part of Nigeria the Mangrove Facies of the Cross River–Calabar area and the Rio del Rey Basin is independent of the Benin Facies and has developed over Upper Cretaceous rocks. It is not clear whether we should think of the Mangrove Swamp facies as part of the Agbada facies in this area. Conditions prevailing in this area now indicate that we are dealing with a high energy destructive estuarine delta facies.

Halophytic red mangroves are the main plants in the Mangrove Swamp Environment and the swamps develop wherever the tide pushes salt water inland through the barrier beaches, creeks and estuaries. Main channels or creeks and intercreek flats are the main morphological elements in the swamps. The largest creeks in addition to carrying tidal water also carry river water and are part of the distributary system. The intercreek flats lie between low and high tides.

Deposits in the mangrove swamps have accumulated within five main subenvironments:

Main channel
Gulleys
Inter-creek Flat
Point Bar
Intra Swamp delta

Intra-Swamp deltas form as complexes of islands and mudflats in the widest parts of the larger channels (Allen 1965, p. 574). Sediments entering the mangrove swamp environment are polygenetic. Suspended fines enter the system both from the sea and fresh from the rivers. Bed sands are derived either from river sand or erosion of older sediments from creek beds. Tidal and river currents are often strong enough to transport sand. At slack tide clay and silt flocculate and once deposited such floccules are difficult to move again. Sands give way rapidly to silts and clays and interlayered sand, silt, clay and plant debris form. The main characteristics of the sediments being laid down in the mangrove swamp environments are summarized in Figure 119. In the intercreek flats and intra swamp deltas the deposits consist of mainly uniform fine, organic rich clayey silts and silty clays with roots and root mottles. The deposits of the Mangrove Swamp Environment are by implication part of the Agbada facies of Short and Stauble (1967) and Weber (1971).

2.5.5.11.4 Beach environment and deposits

Barrier Bar Deposits forming part of the beach ridge barrier island complex rim the seaward margin of the present day Niger Delta for about 300 miles. Beachridge-barrier islands (Allen 1965) extend westwards into the Lagos area where they were described by Webb (1958). The Barrier Bar Complexes consist of the modern active beach and sand ridges developed in relation to older strand lines. Allen (1965) recognized 20 major barrier islands in the delta area separated by deep tidal ebb and flow channels. It is not clear how extensive these deposits are but Allen (Unpublished MS University of Ibadan Library)

Figure 119
Summary of Characteristics of Sedimentary Environments and Lithofacies of Present Day Niger Delta Complex

	Environment	Environmental Characteristics	Lithofacies
BENIN FACIES	Lower floodplain	Strong currents in channels, mainly in lower flow regime. Meander migration and channel cut-off—lateral sedimentations. Periodic flooding of topstratum levees and backswamps—vertical sedimentation. Abundant plant growth.	*Channels and point bars*: Mainly cross-stratified f. to v.c. sand. Layered sands and silts on bar tops and shoals. *Levees*: Layered v.f. sands and silts grading out to silty clays and clayey silts. Mottles. *Backswamps*: mainly silty clays and silts, some v.f. sand layers. Drifted and autonomous plant debris. Root mottling. *Cut-off channels*: similar to backswamps.
AGBADA FACIES	Mangrove swamp	Strong reversing tidal currents mostly in lower flow regime. Meander migration with lateral sedimentation. Diurnal flooding of inter-creek flats with vertical deposition.	*Channels and point-bars*: Mainly layered cross-stratified f. to v.c. sand and organic-rich silty clay. Abundant drifted plant debris. *Inter-creek flats and intro-swamp deltas*: Mainly uniform fine, organic-rich clayey silts and silty clays with roots and root mottles
	Beach	Strong wave attack on active beaches with swash-backwash flows reaching upper flow regime. Longshore currents diverging from delta tip. Soil formation and plant growth on beach ridges.	*Delta tip*: Mainly evenly laminated, clean, f. to m. sand. Primary current lineation. Very rare small ripples. No mottling. Shell and plant debris very rare. *Delta flanks*: Mainly v.f. and f. sand with other features as for delta tip. *Beach ridges*: Lamination destroyed through plant growth on high parts. Soils.
	River mouth bar	Very strong wave action and reversing tidal currents. Longshore currents. Energy conditions decrease inland and seaward from bar crests, with increase in depth.	*Crests*: Mainly clean, v.f. to m. sand with even lamination, cross-stratification, or cut-and-fill. Shell and plant debris rare. *Bar flanks*: Layered v.f. sand, clean v.c. silt and clayey silt. Drifted plant remains sometimes in thick layers. Rare mottles.
	Delta-front platform	Strong to moderate wave and tidal current action. Longshore currents. Guinea Current. Rip currents. Energy conditions decrease from shoreface to outer edge.	*Delta tip*: On inner platform uniform coarse v.f. sand, and v.c. silt with even laminations. Mottles few or absent. On outer platform layered v.f. sand, v.c. silt, clayey silt, and silty clay with plant debris. *Delta flanks*: On inner platform uniform coarse v.f. sand and v.c. silt. On outer platform layered v.c. silt, clayey silt, and silty clay with plant debris. Some uniform fine clayey silts.
	Pro-delta slope	Moderate to weak wave and tidal current action, decreasing in strength from shallow to deep water. Guinea Current.	*Delta tip*: Layered v.f. sand, v.c. silt, clayey silt; and silty clay. Coarser layers with even lamination, cross-stratification. Plant debris and mica flakes. Common to abundant mottles. *Delta flanks*: Layered v.c. silt, clayey silt and silty clay in shallower parts. Uniform fine clayey silts and silty clays in deeper areas. Abundant mottles.
AKATA FACIES	Open shelf	Weak wave and tidal current action. Deep ocean currents of unknown strength flowing northward over shelf edge.	*Delta tip*: Rare layered v.c. silt, clayey silt, and silty clay. Mainly uniform fine clayey silt and silty clay. Abundant mottles and pelagic foraminifera. *Delta flanks*: Mainly uniform fine silty clays. Abundant mottles and pelagic foraminifera.
	Non-depositional	Weak wave and tidal current action in deeper parts. Strong to moderate action inshore. No, or very slow, deposition of suspended fines. Abundant benthos. Organic debris concentrated.	Mainly mottled v.f. to v.c. quartz sands largely out of equilibrium with prevailing current conditions. Shell debris. Glauconite, foraminifera, and clay-silt increase from shallow to deep water. Partly Late Pleistocene in age. Deposit interpreted as of strand-plain origin.

Based on Allen (1965).

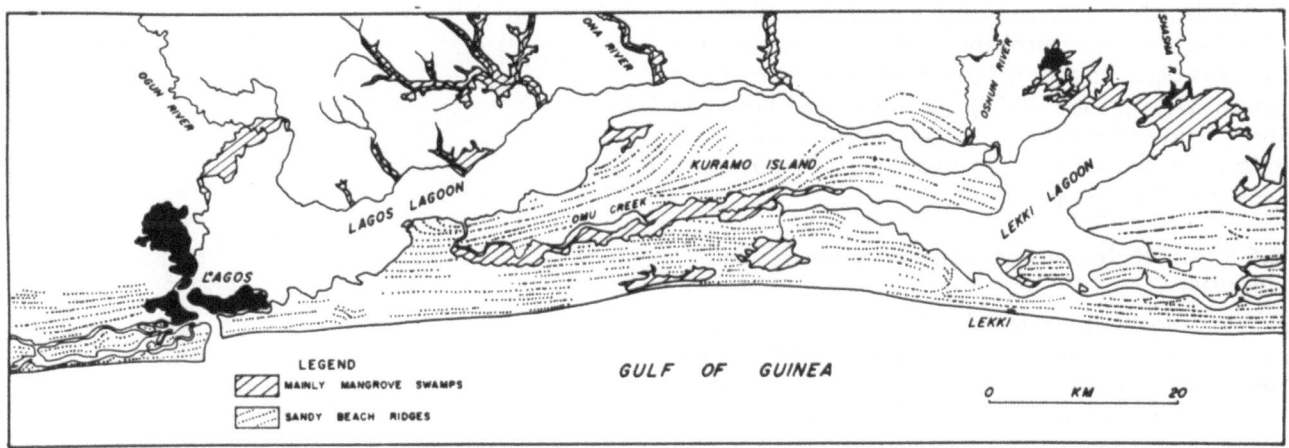

Figure 120. Map showing main environments of deposition in the Lagos and Lekki Lagoon areas, Nigeria. Based on Allen (1965) and aerial photographs. Reproduced with permission of the American Association of Petroleum Geologists.

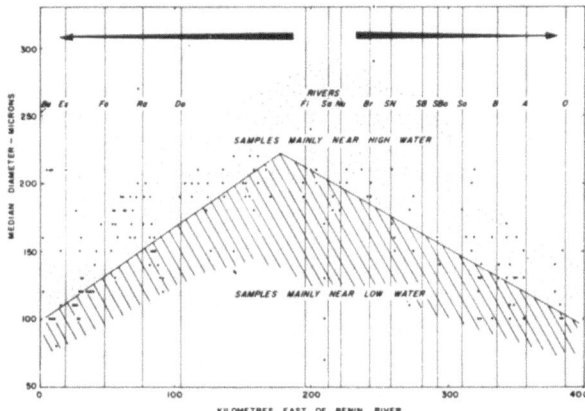

Figure 121. Graph showing across-delta grading of Present Day Beach Sands, Niger Delta Complex. Arrows show the directions of longshore currents. Compiled from data listed by NEDECO (1954 and 1961). Based on Allen (1965) and aerial photographs. Reproduced with permission of the American Association of Petroleum Geologists.

estimated that they certainly exceed 694 square miles in extent. The barrier islands range from a few hundred yards wide to 7 miles wide. Length varies from 3 to 23 miles and averages 11 miles. Seaward margins of the islands are straight whereas the inland margins are irregular and eroded by meandering tidal creeks. Ridges stand around 6–12 ft high and are fixed by palms, scrub and grasses (Allen 1965) and the high atmospheric and surface humidity. Ridges are of two types (Webb 1968):

1. Small ridges standing 3 ft above adjacent troughs extending for hundreds of yards along the coast. Crest to crest distance is small ranging from 80 to 220 yards and averaging about 150 yards.
2. Larger ridges consisting of smaller areas. Ponds, lagoons etc. occur within the troughs.

Wind blown sand plays little role in forming the beach ridges largely because of high humidity, heavy rainfall and the dense vegetation which clothes the ridges.

Sand which forms the beach ridges can be transported long distances, and east of Lagos, for instance, the beach is composed of sand transported for over 75 miles by the west-east setting current. Between the Benin River and the Calabar River the sand forming the beaches is mainly river derived sand transported by the diverging current systems (Figure 138). Between Lagos Lagoon and the Benin River the beach is predominantly muddy because little sand is transported along the shore in this area.

Tidal ranges are narrow and the horizontal distance between high and low marks is typically 50–150 yards. Slopes are typically 1 in 100 to 1 in 50. In areas where beaches are eroding slopes increase to 1 in 5 to 1 in 10 (Allen 1965).

The present day beaches are characterized by well sorted fine quartz sand as are many of the beach ridges. Medium sand is less common than fine sand and very fine sand is less abundant than fine sand. Mica flakes are conspicuous by their absence from beach sands but opaques and heavy minerals are more abundant. There is little shell or plant debris because of the destructive high energy environment. Data analysed by NEDECO (1954 and 1961) on grading is presented in Figure 121. Median diameters of 200 microns are much more common on the delta tip beaches than on flanks. Beach structures demonstrating the high energy and swashback features are conspicuous.

Allen (1965) (Figure 119) summarized the environmental characteristics:

Strong wave attack on beaches with swash-backwash flows reaching upper flow regime. Long shore currents diverging from delta tip. Soil formation and plant growth on beach ridges.

Allen (1965) also summarized beach lithofacies data:

Delta tip: Mainly evenly laminated, clean, f to m sand. Primary current lineation. Very rare small ripples. Shell and plant debris very rare.
Beach Ridge: Lamination destroyed through plant growth on high parts. Grits.

The Beach and Beach Ridge deposits are obviously of interest to the petroleum geologist because such deposits occur in the Agbada Formation as reservoir sands. Dessauvagie (1974) attempted to show the distribution of some of the larger abandoned beach ridges for the subaerial section of the Niger Delta Complex (Figure 9).

2.5.5.11.5 River mouth bar environment and deposits

These deposits are included within the Agbada facies of Short and Stauble (1967) and Weber (1971) and also form reservoirs within the Agbada Formation.

River Mouth Bars develop at the seaward entrance of channels (Figure 122). The high physical energy environment prevailing at river mouths (tidal and longshore currents) produces well sorted sand deposits. Maximum energy conditions prevail on bar crests and sands are chiefly fine to medium grained. Energy decreases landwards and to the deep. Shell and plant debris is rare. Silts appear on the bar flanks and layered deposits occur near bar toes.

The bar environment is characterized by very strong wave action and reversing tidal currents, and by longshore currents. Allen (1965) described the River Mouth Bar Facies as:

> *Crests*: Mainly clean, vf to m sand with even laminations cross stratification, cut and fill. Shell and plant debris rare.
> *Bar Flanks*: Layered vf sand, clean vc silt and clayey silt. Drifted plant remains sometimes in thick layers. Rare mottles.

Representative river mouth bar shapes from the present day Niger Delta are shown in Figure 122. Shapes range from almost straight in small river mouth with weak tidal flows to narrowly U-shaped in large rivers with high discharges. Lengths from river entrances to bar toe range from 2 to 11 miles and average 6 miles, width averages 8 miles ranging from 3 to 16 miles. The Bonny Bar is the largest of the bars in the present day delta. According to Allen (Unpublished MS University of Ibadan Library) River Mouth Bars exceed 617 square miles in extent. The River Mouth Bar deposits form part of the Agbada Facies of Short and Stauble (1967) and Weber (1971).

2.5.5.11.6. Delta front platform environment and deposits

Strong to moderate wave and tidal current action still prevails in this environment and sediments are distributed mainly by longshore currents either assisted or opposed by the deeper Guinea current (Allen 1968) (Figure 138). Energy conditions decrease outwards from the shore.

The Delta Front Platform is 6–10 miles wide off the nose of the delta and diminishes to 5 miles on the flanks. Off the estuaries of the Calabar River the delta platform is about 12 miles wide. The slope of the platform is between 1 in 2000 and 1 in 4000 and is bounded by a marked steepening on the seaward

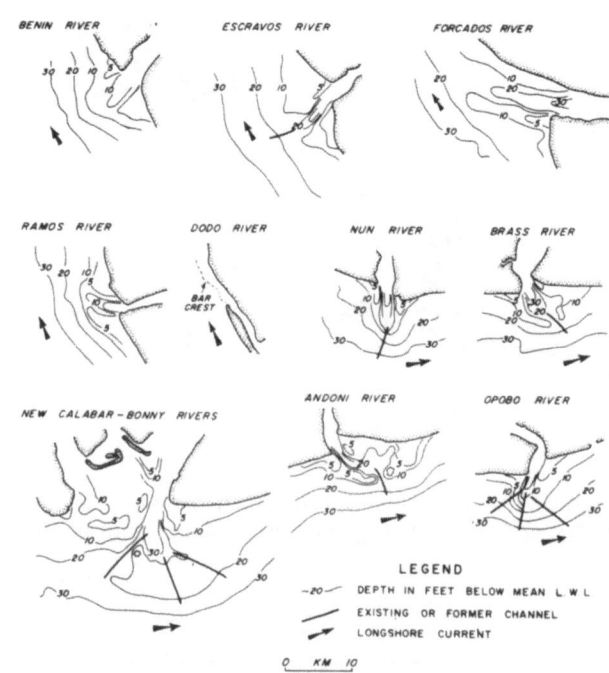

Figure 122. Series of maps showing morphology, current directions etc. of mouth bars (representative) from the Present Day Niger Delta. Based on Allen (1965). Reprinted with permission of the American Association of Petroleum Geologists.

side. A break of slope lying 1–3 miles from the beach limits it on the beach side. Allen (Unpublished MS University of Ibadan Library) estimated that this environment exceeds 1737 square miles. Sediments on the Delta Front Platform include (Allen 1965):

> *Delta tip*: On inner platform uniform coarse vf sand and vc silt with laminations. Mottles few or absent.
> *Delta flanks*: On inner platform uniform coarse vf sand and vc silt. On outer platform layered vc silt, clayey silt and silty clay with plant debris. Some uniformed fine clayey silts.

Outcrops of Older Sands break up the continuity of Delta Front Platform deposits, and provide local sand sources. The outer limit of Delta Front Platform deposits is taken at 30–50 and in Weber's classification (1971) these deposits which would be referred to as Barrier Foot–Fluvio Marine Deposits (30–100 ft water depth).

2.5.5.11.7. Pro-Delta Slope deposits

The Pro-Delta Slope is characterized by gradients of 1 in 1500. The Upper units of this zone occur between 24–36 ft on the delta flanks and 32–40 ft off the delta tip. The lower limit is usually taken at the 132 ft line outside of which the bottom gradient is between 1 in 800 and 1 in 400. This is the average depth of the horizontal thermocline separating the warm Guinea Current surface water from cooler central–south Atlantic Water (Allen and Wells 1962; Allen 1965). The Pro-Delta Slope Environment is dominated by the Guinea Current and ranges in width from 9 to 25 miles (Allen 1965). Allen (Unpublished MS University of Ibadan Library) estimated the extent of the Pro-Delta Slope covers over 3011 square miles.

There is moderate to weak wave and tidal current action over the Pro-Delta Slope and deposits, which according to Allen (1965) include:

Delta tip: Layered vf sand and vc silt, clayey silt and silty clay. Coarser layers with even lamination, cross stratification. Plant debris and mica flakes.
Delta flanks: Layered vc silt, clayey silt and silty clay in shallower parts. Uniform fine clayey silts and silty clays. Abundant mottles.

We have already commented on the arbitrary nature of the boundary between the Agbada and Akata facies and in present day depositional terms it must lie on the continental shelf and perhaps include the deposits on the Lower part of the Pro-Delta Slope.

2.5.5.11.8. Open Shelf Environment and Deposits

The open shelf environment extends from Guinea Current thermocline (average 132 ft in depth) to the shelf break which ranges from 468 to 648 ft off the delta tip to around 300 ft on the flanks and according to Allen (Unpublished MS, University of Ibadan) extends over 3861 square miles. Deposits laid down in this environment and on the open shelf and on lower Pro-Delta Slope belong to the Akata Facies.

2.5.5.11.9. Older Sands

The Older Sands constitute the main deposits in the Non-Depositional areas (Figure 138) and were laid down in the Late Pleistocene and Early Holocene. They are sheet deposits of shelf wide extent. Allen (unpublished MS, University of Ibadan) gave a figure of 1969 square miles for Older Sands outcrop subsea. Downcutting during the Late Wisconsin low of sea level caused rivers to cut deeply into the Late Pliocene and Early Pleistocene Agbada and Benin Facies which had developed in relation to a shore situated 25–30 miles seaward of the Present Day Shore. Transgressive deposition took place as sea level rose and regression developed as a balance became established between sea level fluctuation, sediment supply and subsidence (both local and regional). As sea level rose alluvium was first deposited in the river valleys and as the shelf became inundated sheets of beach and offshore shoal deposits developed as the sea overspread the delta complex. The Older Sands were formed during this transgression. They consist of coarse grained sands bearing shallow benthonic foraminifera. As the sea deepened, terrigenous muds with foraminifera accumulated over the sands. Terrace and ridge systems are developed on the upper surface of the Older Sands and Allen (1965) suggests that the balances between sea level net rise, subsidence and sediment supply enabled two main barrier island complexes to develop seaward of the Present Day Shore. Allen (1965) refers to the "Younger Suite", the sands, silts and clays which form the modern Niger Delta Complex. The contact between the "Younger Suite" and "Older Sands" of Allen (1965) is time transgressive.

Thickness and distribution of the Late Quaternary Younger Suite sediments of the Niger Delta are shown in Figure 123. Allen (1965) has estimated that the Late Quaternary portion of the Niger Delta Complex has a minimum volume of 900 cubic kilometres.

The most recent study on the Late Quaternary Niger Delta Complex was made by Oomkens (1974) who studied cores from three holes drilled largely within the Lower Deltaic Plain environment of the Present Day delta (Figures 124–134). The Post Glacial deltaic sediments were divided into three units (Oomkens 1974):

YOUNGEST
3. The offlap complex of fluviomarine and coastal deposits which consist of a lower member of marine clay and silt and an upper member of tidal channel and coastal barrier sand.
2. An onlapping complex of lower coastal plain deposits which consists of fine grained lagoonal and mangrove deposits and an upper member of tidal channel and coastal barrier sands.
1. Alluvial valley – fill sands and conglomerates.
OLDEST

Unit 1 and Unit 2 were deposited during the Post – Glacial rise in sea level. Unit 3 indicates that deltaic progradation started as soon as the rapid postglacial rise in sea level (Flandrian Transgression) has slowed down. The offlap complex (Unit 2) is more than 100 ft thick.

The lithologies and depositional sequences penetrated in the fourteen coreholes drilled are shown in Figures 124–134.

In other sections at this point in formational description we have included a discussion of oil and gas occurrences and commented upon the petroleum prospects of formation. However, because of the close association of oil and gas occurrences with rollover anticlines (almost all the 150 or so oil fields discovered in the delta province are associated with synsedimentary structures — rollover anticlines, growth faults, mud diapirs etc.). We have reserved discussion of these topics until after we have dealt with the synsedimentary growth fault–rollover structures in Chapter 3, Structure.

2.5.5.12. Present day Niger Delta and Niger-Benue Drainage System Basic Environmental Data

2.5.5.12.1. General

Having considered the stratigraphic details of the formations which make up the Niger Delta Complex and the present day deposits we must now look at a miscellany of environmental data which also bear on origin of the delta complex. As we have pointed out the Present Day–Late Pleistocene deposits sit on the top of a great pile of mainly deltaic sediments which have accumulated since Late Cretaceous time

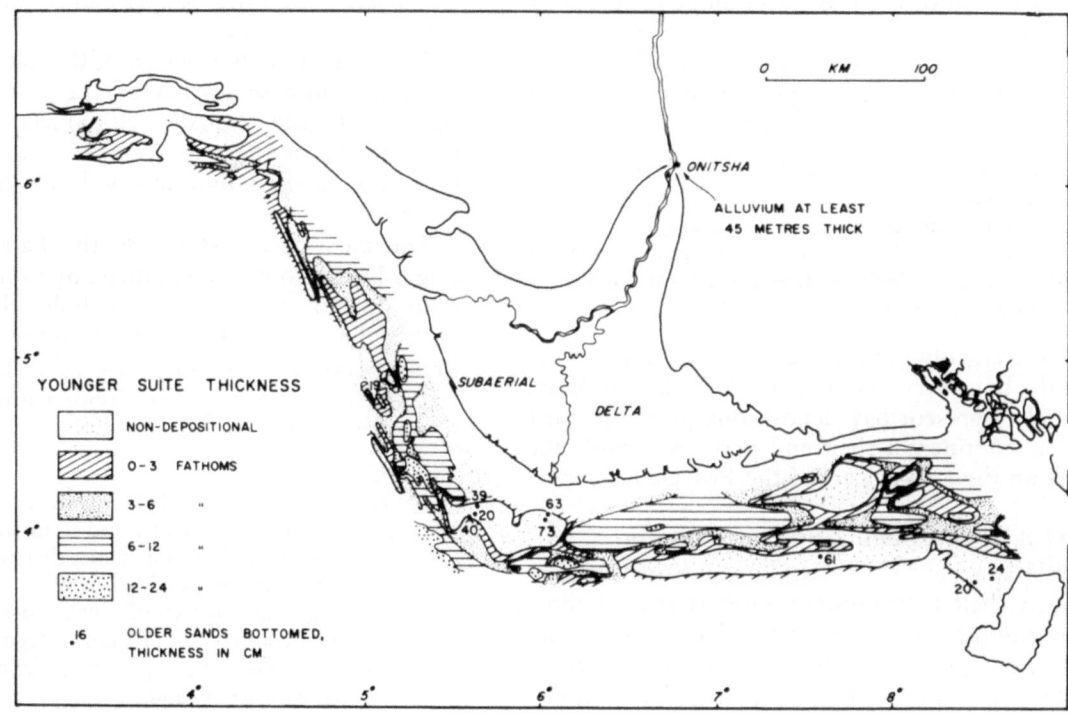

Figure 123. Map showing the thickness of Late Quaternary sediments of the Niger Delta Complex. Distribution of Younger Suite based on echograms and Older Sands based on cores. Based on Allen (1965). Reprinted with permission of the American Association of Petroleum Geologists.

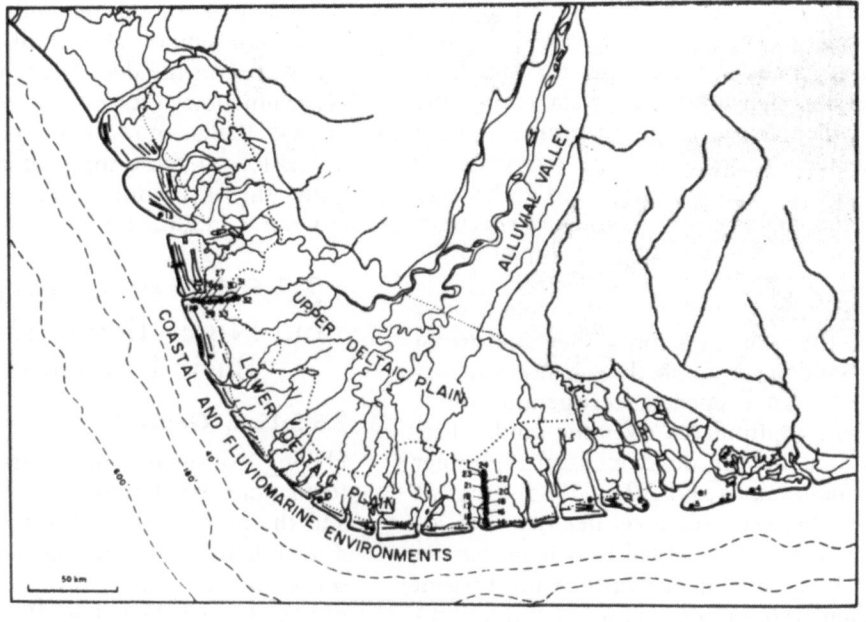

Figure 124. Generalized map Late Quaternary Niger Delta Complex Core hole location map for Figures 125–134. Based on Oomkens (1974).

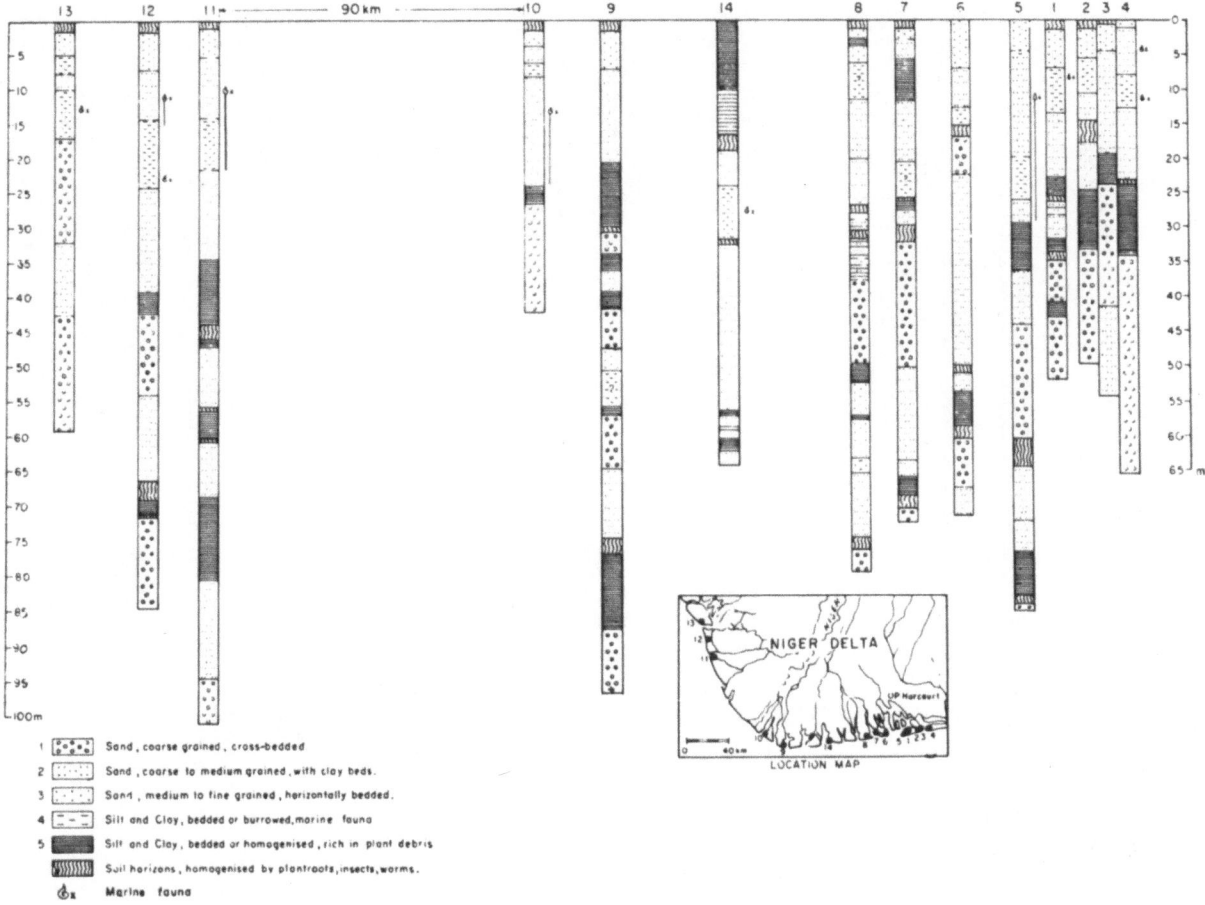

Figure 125. Sections showing the distribution of lithofacies encountered in fourteen coreholes, Late Quaternary Niger Delta Complex. Based on Oomkens (1974).

within the structural constraints defined basically by a complex RFr triple junction system (Figures 7, 22, 224B etc.).

The Present Day sediment distribution pattern and the general physiography of the Present Day–Late Pleistocene Delta have developed because of several major factors which include:

1. transport;
2. mode and ultimate deposition of sediment load;
3. shape and growth of the delta;
4. climate and climatic fluctuation;
5. delta dynamics etc.

Some of these factors are discussed in this section; others are discussed elsewhere.

Sediment Supply and Dispersal: The Niger River enters the delta from the north some 145 miles below its confluence with the Benue River. The Benue is a major tributary and as far as sediment source is concerned more important than the Niger itself; indeed if we consider provenance of the sediments which make up the Niger Delta Complex, it might better be called the Benue Delta Complex.

Sediments entering the Present Day Delta come mostly from the coarsely crystalline Basement Com-

plex and from Cretaceous and Cenozoic basement derived sediment which crops out in the catchments of the Benue and Niger rivers. Overall this situation has not changed significantly since the early Cenozoic, except that the input of volcanically derived sediments into the Benue drainage was significantly higher in the Neogene than in the Palaeogene. The volcanically active Cameroun Zone was initiated in the Miocene (Hedberg 1968; Whiteman 1973; Burke and Whiteman 1973).

The River Niger rises in the Guinea Highlands and first flows northeastwards into the so-called Inland Delta in the Mopti-Tombouctou area. It then swings southeastwards into Nigeria flowing through the Bida Basin and then down to Lokoja where it joins the Benue River (Figure 4).

Important tributaries such as the Sokoto, Zamfoura and the Kaduna join the Niger above Lokoja, and drain the extensive upland area of Basement Complex centred on Kaduna. The Niger breaks through an important east-west watershed at Idah before entering the Upper Deltaic Plain. The course of the Niger above Lokoja is thought to be a relatively new feature which evolved in the Plio-Pleistocene.

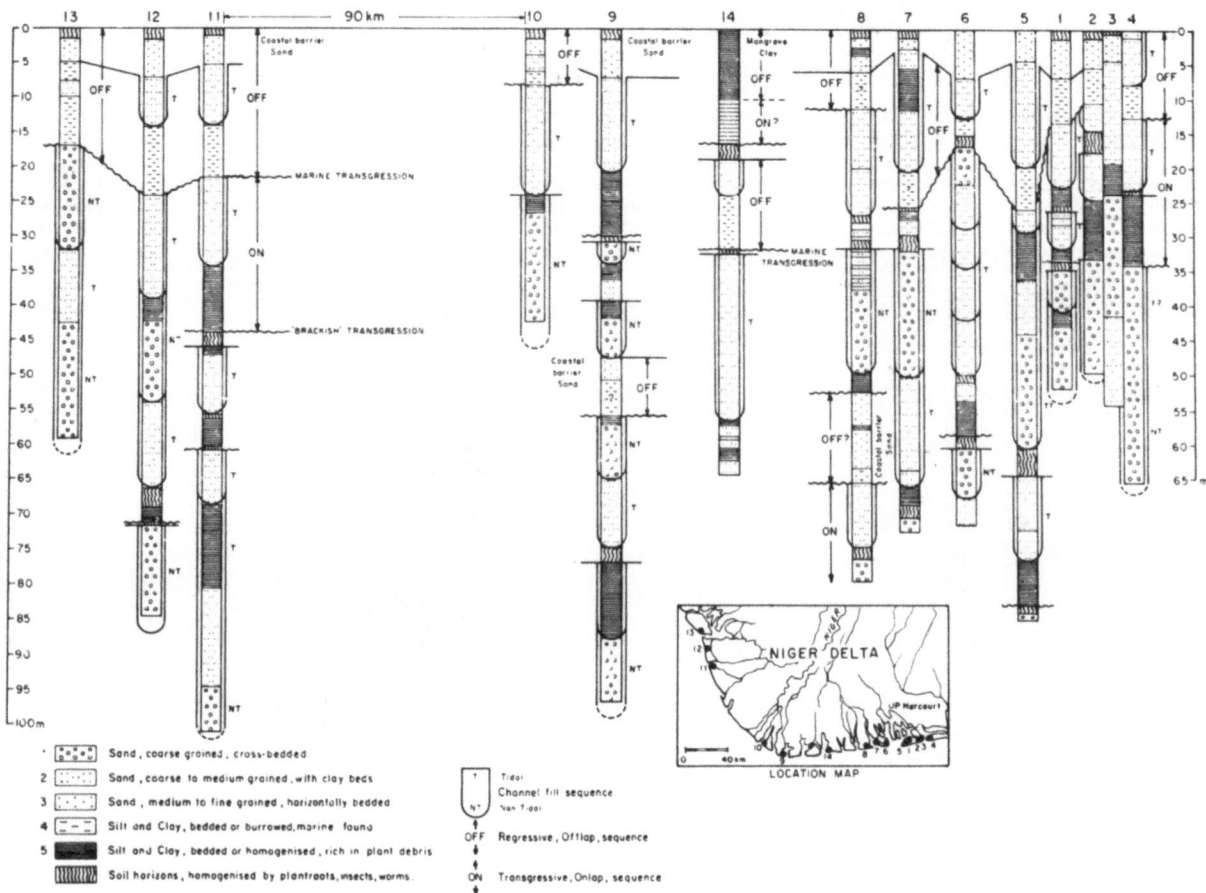

Figure 126. Deposits encountered in fourteen coreholes in Late Quaternary Niger Delta Complex classed into Tidal Channel Fill Sequences, Regressive Offlap Sequences and Transgressive Onlap Sequences. Based on Oomkens (1974).

In Plio-Pleistocene times Upper Niger ran into the Inland Delta (Uvroy 1942; Voute 1962; Burke, Durotoye and Whiteman 1971), a region of interior drainage situated on the northern side of the savannah and on the present day southern margin of the Sahara. Large lake systems existed in this area and in the Chad-Sudan area to the east, during the Pleistocene. Following on the great 20 000 YBP dry phase, when the Saharan edge extended into southern Nigeria and reached the coast near Accra (Burke, Durotoye and Whiteman 1971), extensive lakes developed about 10 000 YBP in the Inland Delta area. Spill water overtopped the sill at Taoussa and the Upper Niger drainage was captured by the Lower Niger drainage which, if we are to judge by the extent of the pediplains in the Bida Basin, must have existed for a considerable part of Cenozoic time (Figure 4).

Climate in Nigeria is dominated by movements of the Intertropical Front and the westerly directed descending (and therefore warming) jet stream as the Intertropical Front migrates back and forth. The rainy season away from the coast and the high ground of the Cameroun–Niger frontier area, is mainly from April to October and southwesterlies bring rainfall ranging in amount from 60 inches in the west to 400 inches in the east.

The monsoonal rains of the Guinea Highlands feed the Upper Niger but a large part of the water evaporates and is lost in the Inland Delta area. Southwesterly monsoon winds dominate the Niger Delta region and in the east some 400 inches of rain falls. The amount decreases rapidly to the north and away from the high ground. At Onitsha and Benin areas some 75 inches fall. About 150–170 inches fall in the delta area proper. Humidity is high nearly all the year around and the combination of high humidity and temperature make for unpleasant weather conditions but stabilizes subaerial deposits.

South of the Intertropical Convergence Zone (ICZ) or "Intertropical Front" humid Equatorial Maritime air dominates. In this part of Africa rain-bearing south westerly winds prevail. To the north lies the dry, dusty extremely low humidity Saharan air. Occasionally the dry harmattan wind reaches the coast but this is only for limited periods in January

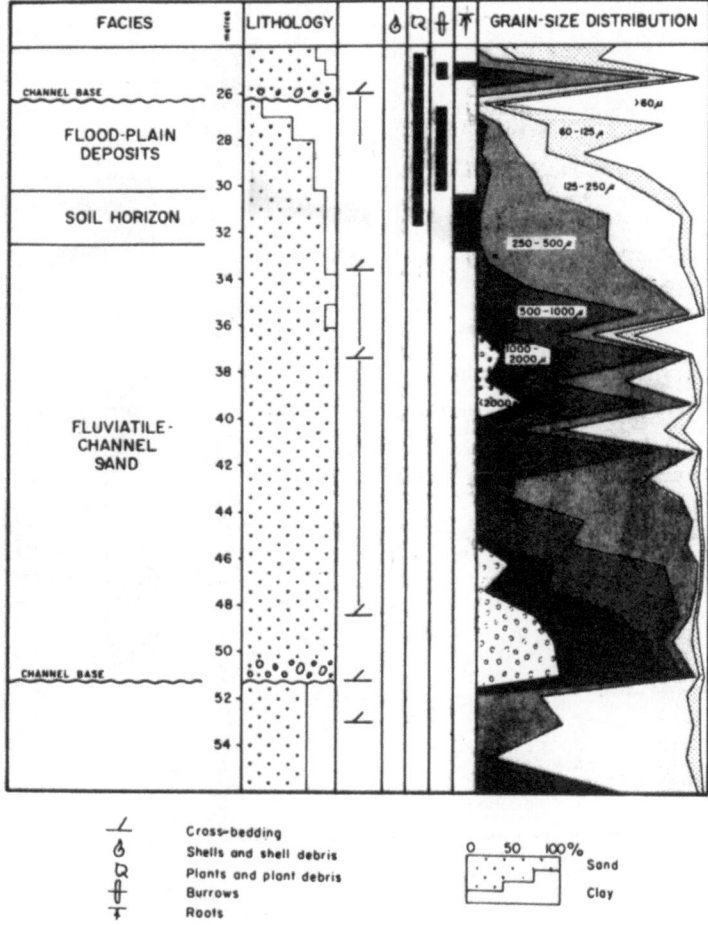

Figure 127. Section showing Facies, Lithology and Grain Size Distribution of a Non-tidal Channel Fill sequence, Corehole 7 (Figure 126). Based on Oomkens (1974).

and February. The rainy season varies in length and duration and time of onset in Nigeria. Temperatures in the delta area are uniform and range from 68°F at night to over 85°F during the day.

2.5.5.12.2. Niger–Benue Sediments

Sediments transported by the River Niger water derived from these systems and reaching the delta are derived mainly from erosion of predominantly acid Basement Complex metamorphic rocks and acid intrusives; from Palaeozoic rocks in Niger and from Cretaceous to Pleistocene sediments elsewhere. At the confluence of the Niger and Benue at Lokoja (Figure 4), the river carries a bed load consisting mainly of medium to coarse sand totalling around 0.3×10^6 cubic metres annually. Suspended sand, silt and clay totals about 4.6×10^6 cubic metres annually. Sand grade and finer materials are characterized by hornblende epidote, sillimanite and zircon.

The total length of the Niger from its source to the sea is around 2562 miles, making it one of the world's eleven longest rivers. A total drainage area of more than 386 100 square miles places the Niger in seventh place after the Amazon, Mississippi, Congo, Nile, Parana and Ganges.

Although the Benue River has a smaller drainage basin than the Niger River, it is a more important sediment source compared with the Niger. From the north important tributaries drain the Jos Plateau and the great area of Basement Complex centred on Kaduna. The tributaries of the Benue draining from the east and south (Figures 4 and 9) such as the Foro, Taraba, Dongu, Katsina Ala etc. rise in the high ground high rainfall area of the Nigeria–Cameroun frontier where heights exceed 3280 ft over wide areas and the highest peaks attain over 8000 ft.

The Benue Catchment is dominated by Basement Complex rocks, thick sandy Cretaceous sediments and predominantly basic volcanics occur within the valley in the Biu and Longuda areas and the much more extensive volcanic outcrop in the Cameroun – Bamenda–Adamawa Mountains. The Benue sediments are characterized by hornblendes, epidote, zircon, garnet, staurolite and sillimanite. The annual bed load discharge (mostly coarse sand) is around 0.6 $\times 10^6$ cubic metres and the suspended load discharge

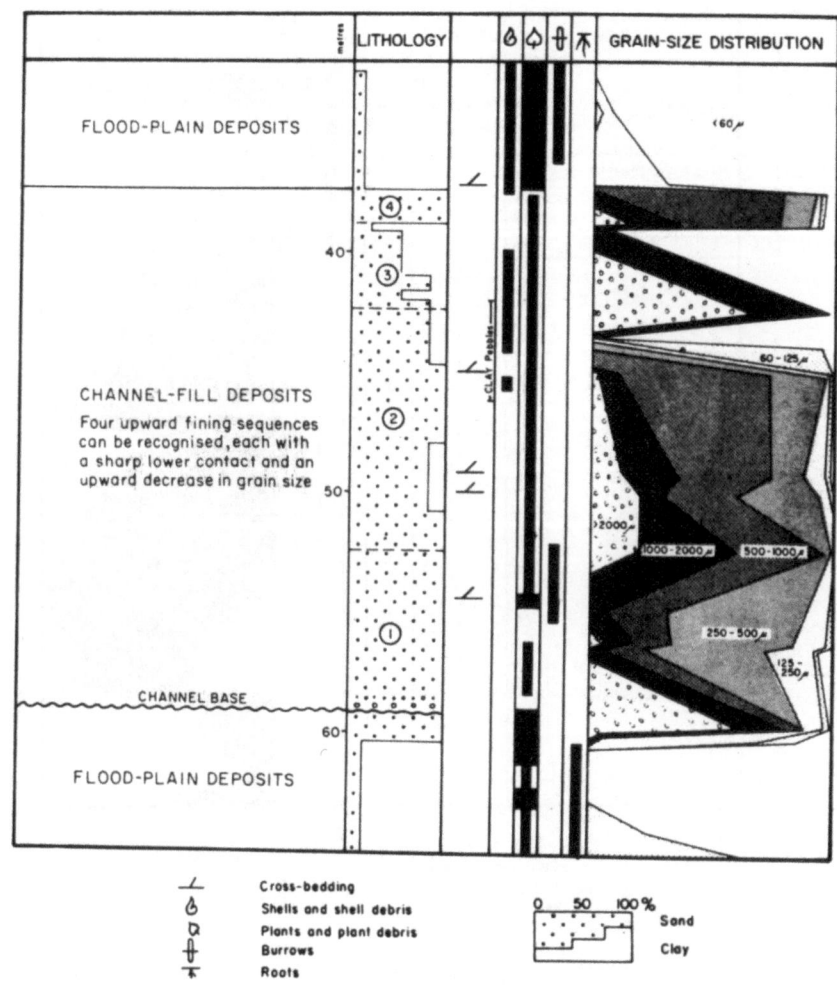

Figure 128. Section showing a Non-Tidal Channel Fill Sequence showing four upward fining sequences, Corehole 5 (Figure 126). Based on Oomkens (1974).

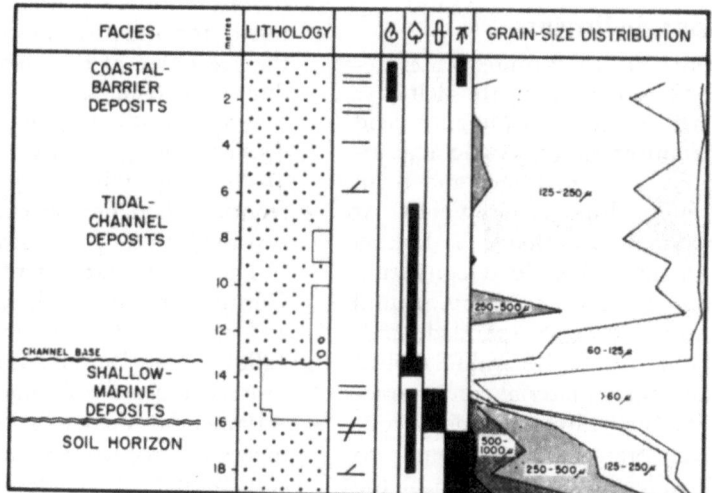

Figure 129. Section showing Coastal Channel Fill Sequence, Corehole 6, (Figure 126). Based on Oomkens (1974).

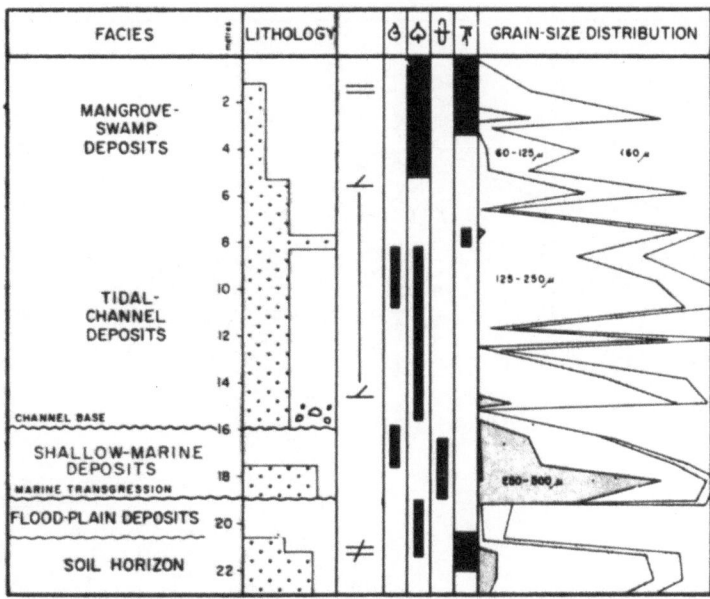

Figure 130. Section showing Facies etc. Inshore Tidal Channel Sequence, Corehole 26 (Figure 126). Based on Oomkens (1974).

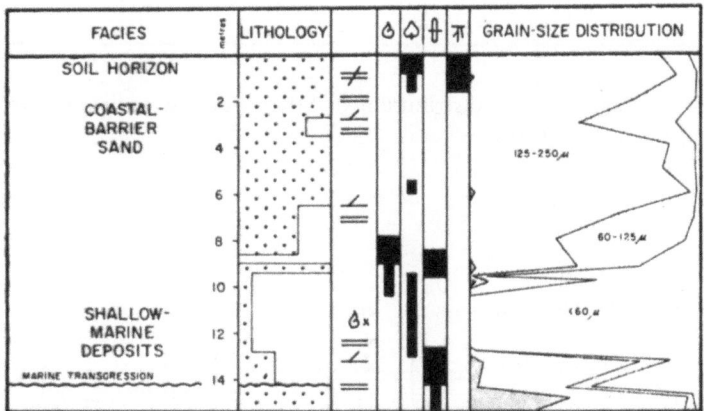

Figure 131. Section showing Facies etc. Regressive Sequence with a Coastal Barrier Sand Member, Corehole 1 (Figure 126). Based on Oomkens (1974).

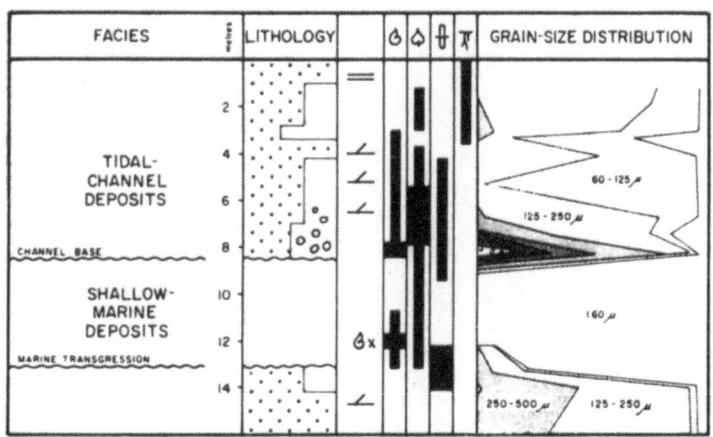

Figure 132. Section showing Facies etc. Regressive Sequence with Tidal Channel Fill Sand Member, Corehole 4 (Figure 126) Based on Oomkens (1974).

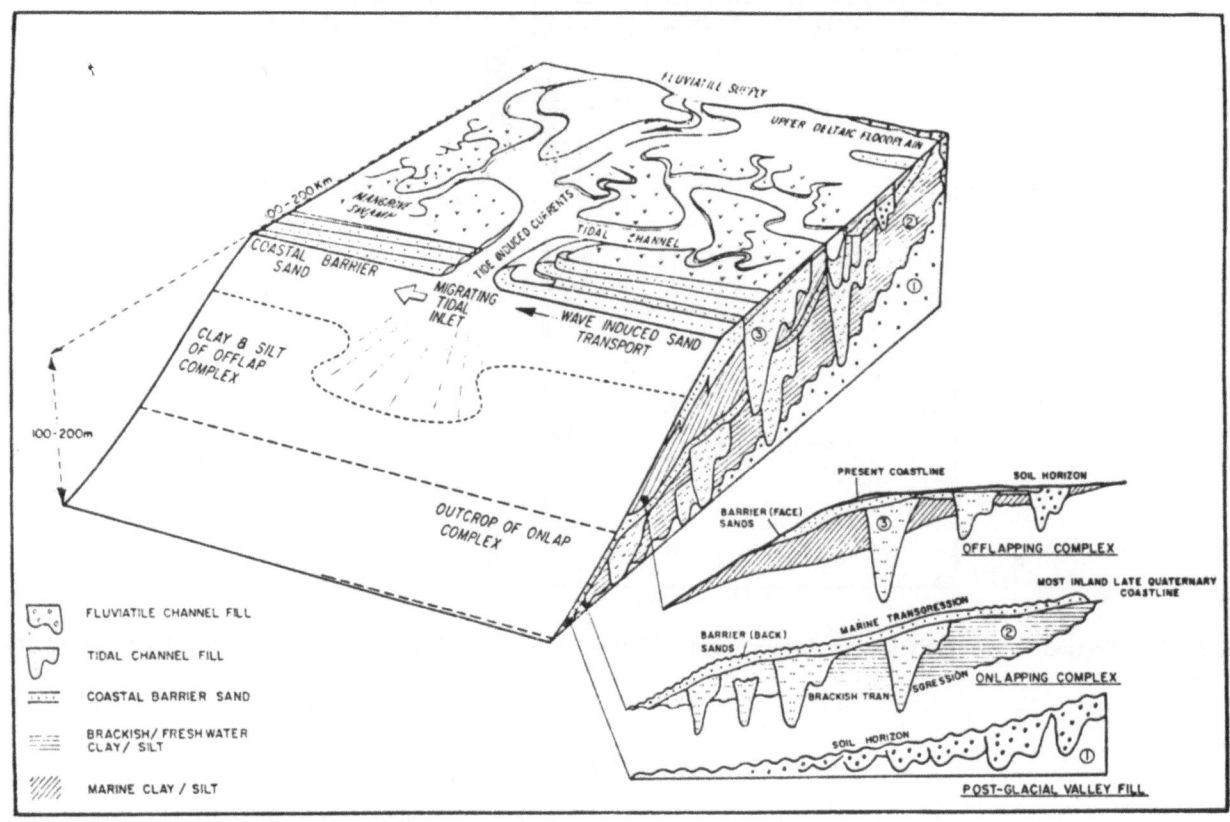

Figure 133. Physiographic diagram showing distribution of lithofacies and environments in the Late Quaternary Niger Delta Complex. Based on Oomkens (1974).

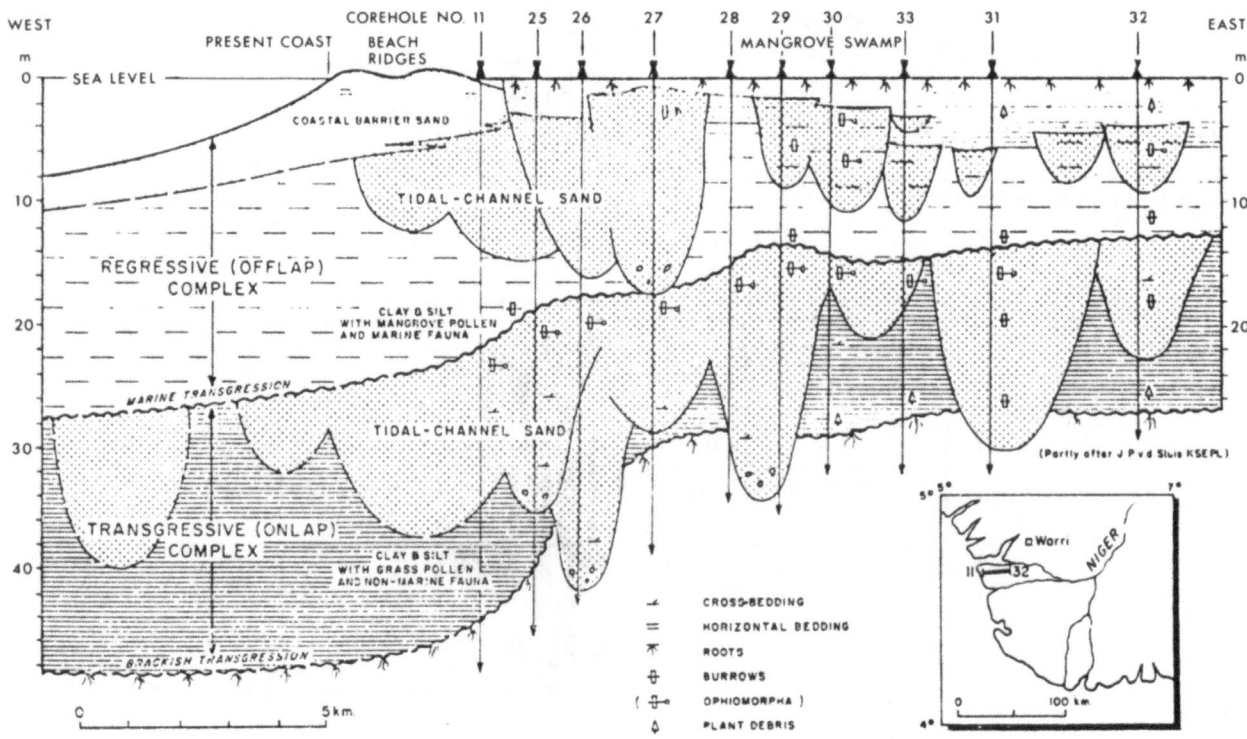

Figure 134. Sections showing transgressive and regressive sequences in the Late Quaternary Niger Delta Complex. Based on Oomkens (1974).

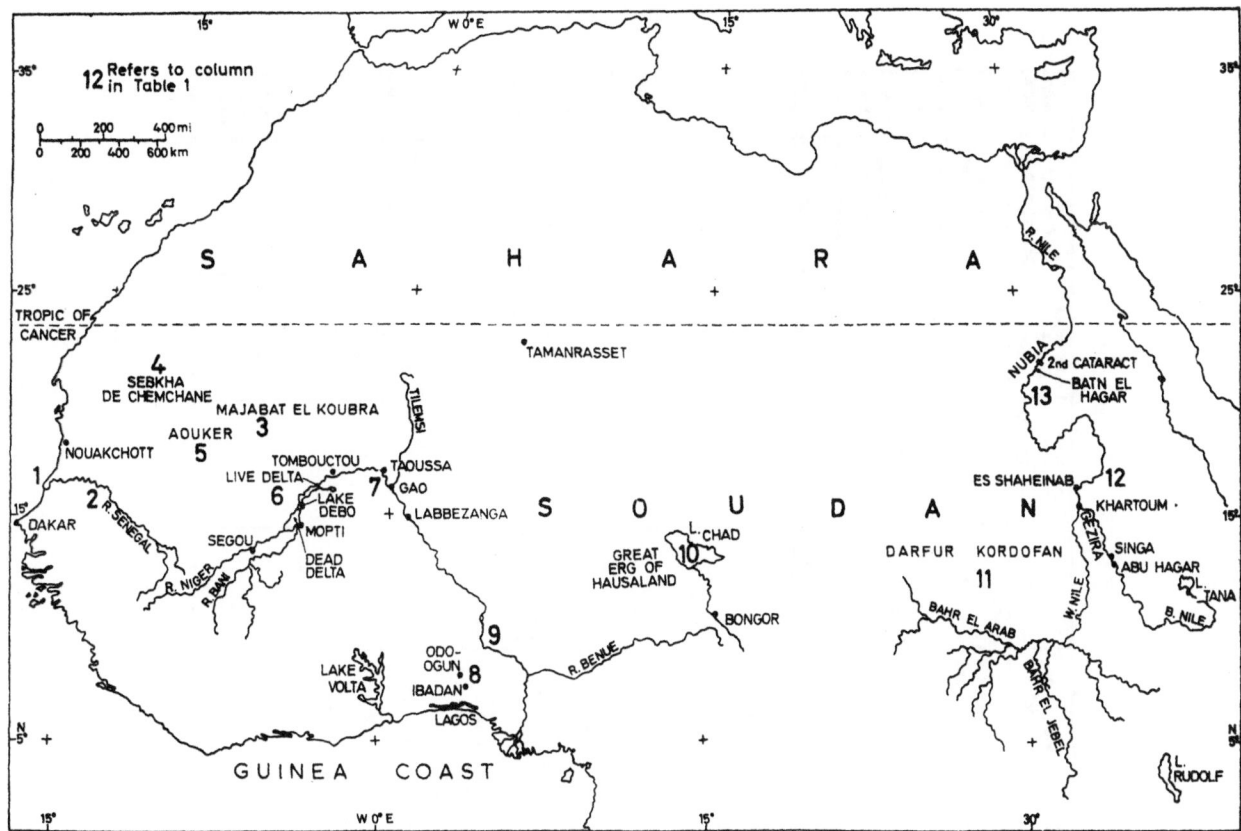

Figure 135. Map showing the distribution of Dry Phase Deposits 20 000 years ago for West and Northeastern Africa shown in Figure 136. Based on Burke, Durotoye and Whiteman (1971).

amounts to about 11×10^6 cubic metres, clearly making the Benue catchment a much more important sediment source than the River Niger (Figure 139). Of the combined Niger-Benue sand-size load, around 65% is retained within the delta as Benin Sand facies and only some 35% reaches the sea. Of this 35% very little apparently escapes into deep water and is retained within the upper part of the delta complex.

West of the Niger Delta Complex, rivers situated as far west as the Volta River of Ghana contribute sediment. Some of this material eventually becomes incorporated in the delta system—because of the easterly setting longshore currents which prevail as far east as the Mahin Beach in Nigeria (Figure 138). These rivers contribute staurolite, zircon, epidote, kyanite, sillimanite and green hornblende as heavy minerals. Probably their major contribution is in fines.

East of the Present Day Niger Delta the Cross River (most important sediment source), the Bonny, Imo, and Kwa Ibo rivers contribute sediment to the delta complex. The Cross River Catchment includes Basement Complex formations, Cretaceous and Tertiary sediments and Cenozoic volcanics and mineral assemblages include brown hornblende and regional metamorphic minerals. The Imo and Kwa Ibo rivers largely drain areas of Tertiary sands and yield assem-

blages marked by kyanite, zircon, with uncommon tourmaline and hornblende (Allen 1965).

2.5.5.12.3. Distributive Agents, River Flows

NEDECO (1959 and 1961) established that the annual discharge of freshwater into the delta is about 200×10^9 cubic metres and these waters carry about 1×10^6 cubic metres of sand and some 17×10^6 cubic metres of silt and clay. Swift flows in the distributaries carry these sediments unequally through the delta proper from a centre near Aboh (Figure 138).

The annual water discharge of the Niger River ranges from 170×10^9 to 300×10^9 cubic metres per annum or from 5420 cubic metres per second to 940 cubic metres per second. These figures show that the overall discharge is about two fifths that of the Mississippi River. At Onitsha (1925–1957) the combined Benue-Niger River carried on an annual average 19×10^6 cubic metres of sediment. Of this sediment load about one part of sand was transported along the river bed, two parts of sand fraction were carried in suspension along with sixteen parts of silt and clay in suspension (NEDECO 1959).

The amount of sediment transported and its dispersal pattern will depend on the river regime and whether or not the rivers are in flood. Peak water and sediment discharge takes place from September

TIME IN YBP	LOWER SENEGAL VALLEY (1)	UPPER SENEGAL VALLEY (2)	MAJABAT AL KOUBRA (3)	SEBKHA DE CHEMCHANE (4)	AOUKER (5)	MID NIGER SEGOU-TOMBOUCTOU (6)	MID NIGER TOMBOCTOU LABEZZANGA (7)	ODO OGUN IBADAN (8)	MID NIGER JEBBA (9)	LAKE CHAD (10)	DARFUR-KORDOFAN-KHARTOUM-NORTHERN PROVS (11)	NILE VALLEY KHARTOUM AREA (12)	NILE VALLEY BATN EL HAGAR (13)	REGIONAL CLIMATE
	SPIT AT RIVER MOUTH 1620 YBP, INTERDUNE LAGOONS 2470 YBP, COASTAL DUNES (YELLOW)	DELTA FORMATION	WHITE SHIFTING DUNES	DUNE FORMATION	DUNE FORMATION / GULLYING	LAKE DEBO SHRANK	LEVEES & DEFLATION	DEPOSITS ON SLOPES OF RELIVE-NAKED VALLEYS	SMALL STREAMS DOWN CUTTING / BUSSA JEBBA RIVER DOWN CUTTING	DUNES: 1100; FALLING LAKE LEVELS / GULLYING	DRY PHASE; WET PHASE; DRY PHASE; LAKES BETWEEN DUNES	NILE SILTS AND WADI DEPOSITS	MODERN ARID PHASE; MID HOLOCENE TRANSITION PHASE 7200; NILE FLOW 9230 > 2 X PRESENT	DRY / WET / 5000 / DRY / MAJOR WET / 9000
5000	* NOUAK-CHOTTIEN BEACH 5500 YBP; REDDENING OF GREAT DUNES	RIVER SANDS DEPOSITED; REDDENING OF CLAYS BEHIND DUNE BLOCK	REDDENING OF GREAT DUNES	GREEN GYPSIFEROUS * 7320 YBP; * 9230 YBP CLAYS	BLOWN SAND INFILLED RAVINES; REDDENING OF GREAT DUNES / GULLYING; 15 M WADI TERRACE	LARGE LAKE DEBO FED BY NIGER BANI; GREY BROWN YELLOW DUNES ON OLD FLOOD PLAINS; REDDENING OF GREAT DUNES; NIGER FLOWS THROUGH MOPTI	LOWER TERRACE OF NIGER; DUNES OF RECENT ERG N-S TREND; REDDENING OF GREAT DUNES; NIGER FLOWS OVER TAOUSSA; TILEMSI FLOWS; HIGHER TERRACE FROM TOMBOCTOU-LABEZZANGA	LOAMY SAND	LOAMY SAND	DUNES: DISCHARGED (CALC? VIA LAKE BED); C14 ON *SPILLWAY *CALC VIA BOKOR DIATOMITE = 320 M	REDDENING OF DUNES / GULLYING; ALTERNATING WET & DRY CONDITIONS	NILE SILTS AND WADI DEPOSITS; KHOR UMAR GASTROPOD SITE 7450; 382 M LAKE 8700; WHITE NILE; 382 M LAKE 11300; 396 M LAKE ?	7200; 9230; 14000	10000
10000	FORMATION OF GREAT DUNES; BLOCKING RIVER SENEGAL	RIVER CUTS THROUGH DUNES; FORMATION OF GREAT DUNES; CLAYS DEPOSITED BEHIND DUNES	FORMATION OF GREAT DUNES	FORMATION OF GREAT DUNES	FORMATION OF GREAT DUNES C 50 M HIGH	FORMATION OF GREAT DUNES; NIGER BLOCKED N. OF SEGOU	FORMATION OF GREAT DUNES; NIGER DID NOT FLOW	UNCEMENTED PEDIMENT GRAVEL	UNCEMENTED PEDIMENT GRAVEL; 25 M BELOW SEA LEVEL; BURIED CHANNEL	FORMATION OF GREAT DUNES; GREAT ERG OF HAUSALAND; LAKE DRY	FORMATION OF GREAT DUNES; QOZ OF KORDOFAN	UPPER CLAY MEMBER GEZIRA FM EQUIVALENT; SEBILIAN SILTS	SEBILIAN SILTS	MAJOR DRY / 20000
15000										RISING LAKE LEVELS				
20000	VALLEY CUTTING PHASE	GRAVELS DEPOSITED BY RIVER				NORTH OF SEGOU NIGER & BANI FLOW INTO DEAD INLAND DELTA		CEMENTATION OF OLDER PEDIMENT GRAVEL	CEMENTATION OF OLDER PEDIMENT GRAVEL			MIDDLE SANDS AND GRAVEL MEMBER GEZIRA FORMATION		25000
25000	* 31000 YBP INCHIRIEN SUPERIEUR * 32000 YBP													30000 YBP WET
30000														

KEY

(hatched)	Wet conditions
(dotted)	Dry conditions
(blank)	Uncertain
12	Column numbers as used in text
*	Generalized radiocarbon dates

Figure 136. Correlations of environments, deposits and climate West and Northeastern Africa over the last 30 years. Based on Burke, Durotoye and Whiteman (1971).

154

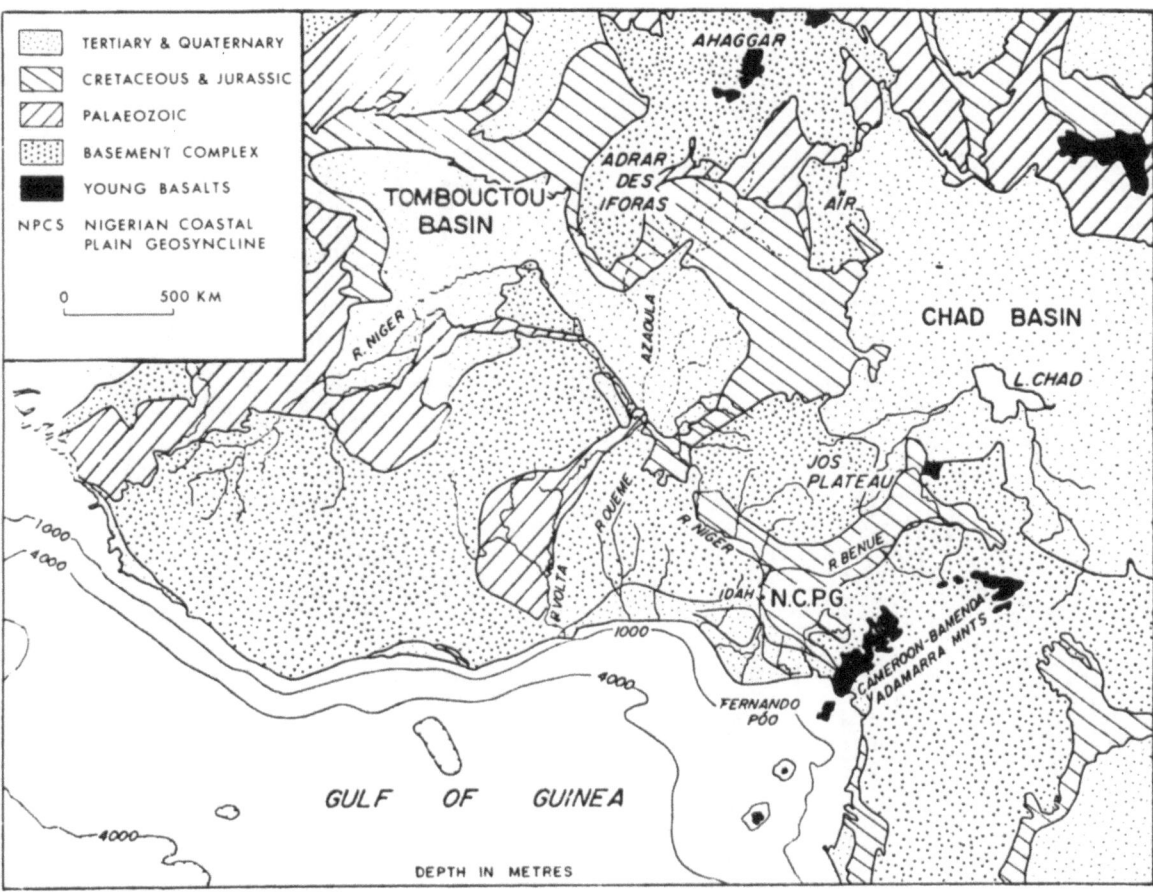

Figure 137. General geological map showing the main rivers draining into the Niger-Benue Catchment and supplying sediments to the Niger Delta Complex. Based on Allen (1965). Reprinted with permission of the American Association of Petroleum Geologists.

to October and from December to May discharge is low (NEDECO 1961). Low mean flow velocities (37 – 82 cm/sec) do not transport gravel (Allen 1965). During the flood, levees are overtopped, crevasses develop and diversion channels are occupied.

Some water enters the independent rivers marginal to the main system in the Lower Floodplain during the flood and water escapes from flooded bottom lands into the Ase and Orashi Rivers in the Upper Floodplain channels (Figure 9 etc.).

In the Lower Floodplain the Niger divides into numerous distributaries (Figure 138) extending south to the Mangrove Swamps. Dispersal occurs through an arc of 85° and directions ranging between south and west. Debris of sand grade and finer is transported. The Nun distributary passes about 26% of the fresh water to the sea and the Ramos and Forcados exits pass about 29 and 15% of the fresh Niger water, respectively. The remaining 30% of the fresh water is discharged via 19 exits (NEDECO 1959). The proportions of water and sediment discharged by major distributaries of the Niger Delta are shown in Figure 140.

2.5.5.12.4. Distributive Agents, Tidal Flows

Within the intertidal mangrove belt and in shore waters reversible tidal currents disperse sediment. Tides are diurnal and approach the shore from the south-southwest through more than 20 major channels. Tidal ranges are from 1m at Lagos, 1–3 m at the Forcados River to 2.8 m on the Calabar River where a high energy, destructive estuarine delta has developed. In the tidal range 1–2 m the tidal capacity of the swamp is about 480×10^6 cubic metres and the discharge from it of salt water greatly exceeds that of freshwater during each tidal cycle (Allen unpublished MS Ibadan University Library). Expressed differently the swamp belt and associated provinces have a collective tidal capacity of around 3000 times the fresh water influx during each tidal cycle. During the flood sediment-laden river water is backed up by tidal currents and forced up delta and into marginal swamps. This returns some fines and some sand from the shore areas and redistributes them through the swamp. Sediment laden waters ponded back during flood tide is released during the ebb and turbid water can be seen as far as 10 miles from the shore.

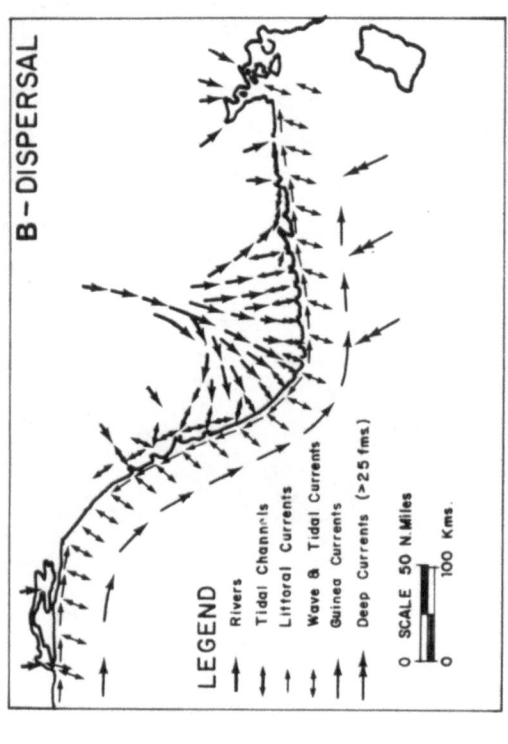

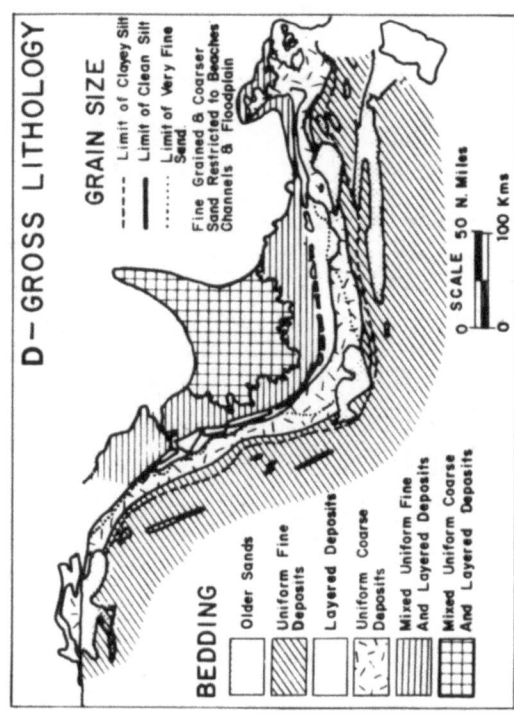

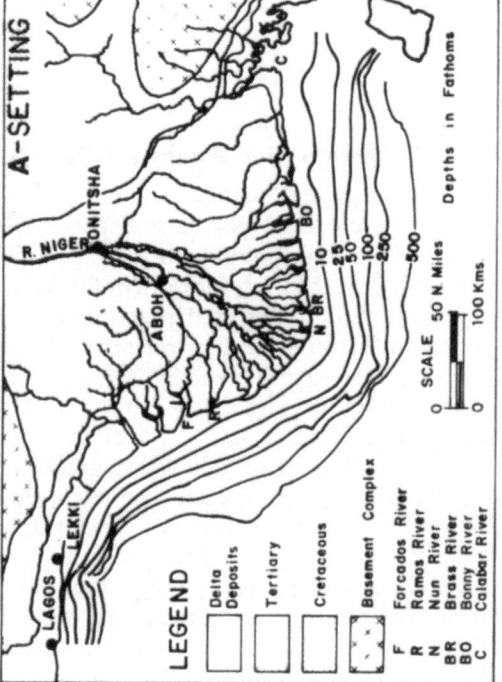

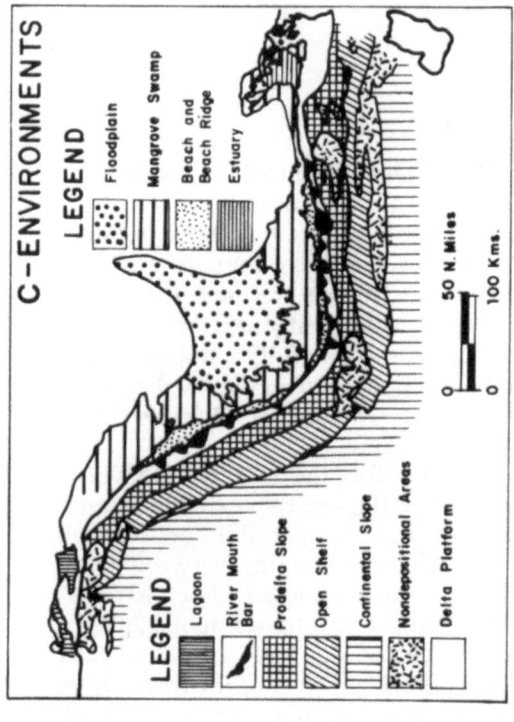

Figure 138. (A–D) Sketch maps showing important environmental and sedimentological features Niger Delta Complex. (A) General Geological Setting; (B) River and Ocean Dispersal Systems; (C) Main environments of Deposition; (D) Gross Lithology. Based on Allen (1965). Reprinted with permission of the American Association of Petroleum Geologists.

Figure 139
Niger and Benue Sediment Discharge

	Benue x 10⁶ m³	Niger x 10⁶ m³
Annual Bed Load Discharge	0.6	0.31
Annual Suspended Load Discharge	11	4.6
Combined Sediment Load at Onitsha		19

NB Around 35% of combined sand load reaches the sea
Around 65% of combined sand load retained in Delta Benin Facies.

Based on NEDECO (1959) and Allen (1965). Reprinted with permission of the American Association of Petroleum Geologists.

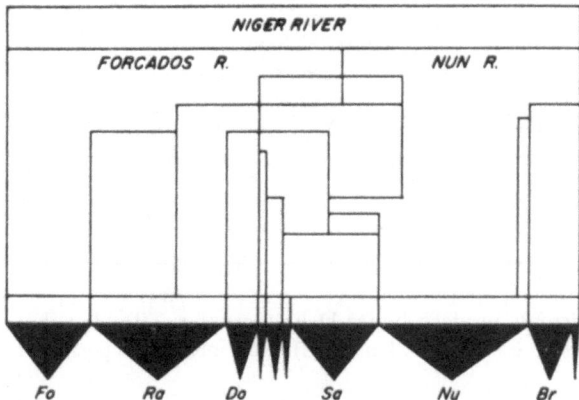

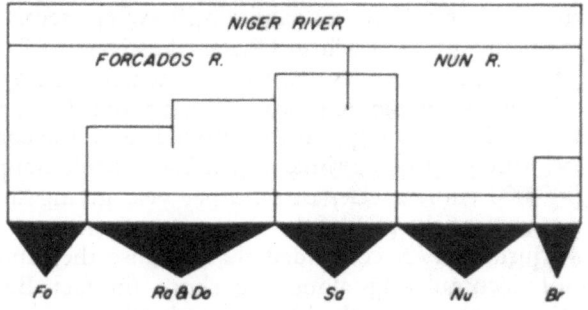

Figure 140. Division of water and sediment discharged by major distributaries of the River Niger. Discharge of the Niger River above the Forcados – Nun bifurcation is taken as 100%, the width of each rectangle expresses the proportion of that discharge carried by a particular stream. Key to rivers is on Figure 138A. (Allen 1965). Reprinted with permission of the American Association of Petroleum Geologists.

Tidal currents are strongest over shallow mouth bars and during ebb tides velocities range from 60 to 280 cm/sec (Allen 1965).

Depth limits the active and dispersive powers of tides (Figure 141) but on the outer shelf, currents appear to be strong enough to keep silt and clay suspended at least for peak flow periods. Once the fines are deposited currents do not easily erode these grades and there is a net sedimentary gain if the area is subsiding.

2.5.5.12.5. Distributive Agents

Wind-wave Currents: Basic data on wind and tidal currents are presented in Figures 138 and 141. Wave action is high in the Gulf of Guinea and prevailing southwesterly winds because of their long oceanic fetch, generate large swells. As far as the Palaeogene and Neogene Niger Delta Complex is concerned it should not be forgotten that in Early Palaeogene times the proto-Atlantic Ocean was at least 1400 miles wide from the eastern tip of South America to Onitsha, Nigeria and so wave power may well have been considerable.

Maximum orbital velocities decrease into deeper water and in depths of around 50 ft the maximum velocity is around 50 cm/sec. Such velocities will remove, either partially or wholly, whatever silt or clay was deposited at slack water. At 120 ft depth the maximum velocity falls to about 25 cm/sec and cohesive clay sediments probably remain unscoured (Allen 1965). At greater depths wave generated currents probably do very little except temporarily inhibit settling of clay and silt.

Wave generated currents diverge from the delta nose and currents set eastwards from Lagos to form a convergent current system along the Mahin Beach (Figure 142). This system may feed the Avon and Mahin Canyons (Burke 1972). Longshore currents have velocities of 18 – 41 cm/sec in the Bonny Bar area and therefore are probably able to keep sand on the move and silt and clay in suspension. Wave generated rip currents extract sediment from the shore area and move it away from the area for a quarter of a mile or so.

Dispersal Agents – Oceanic Currents: Two major bodies of water have been identified in the shallow part of the Guinea Gulf (Allen and Wells 1962):

1. An upper warm Guinea Current; and
2. Lower cool South Atlantic water

The thermocline intersects the continental shelf at 132 ft. The Guinea Current moves eastward across the delta front at speeds of up to 30 cm/sec and in contact with the sea bed. The current appears to be unable to move sand but outside the wave generated inshore zone is mainly responsible for dispersing suspended fines.

2.5.5.12.6. Distributive Agents: Longshore Drift and the Niger Delta Submarine Canyons

Burke (1972) drew attention to an intriguing consequence of divergent, southwesterly wind generated

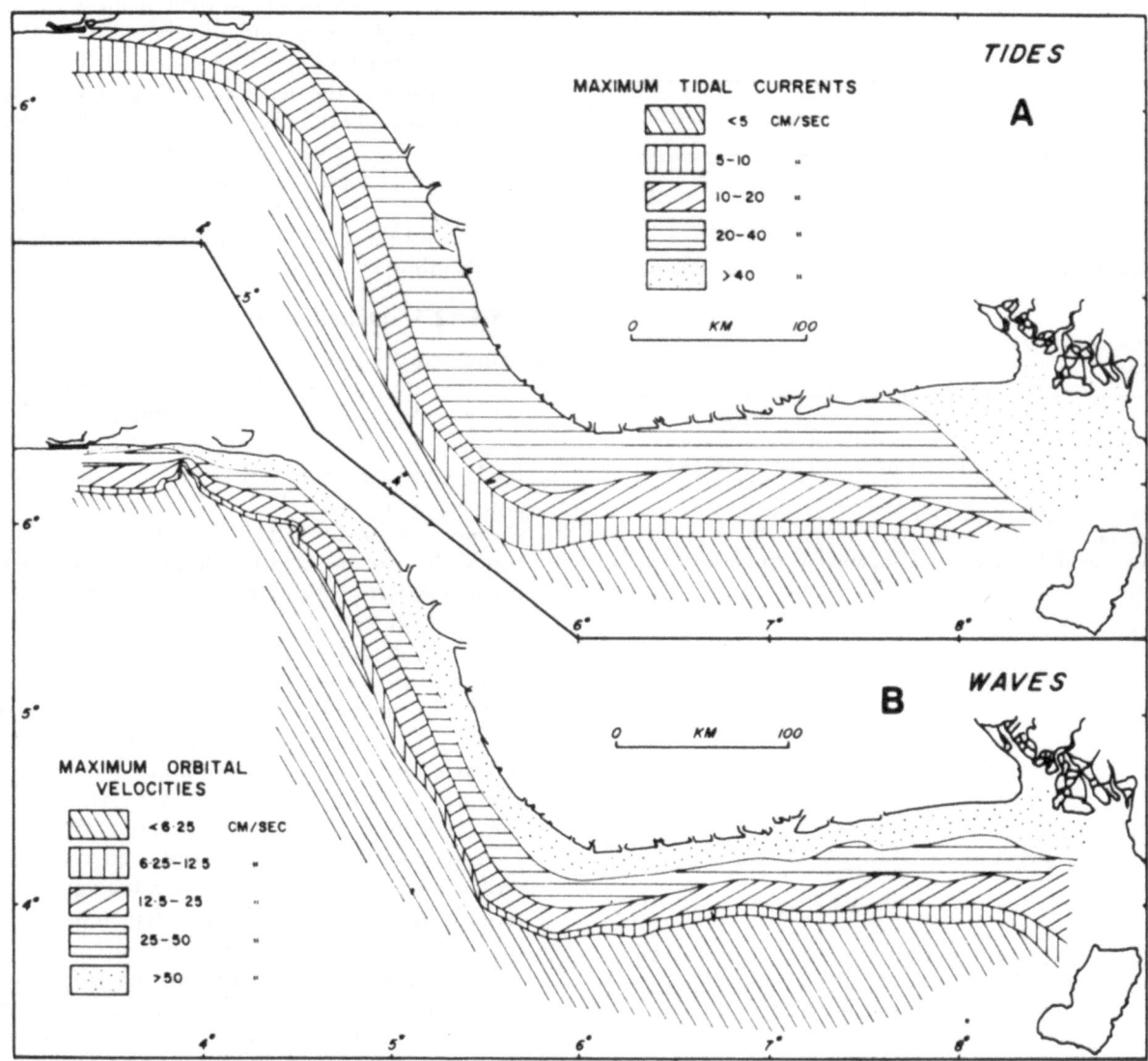

Figure 141. (A) Map showing distribution of maximum tidal currents modern Niger Delta Complex. (B) Map showing maximum orbital velocities of waves modern Niger Delta Complex. Based on Allen (1965). Reprinted with permission of the American Association of Petroleum Geologists.

longshore drift, along the front of the Present Day Niger Delta. He pointed out that large quantities of sand, silt and clay are transported away from the nose of the delta; where they meet opposing sediment transporting current systems and where, because of lack of accumulating shallow sediments, large quantities of sediments must be extracted from the shallow part of the delta to feed two major submarine canyons and their deep subaqueous deltas.

Using NEDECO data (1961, Table VII 5–3) Burke (1972) calculated that northerly directed drift increases northwards from the delta nose as successive distributaries contribute so that between the Forcados and Ramos distributaries the drift exceeds 50 000 cubic metres per year (Figures 9 and 142). A sand beach extends up the coast to the Benin Mouth but north of the mouth the Mahin Shore is muddy, despite the fact that more than 1 000 000 cubic metres of sand etc. are carried annually towards Mahin (Fig-

ure 143). Similarly, longshore drift sweeps sediment from the Volta Beach in Ghana, the rivers of Togo and Benin (Dahomey) past Lagos, where the Nigerian Rivers contribute sandy sediment, but this sandy material does not appear to continue along the coast. NEDECO (1961) estimated that Lagos Mole trapped 500 000 cubic metres of sand per year giving some measure of the size of the easterly sand drift.

Burke (1972) concluded that because the sand is not accumulating along the shore (in fact Beach Ridges are rare on the Mahin Beach which is muddy as we have said) therefore the sand etc. must be transported and deposited elsewhere. The Avon's and Mahin Canyons are located in appropriate positions so that sand can be channelled down to fans at the delta complex foot. Other canyons exist along this coast (Figure 142). Rivers which may have fed into Avon's Canyon during the Pleistocene lows in sea level are shown in Figure 148.

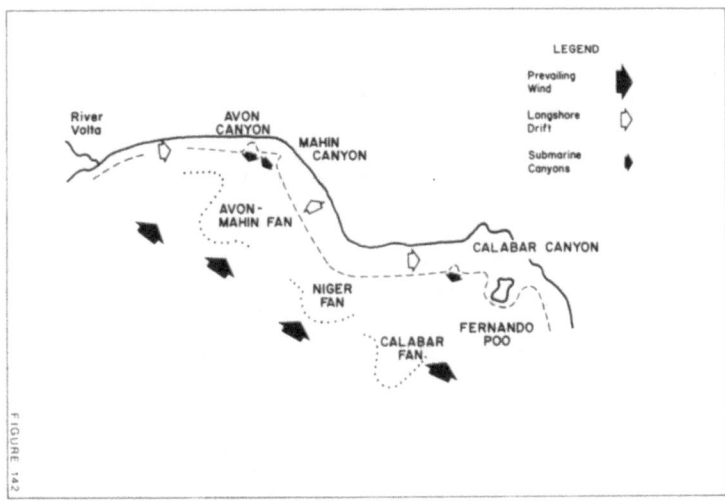

Figure 142. Sketch map showing the positions of submarine canyons, directions of longshore drift and prevailing winds for the Eastern Gulf of Guinea and the Present Day Niger Delta Complex. Based on Burke (1972). Reprinted with permission of the American Association of Petroleum Geologists.

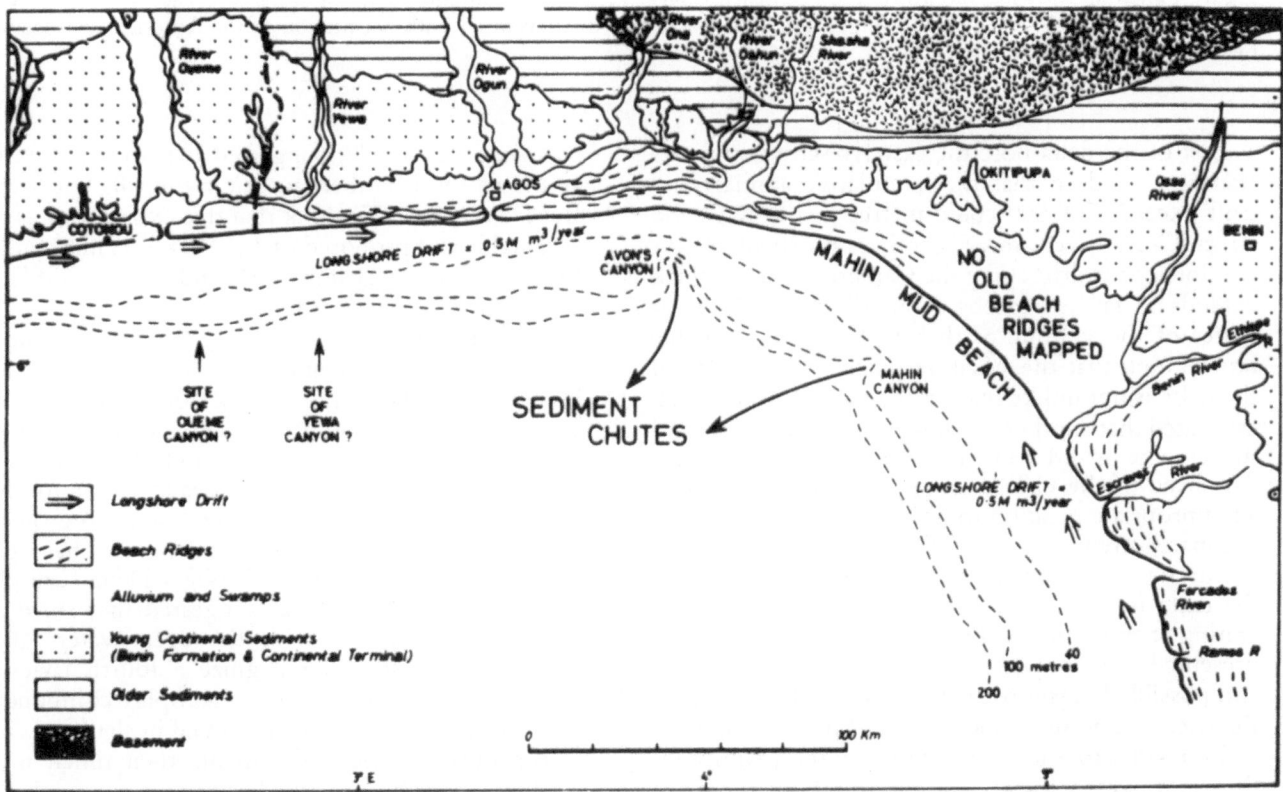

Figure 143. Map showing the distribution of Superficial Deposits from Cotonou (Benin) to the Ramos River Nigeria, sites of canyon heads and sediment chutes. Mahin Mud Beach exists because of converging longshore drift sweeps coarse sediments into the Avons' and Mahon canyon sediment chutes. Based on Burke (1972). Reprinted with permission of the American Association of Petroleum Geologists.

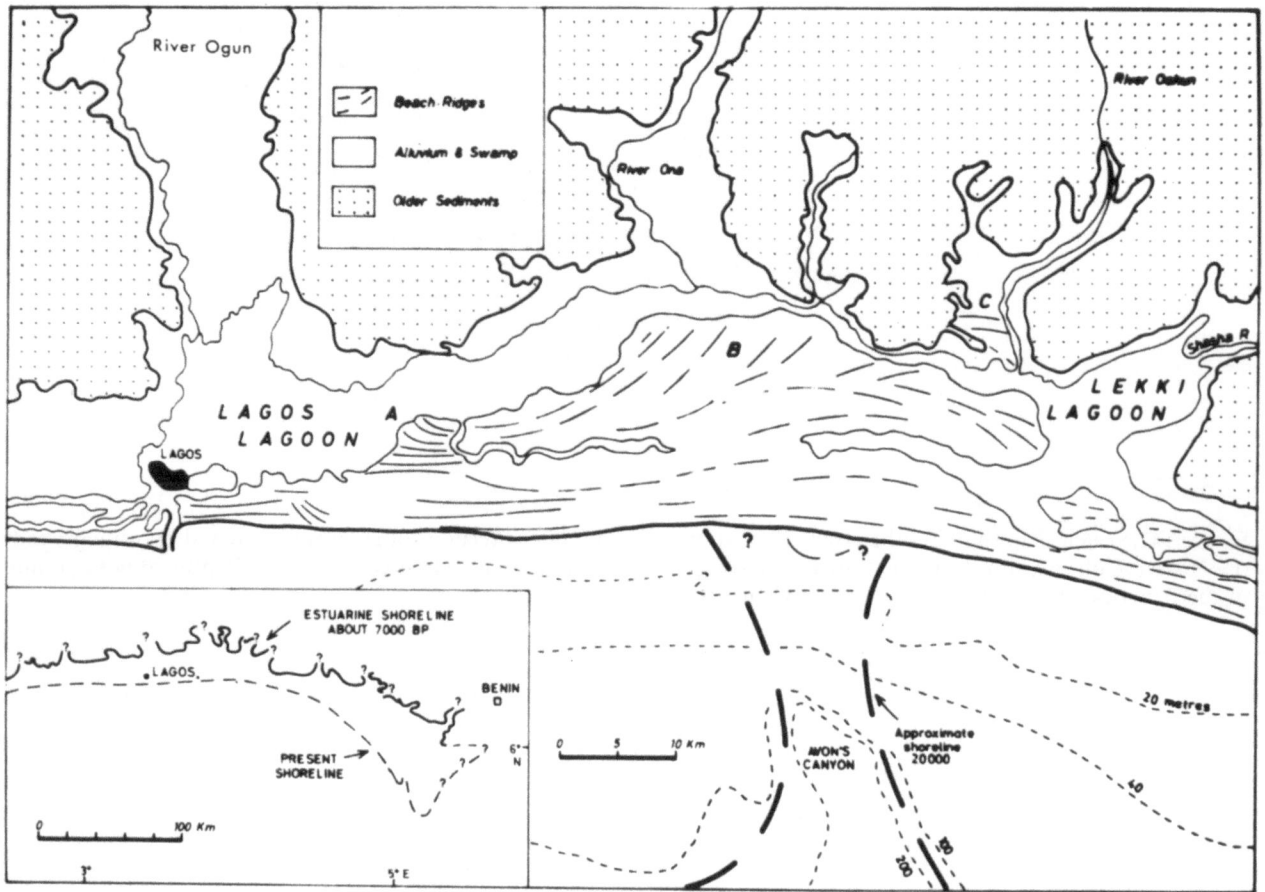

Figure 144. Map showing the distribution of beach ridges, Pleistocene shorelines and the possible position of the head of Avons' Canyon at 20 000 year BP stage. Based on Burke (1972). Reprinted with permission of the American Association of Petroleum Geologists.

East of the delta nose the easterly setting currents transport sand so that the cumulative annual sand drift exceeds 500 000 cubic metres in the Kwa Ibo River–Cross River area. Figures are not available for the northward drift along the Cameroun coast which extends as far as Souleba Point (Pugh 1954). Estimates of the volume of this drift are not available but arguing that the situation on the eastern side of the delta is not unlike that on the west, Burke (1972) estimated that another 1 000 000 cubic metres of sand must be removed from the shallow water system. The head of the Calabar Canyon (Allen 1964) is in an appropriate position to channel this sand into the Calabar Canyon and Fan (Figure 145).

Burke (1972) argued that the Niger Fan in front of the delta nose was fed by a submarine canyon operating at times of low Pleistocene sea level. He also presented a set of maps showing delta shore positions and possible longshore drift for Medial Eocene, Late Eocene, Oligocene, Miocene, Late Miocene-Pliocene and Plio-Pleistocene times, drawn on the assumption that southwesterlies prevailed throughout these intervals of time (Figure 147). The delta corners developed under a high energy lobate–arcuate system were probably always sites where opposing currents met and where canyon chutes existed.

Also Burke (1972) suggested that the Afam Clay and Kwa Ibo Members were canyon fills but Weber and Daukoru (1975) think that this explanation is not acceptable and proposed that the deposits infill gullies formed during transgressions and are not infills at the heads of canyon systems.

The sand which escapes being trapped and incorporated in (1) the Benin Facies in the Upper and Lower Flood Plain environments and (2) in the Mangrove Swamp Barrier Ridge and Barrier Flood environments, is then transported by longshore currents to canyon heads and cascades down to form fans at the foot of the continental slope and on the rise. Burke (1972) thinks that this process has operated since the initiation of the Niger Delta Complex in the Palaeogene and has suggested that since the deposits interdigitate with the deep water Akata Shale that we should recognize a fourth facies of canyon and fan sand as a delta complex component.

If elongate deltas were developed in shallow water in the pre-Miocene embayments then much more sand may have been available on the continental shelf and eventually may have drifted and may have been transported into marginal canyon systems. The lobate–arcuate high energy constructive delta of the post-Oligocene–Present Day holds nearly all sand

160

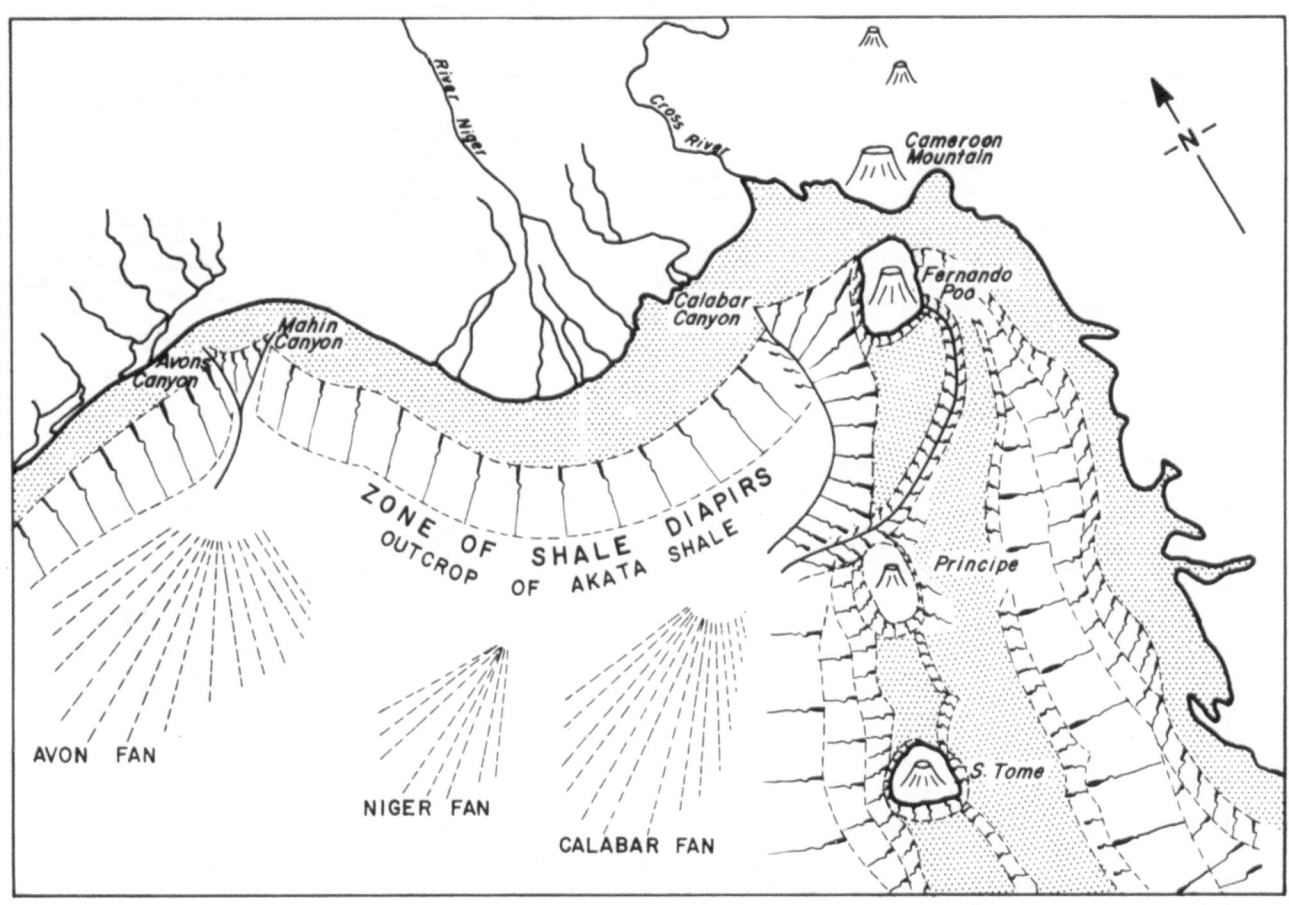

Figure 145. Physiographic sketch map of the Present Day Niger Delta Complex. Based on Burke (1972). Reprinted with permission of the American Association of Petroleum Geologists.

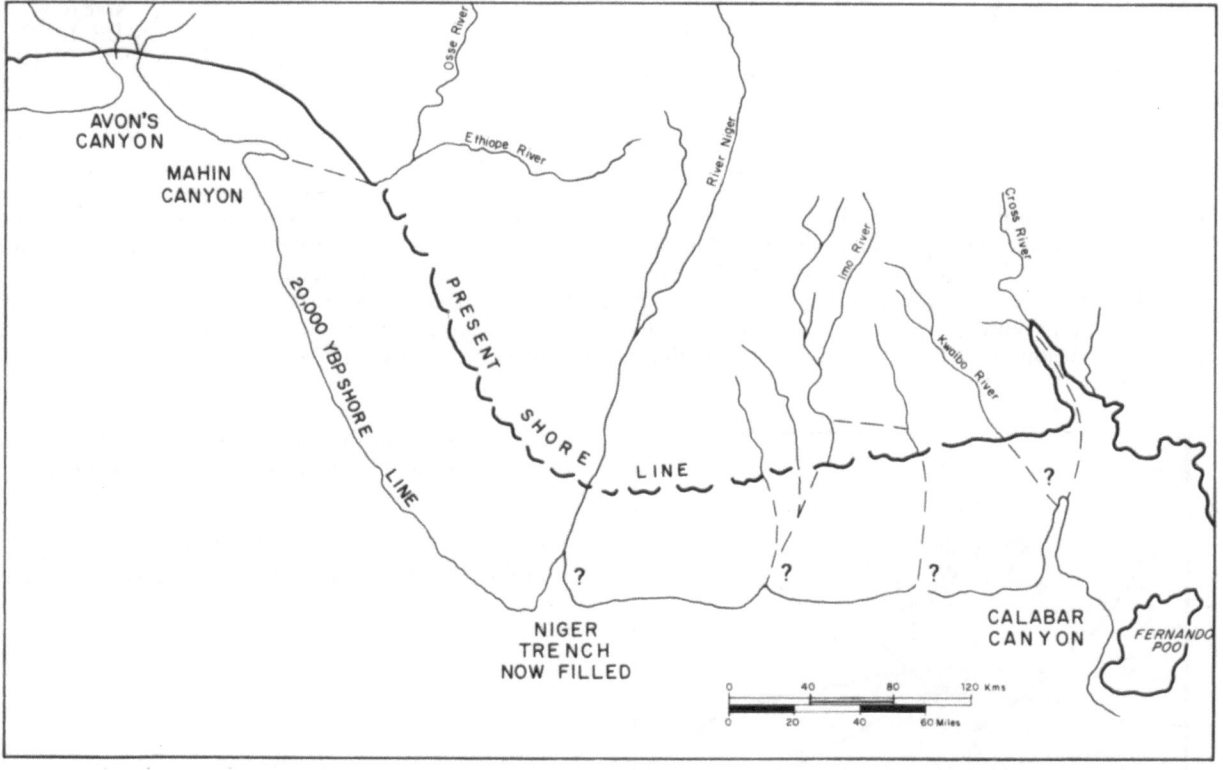

Figure 146. Sketch map showing the positions of the Present Day and 20 000 YBP shorelines, channel and canyon positions, Niger Delta Complex. Based on Burke (1972). Reprinted with permission of the American Association of Petroleum Geologists.

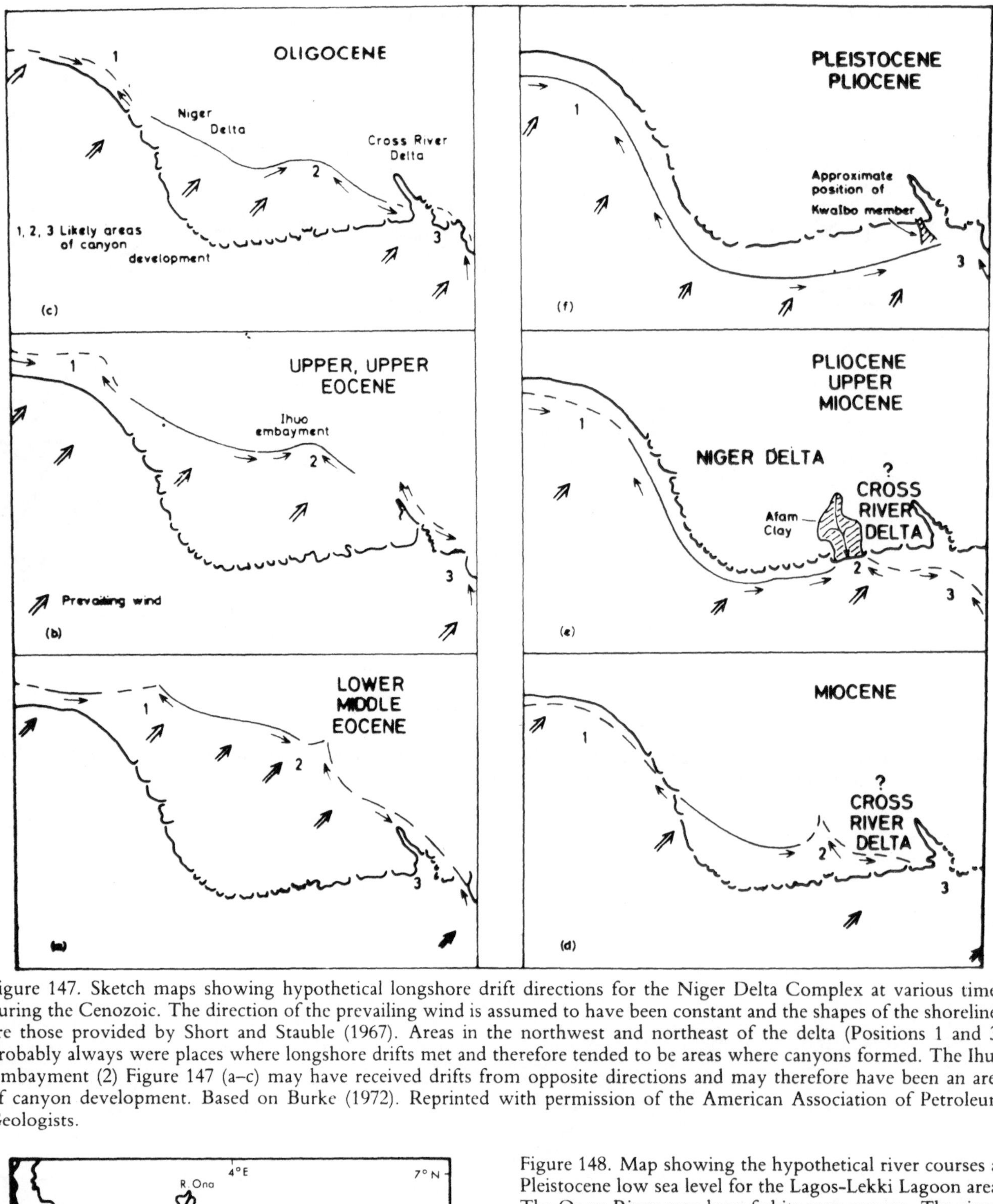

OLIGOCENE

Niger Delta

Cross River Delta

1, 2, 3 Likely areas of canyon development

(c)

PLEISTOCENE PLIOCENE

Approximate position of Kwalbo member

(f)

UPPER, UPPER EOCENE

Ihuo embayment

Prevailing wind

(b)

PLIOCENE UPPER MIOCENE

NIGER DELTA

CROSS RIVER DELTA

Afam Clay

(e)

LOWER MIDDLE EOCENE

(a)

MIOCENE

CROSS RIVER DELTA

(d)

Figure 147. Sketch maps showing hypothetical longshore drift directions for the Niger Delta Complex at various times during the Cenozoic. The direction of the prevailing wind is assumed to have been constant and the shapes of the shorelines are those provided by Short and Stauble (1967). Areas in the northwest and northeast of the delta (Positions 1 and 3) probably always were places where longshore drifts met and therefore tended to be areas where canyons formed. The Ihuo Embayment (2) Figure 147 (a–c) may have received drifts from opposite directions and may therefore have been an area of canyon development. Based on Burke (1972). Reprinted with permission of the American Association of Petroleum Geologists.

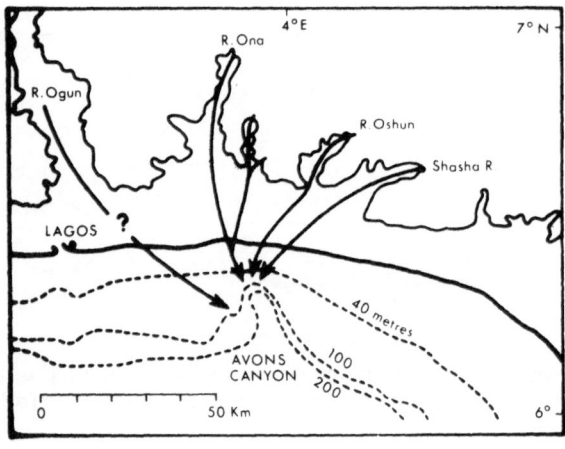

Figure 148. Map showing the hypothetical river courses at Pleistocene low sea level for the Lagos-Lekki Lagoon area. The Ogun River may have fed its own canyon. The rivers Ona, Oshun and Shasa probably fed into the Avon's Canyon. Based on Burke (1972). Reprinted with permission of the American Association of Petroleum Geologists.

within the Barrier Ridge–Mangrove Swamp "baffles" in the Upper and Lower Flood Plain environments. Some 65% of the sand transported by the Niger is retained, largely within the subaerial delta and has accumulated as Benin Formation (maximum thickness over 10 000 ft) and the remainder is held largely within the Barrier Ridge System (Figures 95 and 96).

2.5.5.13. Size, Shape etc. of the Niger Delta Complex and the Present Day Niger Delta System

The Cenozoic Niger Delta Complex is much larger than the Present Day Niger Delta and occupies around 30 000 square miles (onshore) limited by the outcrop of the Palaeocene Imo Shale Formation, itself considered to be an up dip portion of the marine Agbada Facies, and by the Present Day shore. (Figures 9 and 149).

The offshore deep water part of the complex, extending as far out into the Gulf of Guinea as the lower part of the continental slope and the Nigerian Escarpment (Figure 149), exceeds 30 000 square miles and fine grained sediments forming the lower part of the complex extend out towards the Guinea Abyssal Plain over an area which exceeds 36 000 square miles on the lower part of the continental rise. Bathymetric maps show the general subsea form of the delta bulge at the eastern and of the Guinea Gulf (Figures 149 etc.)

Taken together the subaerial and subaqueous portions of the Niger Delta Complex delimited onshore by the Palaeocene Imo Shale/Cretaceous contact and in the offshore by the Cameroun Volcanic Zone on the southeast, the Okitipupa High and the eastern part of the Ivory Coast Embayment on the north and the Guinea Abyssal Plain on the southwest exceed 100 000 square miles.

The "currently" accessible prospective area *offshore* (0 –600ft) is around 18 000 square miles compared with the *onshore* area of the delta complex where it is considerably less than the 30 000 square miles enclosed within the Palaeocene/Cretaceous and present day shore limits.

The Late Quaternary–Present Day Niger Delta is much smaller than the onshore area occupied by the Niger Delta Complex as defined above. The Present Day subaerial delta limits extend from the eastern end of the Lagos–Lekki Lagoon Complex at the mouth of the Siluko River (Figures 145 and 157), via a series of sinuous arc shaped boundaries to Onitsha Gap, then along the valley of the Orashi River to near the mouth of the Imo River and eastwards to the Kwa Ibo Mouth. Within these limits onshore delta environments occupy around 6454 square miles (Figure 151).

Near the Kwa Ibo mouth marine estuarine conditions develop in the Cross River and Rio del Rey areas and extend into Cameroun. These areas may be classed as belonging to High Energy Destructive Estuarine Delta Environment in contrast to the High Energy Constructive Lobate–Arcuate Delta Environment situated to the west (Figure 151). Whether in earlier times constructive deltaic conditions were developed in these areas is not clear but if the Cross River, Imo River and other main rivers were more active in transporting sediments than they are today, then constructive delta complexes may have developed. The presence of growth faults in the offshore of the Cross River Mouth may indicate that sand, silt clay differentiation may have taken place on a large scale in the past.

The amount of sediment discharge by the combined rivers entering the ocean east of the Present Day Delta is not known exactly but it is unlikely to exceed a few per cent of that carried by the Niger-Benue Complex. Tidal ranges are over two times as high as they are at Lagos (3 ft) in the Calabar River area and tidal currents in the Cross River–Rio del Rey–Fernando Po area exceed 40 cm/sec (Figure 141) and are able to keep sand on the move and keep silt and clay particles in permanent suspension, so preventing or retarding Lagoon–Barrier Bar– and Pro-Delta Slope formation. Tidal ranges are likely to have been high in the past because of a restrictive geographic environment at the head of Gulf of Guinea and shallow water in between the island of Fernando Po and the mainland.

Onshore from the mouth of the Benin River to the mouth of the Cross River is about 200 miles and from the head of the delta near Onitsha to the "nose" of the Present Day delta between the Sengana and Brass Rivers is about 200 miles (Figures 9 and 92).

The onshore Niger Delta Complex (base of Palaeocene to Present Day shore) occupies an area of around 30 000 square miles which should be compared with (1) around 6454 square miles from the area covered by Present Day Niger Delta environments; (2) with an area of around 9300 square miles for the *Present Day* Nile Delta (limited by sea level and the Damietta and Rosetta distributary systems) and (3) with an area of around 20 000 square miles for the surface Mississippi Delta System (see Fisher and Brown 1969) (Figure 152).

Clearly the Delta areas mentioned above are not strictly comparable, and the Niger, Mississippi and Nile Delta Complexes as a whole are larger, and extend down the slope and continental rise tailing off in near-abyssal or abyssal depths. In the case of the Nile Delta older portions of the complex lie well to the west of and are disconnected from the Present Day–Pleistocene–Pliocene Miocene part of the complex.

The sediments which constitute the Niger Delta Complex exceed 32 000 ft on the shelf immediately off the mouth of the Brass River and thin both seawards and landwards. On the landward side of the Nigerian Escarpment they still exceed 10 000 ft and the 3200 ft isopach lies some 600 miles out from the Brass River distributary in the Guinea Gulf.

The Late Quaternary Deltaic Section of the Niger Delta Complex has a volume exceeding 900 cubic

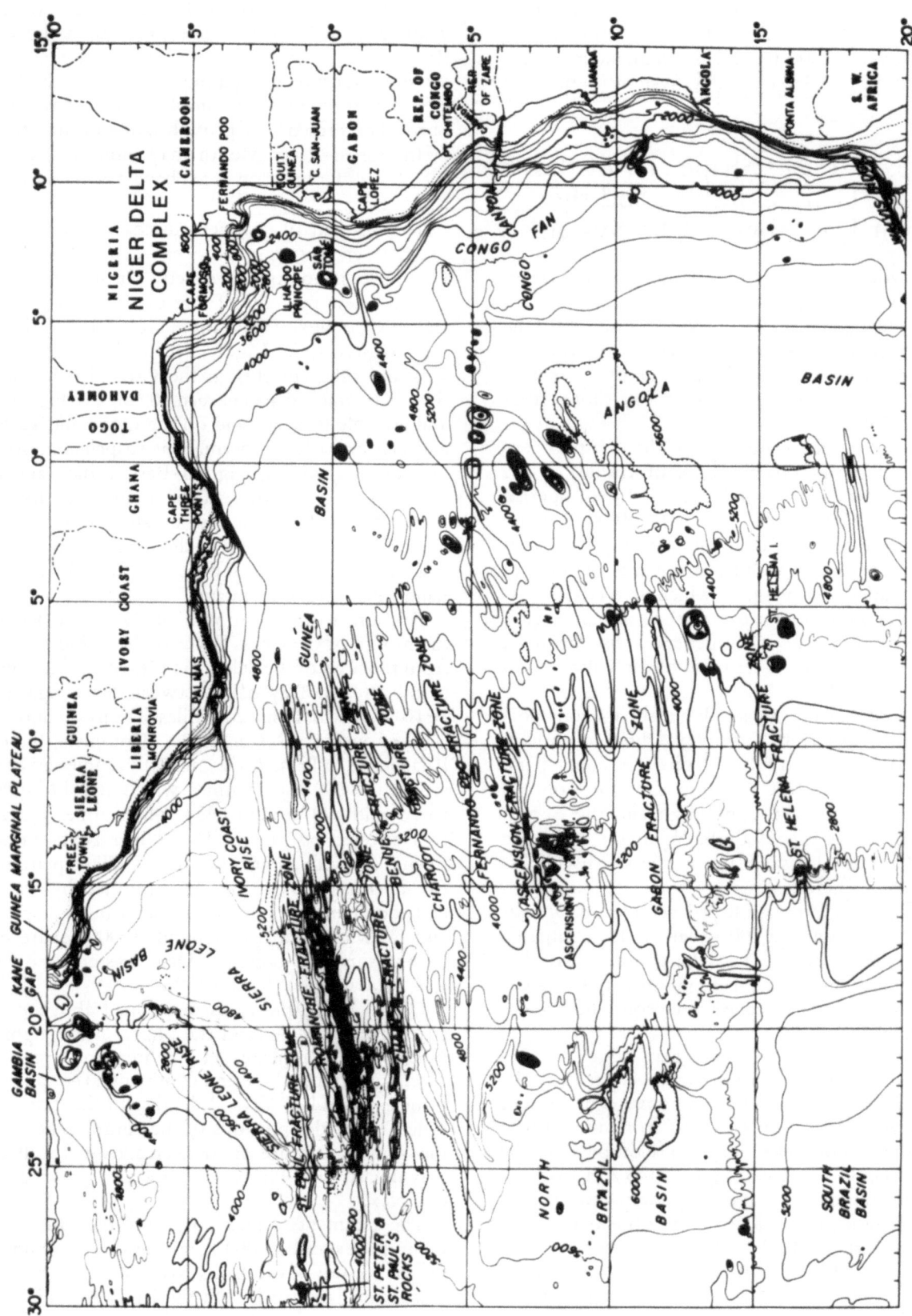

Figure 149. Bathymetric map mainly eastern Equatorial and South Atlantic showing prominent sedimentary bulge of the Niger Delta Complex situated in the eastern Gulf of Guinea. Based on Emery *et al.* (1975). Reprinted with permission of the American Association of Petroleum Geologists.

Figure 150
Estimated Areas Occupied by Selected Environments, Present
Day Niger Delta Complex

	Square miles
Estimated total onshore and off-shore Niger Delta Complex (ino Shale – Abyssal Plain)	100 000
Estimated area onshore part of "Fossil" delta complex	80 000
Upper and Lower Floodplain	3 243
Mangrove swamp	1 900
Barrier Island	694
River mouth bar	617
TOTAL	6 454
Delta Front Platform	1 737
Pro Delta slope	3 011
Open shelf (to shelf break)	3 861
Continental slope (Shelf break to 4200 ft. depth	6 177
TOTAL (Offshore to 4200 ft contour)	14 786

Source: Various, including Allen (1965) and Whiteman (1973)

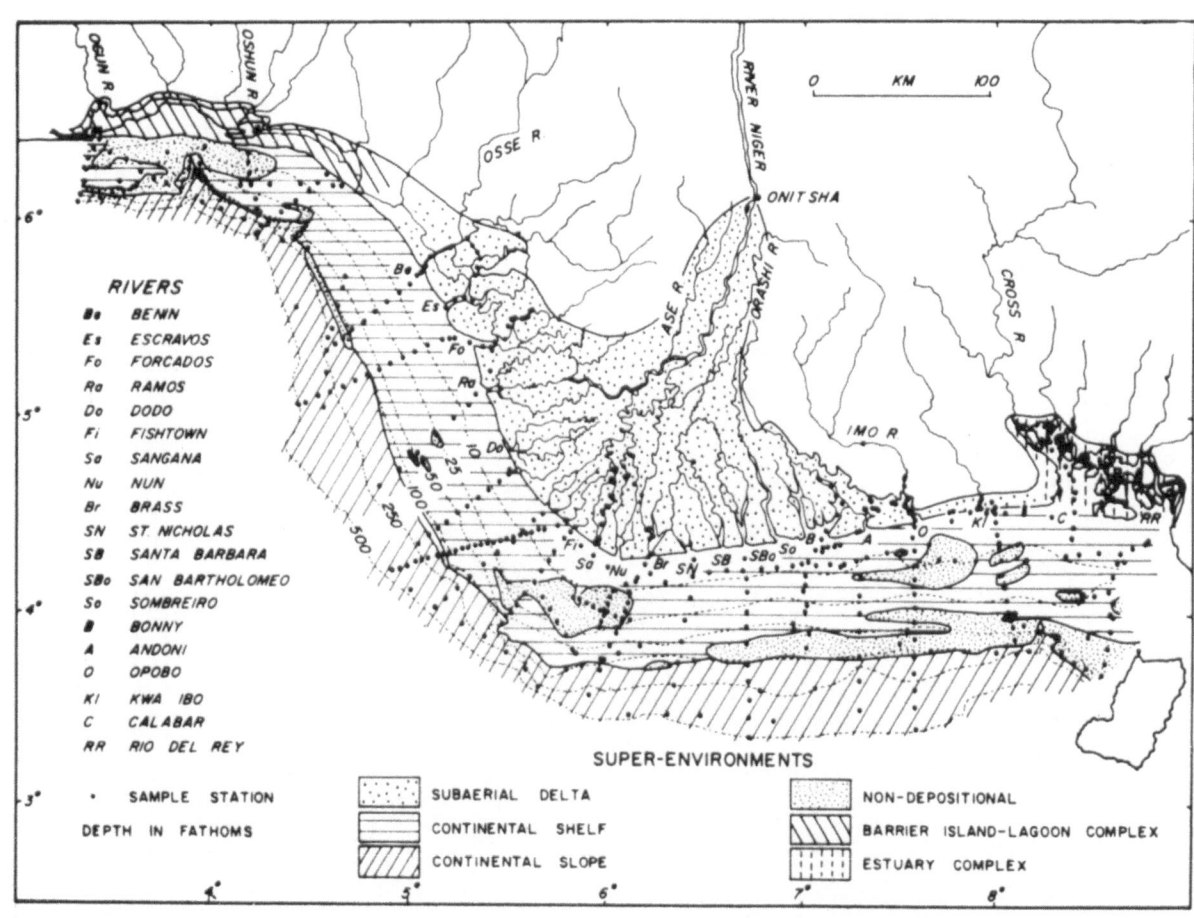

Figure 151. Sketch map of the Present Day Niger Delta Complex, Cross River Estuarine Complex and adjacent areas. List of river names given. Based on Allen (1965). Reprinted with permission of the American Association of Petroleum Geologists.

165